B. C. Anderson · Y. Imanishi (Eds.)

Progress in Pacific Polymer Science

Proceedings of the First Pacific
Polymer Conference
Maui, Hawaii, USA, 12–15 December 1989

Springer-Verlag
Berlin Heidelberg New York
London Paris Tokyo Hong Kong Barcelona

Dr. B. C. Anderson

Central Research & Development Department
E. I. du Pont de Nemours & Company
Experimental Station
P. O. Box 80328
Wilmington, DE 19880-0328
USA

Prof. Dr. Y. Imanishi

Department of Polymer Chemistry
Kyoto University
Kyoto 606
Japan

ISBN-13: 978-3-642-84117-0   e-ISBN-13: 978-3-642-84115-6
DOI: 10.1007/ 978-3-642-84115-6

Offsetprinting: Mercedes-Druck, Berlin; Bookbinding: Lüderitz & Bauer, Berlin
02/3020-543210 – Printed on acid-free paper

Keynote Speakers and Invited Lecturers

1. Higashimura	6. Morris	11. Kim	16. Hill
2. MacLachlan	7. Salamone	12. Ghiggino	17. Napper
3. Vogl	8. Abe	13. Gaudiana	18. Verbicky
4. Saegusa	9. Karim	14. Israel	19. Bailey
5. O'Donnell	10. Yamabe	15. Edmonds	

Acknowledgements

The editors acknowledge considerable support and assistance from the Council of the Pacific Polymer Federation and the first President, Professor Vogl, and the Chairman of the Organizing Committee, J. C. Salamone. Especial thanks are due the Conference Business Manager, Jane C. Vogl. In preparation of the book, Mrs. Donna Weibley and Dr. A. L. Logothetis of the Du Pont Company were truly assistant editors.

Preface

This book is a collection of the addresses of the keynote speakers and invited lecturers as well as manuscripts of a few outstanding papers which were delivered at the First Pacific Polymer Conference organized by the Pacific Polymer Federation in Maui, Hawaii, 12-15 December, 1989.

The First Pacific Polymer Conference covered a wide variety of topics in macromolecular science, demonstrating the emphasis given to polymer research in the Pacific Rim countries. The keynote speakers and invited lecturers are excellent scientists and leaders of effort who covered their fields expertly and in many cases gave their own perspective on the future of polymer science and engineering. A panel discussion on the role of polymers in the arts interested the attendees and emphasized the pervasiveness of polymers in all facets of life. The meeting was attended by over 500 scientists from all over the world. The participants left the meeting with renewed feeling for the importance of polymers in the material sciences and impressed by the progress in polymer research and development.

This book, therefore, provides a wide -angle snapshot of the polymer research as we enter the 1990's. It is a useful book for all scientists interested in polymers and the progress of our science in the countries of the Pacific Rim. We hope that many attendees were stimulated by the meeting and that new ideas and new collaborations will result which will further enrich research, and lead to new useful polymers for all countries.

We would like to express our most sincere appreciation to all the speakers for the excellent talks they presented at the First Pacific Polymer Conference and the high-quality papers they submitted for publication.

B. C. Anderson
Y. Imanishi

Table of Contents

X

Welcoming Addresses

The Pacific Polymer Federation

The Constitution of the Pacific Polymer Federation was signed on October 19, 1987 in Tokyo, Japan at the International House in Roppongi, Tokyo by the Chairman of the Division of Polymer Chemistry, ACS, Ronald K. Eby and the Chairman of the Foreign Affairs Committee, Otto Vogl; by the President of the Society of Polymer Science, Japan, Motowo Takayanagi, and by Vice President for International Affairs, Society of Polymer Science, Japan, Akihiro Abe; by the Chairman of the Polymer Division of the Royal Australian Institute of Chemistry, David J.T. Hill, and by the Chairman of the International Committee James H. O'Donnell.

The FEDERATION was created as an organization to advance and benefit polymer science and technology in the Pacific Basin; it has as its objectives to encourage and facilitate: i.) Interaction between polymer organizations of the Pacific Basin; and ii.) Exchange of scientific knowledge, by participation in national meetings and by visits of polymer scientists of the Pacific. The membership of the FEDERATION is open to all societies and associations of scientists and engineers which have at least part of their activities devoted to polymer science and/or technology.

The duly appointed Council met in the afternoon of October 19, 1987 in Tokyo for its so-called "zero" meeting. O. Vogl was elected the first President of the FEDERATION, T. Saegusa, Vice President, J. O'Donnell, Deputy Vice President and J. Salamone Secretary/Treasurer. The President was charged to set up the FEDERATION, and particularly, to organize the First Pacific Polymer Conference in December 1989 in Maui, Hawaii.

It was very gratifying to see a regional organization involving the Polymer Scientists of the Pacific region come into being. Polymer Science is presently one of the fastest growing scientific and technological disciplines and the Pacific Basin is the fastest growing region scientifically, technologically and economically in the world today. It was only logical, that the organizations of the region found a mechanism of cooperation and also found a mechanism

B. C. Anderson · Y. Imanishi (Eds.)
Progress in Pacific Polymer Science
© Springer-Verlag Berlin Heidelberg 1991

of facilitating the interaction of individual scientists. In the
last year of interaction between the organizations and the Council
members, representing the organizations, a great deal has been accomplished which promises that the original concept, the dream of a
smoothly functioning regional cooperation in polymer science can be
established.

The process of establishment of the FEDERATION began with a
meeting of Saegusa, O'Donnell and Vogl at a meeting in Hawaii in
December 1984; at that time the concept of a regional scientific
organization was developed. It was decided to explore with the
parent organizations (which became the Founding Members) to determine if it were possible to devise a Constitution which could incorporate all the interests of the polymer oriented organizations
of the Pacific Basin. Much of the individual writing was done prior
to and during the U.S.-Japan Seminar in October 1985, and at the
Annual Meeting of the Society of Polymer Science, Japan in May 1986,
at which time all parties were present. In New York, in September
1986, and at the Annual Meeting of the Polymer Division of the Australian Institute of Chemistry in February 1987 the details of the Constitution and the subsequent understandings were worked out and final
agreement to establish the PACIFIC POLYMER FEDERATION was reached.
All these negotiations were done before the benefits of Fax machines
were available and much thought went into the preparation of the doc-
ument. At the same time, the organizations and individuals involved
in the negotiations had time to realize how important the development of the FEDERATION was for Polymer Science and for the interests
of polymer oriented organizations and individuals of the Pacific
region.

At the second meeting of the Council of the FEDERATION in Kyoto
July 31, 1988, 3 applications for admission to the PACIFIC POLYMER
FEDERATION had been received. The Council decided to admit all three
organizations: The Macromolecular Science and Engineering Division
of the Chemical Institute of Canada, the Society of Polymer Science
of Korea and the Polymer and Industrial Section of the Malaysian Institute of Chemistry and invited them to provide one Councillor each
for their admission date, January 1989. The second Council Meeting
was held near Brisbane, Australia on February 6, 1989. Two more organizations were admitted to the FEDERATION: The Polymer Division of
the Chinese Chemical Society of the People's Republic of China and
the Polymer Group of the New Zealand Institute of Chemistry. We now

have applications to join the PACIFIC POLYMER FEDERATION on hand
from the American Physics Society. High Polymer Division and polymer oriented organizations from Singapore and Vietnam.

The FEDERATION has already become a healthy organization; it
functions with 8 committees. has produced the first Newsletter and,
I hope, will have an excellent scientific meeting here in Maui.

I am convinced that the organization will continue to grow vigorously during my term as President of the PACIFIC POLYMER FEDERATION
and during the terms of my successors to an essential part of science
and technology in the Pacific region.

Otto Vcgl, President

Maui, Hawaii, December 12, 1989

Message from the Society of Polymer Science, Japan

As the President of the Society of Polymer Science, Japan, it is a great pleasure for me to congratulate Professor Otto Vogl and the Organizing Committee on successfully holding the First Pacific Polymer Conference in the splendid environment of Maui, Hawaii, U. S. A., December 12-15, 1989. The number of the active participants exceeded 400, including over 140 from Japan. On behalf of the Japanese participants, I wish to thank the organizers and the Local Committee members for their many efforts to prepare and hold this exciting international conference.

The meeting has been successful not only in quantitity but also in quality, covering a variety of important facets of contemporary polymer science and technology, on which interesting and excellent keynote and invited lectures and contributed papers have been presented. Consequently, I do believe that the first conference of the Pacific Polymer Federation has already achieved its primary goal: the promotion of the scientific and human interactions among the polymer scientists in the Pacific and other regions of the world. It is our hope that this success will continue and be extended in the future, with an expanding membership of polymer-related societies.

B. C. Anderson · Y. Imanishi (Eds.)
Progress in Pacific Polymer Science
© Springer-Verlag Berlin Heidelberg 1991

As announced in Maui, the Second Pacific Polymer Conference will be held in Otsu, Japan, from November 26 to 29, 1991. We would like to extend our most cordial invitation for all the polymer scientists and engineers in the world, particularly those in the Pacific Rim Region, to attend at this second conference of the Pacific Polymer Federation in Japan.

T. Higashimura

Toshinobu Higashimura
President
The Society of Polymer
Science, Japan

Message from Governor John Waihee

December 12, 1989

It gives me great pleasure to welcome to Hawaii the participants of the First Pacific Polymer Conference taking place at the Royal Lahina Resort on Maui.

The Pacific Basin offers endless opportunities to scientific and technological disciplines as rapid economic growth continues along with advancements in science and technology. Hawaii plays a very special part in promoting cooperation among the people and cultures of the Pacific and Asia. Our geography, history and people distinguish us from the rest of the world and enable us to play this significant international role.

It is heartening to have such a distinguished group of scientists meet in Hawaii to share information and foster international cooperation. Although you have a busy schedule, I hope you find time to savor the natural wonders of the islands.

The people of Hawaii welcome the opportunity to serve as hosts to you. Please enjoy the wonderful and varied ethnic cuisines, the color and diversity of our cosmopolitan population, and particularly, the scenic splendor of our island state.

JOHN WAIHEE

B. C. Anderson · Y. Imanishi (Eds.)
Progress in Pacific Polymer Science
© Springer-Verlag Berlin Heidelberg 1991

Message from Mayor Hannibal Tavares

OFFICE OF THE MAYOR
COUNTY OF MAUI
WAILUKU, MAUI, HAWAII 96793

Welcome to Maui and to the Royal Lahaina Resort. We are very glad that the Pacific Polymer Federation has selected Maui as the meeting place for its First Pacific Polymer Conference.

Anything worth doing is worth doing well, especially if there is an incentive for a one of a kind reward at stake.

A conference or a vacation on our beautiful island of Maui is worth giving your all because if there is any place on this planet of ours that can truly be described as paradise, it is here, it is here, on Maui.

Time spent in our island paradise will provide you with precious experiences and memories you will treasure all your life.

Maui is nature's inspiration which in most cases we have done much to enhance, particularly at our renowned vacation resorts where comfort, luxury and charm have been blended to produce a breathtaking ambiance which is comple- mented by our balmy breezes, scenic splendors, exotic flowers, glorious sunrises and sunsets, kind weather and friendly people.

In Hawaii, we have long been known as "No Ka Oi", which is Hawaiian for "The Best". I am positive that after exper- iencing a Maui conference and a Maui holiday, you too will be impressed that the time you spent on Maui will indeed have been "the best".

Aloha,

HANNIBAL TAVARES
Mayor, County of Maui

B. C. Anderson · Y. Imanishi (Eds.)
Progress in Pacific Polymer Science
© Springer-Verlag Berlin Heidelberg 1991

The Role of Science in a Changing Society

Norman Hackerman

The Robert A. Welch Foundation; Rice University, Houston, Texas
and The University of Texas at Austin

The word "science" comes from the Latin *scientia* which means knowledge and, by common usage, science has come to mean knowledge of nature. Actually I prefer understanding of nature, its forces and its materials. These are the two principal items of nature which make up the material universe. There are also space and time which I will not deal with tonight. Science is a part of our culture, the definition of which is the state of advancement of civilization. So, science is as much a part of our culture as is art or music or writing or any other intellectual activity.

Science therefore has its value as a part of our culture but it also has a much more important role. A story (apocrypha) about Michael Faraday may help to set the stage for the latter point. Faraday, as most of you know, was a great 19th century scientist, one of the great ones of all time. One of the things he discovered was that if you pass an electric current through a copper wire while the copper wire is between the poles of a magnet the wire will be repelled by the magnetic field, and if you make a loop of the wire and permit it to turn, rejection will cause it to rotate, thus the electric motor. If you do the reverse, i.e. make a copper wire rotate in a magnetic field you produce an electric current, thus an electric generator. Based in part on this work of others, Faraday first demonstrated this phenomenon in the fall of 1821. It is said that he happened to meet Sir Robert Peel at a reception a few days after he announced the discovery at the Royal Society of London. Sir Robert congratulated him on the interesting discovery and wondered what use there was for it. Faraday is said to have replied, "Mr. Prime Minister, I don't know what use there is but I am sure you will find a way to tax it." That story, which may well be fictional, still is indicative of the dichotomy of science as a cultural as well as a useful item.

Let me shift gears now to tell you what most of you already know. There are currently in excess of five (5) billion people living on this planet. You have heard about the space vehicle, Voyager, which recently went by Neptune and is now leaving the solar system after sending back much useful information. That space vehicle was launched in 1976 and in the time it took to traverse the solar system there has been a net increase of one (1) billion people on earth. Why is that important? I assert that the biosphere cannot spontaneously support five (5) billion people, much less the ten (10) billion estimated for the first fifth of the next century. In fact I don't know what it can support spontaneously, but I am sure we passed that population a few centuries ago.

B. C. Anderson · Y. Imanishi (Eds.)
Progress in Pacific Polymer Science
© Springer-Verlag Berlin Heidelberg 1991

If it cannot support five (5) billion people, how come we are all here? We are all here because technology <u>makes</u> the earth support us; it is because of technology that we are all here, albeit some in considerably less comfort. How does technology do this? It does it by making use of our understanding of nature's forces and materials. This understanding of nature's forces and materials, as already noted, is science. We wrest our needs from the planet, that is needs in food, shelter, energy and health plus those things we demand. To repeat, all of us need the basic parts of the first four things, and then we demand a great deal more. That total demand has to be met out of the biosphere and it could not do so without technology. Technology is nothing more than the use of our knowledge about nature's materials and forces for the purposes just stated. That tells you that you could not do the technology without the science. So I propose a simple system which uses the 19th, 20th, and 21st letters of the Roman alphabet S →T →U, science to technology to use.

Science spins and spins without very much aim; it simply produces more understanding of nature; the connection between science and technology is very fluid. Technology on the other hand is related to utilization by a very solid connector. What we need and what we demand determines what technology has to do. Now, this simple system has all sorts of bells and whistles on it with feedback and loops and so on, but basically it is simple. Somebody has to understand nature; somebody has to take this understanding and translate it to technology and somebody has to transfer the technology into use. That says again that while science is valuable to our civilization because it is valuable to our culture, it is necessary to our needs. It is not a question of whether we like it or want it or don't want it, we have to have it. Without it there is no advance in technology and without the advance in technology somebody is going to have to get off the earth.

We have, therefore, a vital requirement. We have been talking about science and talking about technology which means, obviously, talking about scientists and engineers. It is important to understand that the system will not function unless all parts are serviced. Obviously you need scientists and engineers. But you also need designers, constructors, manufacturers, transporters, distributors, record keepers, dispute settlers, and cycle closers. The last is required because there is no likelihood that we can carry out the technology without impacting surroundings. So it is necessary to have people who know how to repair the surroundings.

Every task is important and the system will not function properly if any one link is missing. However, I shall focus attention on the scientists who discover new truths about nature, the scientists who help translate that into technology, the engineers who help translate that into technology, and the engineers who convert the technology to use. It is important to understand that the source of science is people -- interested, educated, and trained. I make a point of the fact that they have to be both educated and trained because individuals train for the present but are educated for the future. They are educated for the future because change is inevitable and since change is inevitable, training alone is insufficient. In spite of the importance of this topic, I am not going to discuss who these people are or should be. That will have to wait for another time. Incidentally, this is a vital problem at the present time for all countries, not just the United States.

I would like, instead, to talk to you about the condition of the field of science. Since the field is so important to us, its condition must be important to us. Whether you like it or not, i.e., whether or not you like what science is, it is vital to your well being.

What is the condition of science in 1989? The answer to that is easy. In spite of all of the complaints about lack of interest over the past two decades, the fact is that the field is in excellent shape. It has never been better. You know that we have now set about mapping and sequencing the human genome. Over a period of the next ten (10) years or so we plan to develop the entire human genome. We have the audacity to believe that we can work out huge, complex systems like those of the global atmosphere and the world ocean. We are beginning to believe we understand something about the interior of the earth. We know we have learned a great deal about our neighbors in space. We are going to put the Hubble telescope up in the spring of 1990 and that instrument is going to look toward the receding edge of the universe, whatever that may mean. We are highly sophisticated in many areas. We can do things in quintillions of a second; thus we can now look at a chemical reaction in a time faster than it takes for bonds to make and break. Femtosecond laser spectroscopy permits us to work on these particles as they form and reform. We clearly are on an exponential learning curve in the field of science. It will continue as long as there are a few people who are interested in delving into nature's forces and materials.

I would like to give you three fairly simple examples of the effect of science on your lives; these are picked more or less at random. The first example is the Voyager, already mentioned earlier. Voyager is a complex, remote system for absorbing information, translating it into numbers, and communicating those numbers back to earth. This package went up thirteen years ago. Its solar cells are still working. Its communication system is still working. It is heading out into space and we will undoubtedly lose it shortly, but for thirteen years it has been delivering information at an unbelievable rate. Will that have any influence on you? Directly no, but clearly it is important.

The second example is much more recent. It appears in three papers that were published in the October 14, 1989, issue of Science. These papers describe the discovery of the gene which is involved in the genetic disease, cystic fibrosis. The gene has been identified with little question about it and should be important, especially in the future, to those who have it and those near them. It is not a widespread disease but just think of what that means in terms of other diseases which are genetic in character; a tremendous advance just a few months ago.

The third item which I ask you to think about is an old one. It started in 1860 and you use it every day; without it your life would be entirely different. I am talking about the lead storage battery. Just think of what it would be like if there were no such rechargeable battery -- no cars except Stanley steamers, no airplanes, no load leveling of communication centers, no capacity to handle surges in electric generation. Of course, there would be some good without batteries since modern warfare would be impossible without the mobility and communication modern conflicts require. That little black box in 130 years has remade civilization.

It is clear that science is not just something that we can like or dislike, it is something that <u>you must have</u>. <u>We have no choice</u>, except for many of us to get off the planet, but that is not much of a choice. I leave you with the proposition that if what I said is right it behooves all of us to make sure that the system functions well. I also tell you that I appreciate your attention. Thank you.

Scientific Papers

Future Aspect of the Science of Matter and Macromolecules

Kenichi Fukui

Institute for Fundamental Chemistry,
Kyoto 606, Japan

INTRODUCTION

We would like to begin by looking at science in general without limiting ourselves to the specialized field of macromolecular chemistry. In intellectual creation required in the future society, the relationship between scientific techniques and man will emerge as the greatest issue. We have to deliberate what, after all, is science and technology to man.

Science with the objective of acquiring knowledge and technology to apply the knowledge to nature are originally two entirely different concepts. However, technology has been utilized by science to accomplish further advancements in science, and advanced science yielded more sophisticated technology. Science and technology have stimulated each other's development in such a manner that it is difficult today to draw a line between the two fields, hence, the term "Science/Technology" denoting the two together.

Science/Technology, which showed very rapid advancements in this century, brought about enormous convenience and welfare to mankind through reducing illness and starvation and allowing more efficient exploitation of resources and energy, only at the cost of exhausting the resources and energy reserve of the earth and contaminating the natural environments. All the more because Science/Technology proved itself to be powerful enough to realize man's long-entertained dreams and hopes one after another in this century, it drove mankind to the pursuit of insatiable desires and to competitions, depriving man of peace of mind. It so to speak had us bite the forbidden fruit. This lies at the bottom of all contemporary issues such as various sorts of conflicts and frictions.

The accelerative progress of science and technology through their interrelations resulted in extreme maldistribution of

B. C. Anderson · Y. Imanishi (Eds.)
Progress in Pacific Polymer Science
© Springer-Verlag Berlin Heidelberg 1991

Science/Technology on the global scale. While industrialization has been carried to extremes in some regions, other regions are still left out of the benefits of scientific civilization.

This brief review of the history of modern science shows that Science/Technology has almost always had both advantages and disadvantages.

MACROMOLECULAR SCIENCE AS A BRANCH OF MATERIALS SCIENCE

Let us view the above general discussion from the standpoint of science related to materials, especially macromolecular science.

"Materials science", as we call all sciences dealing with materials together, was baptized at the onset of the 20th century first with the quantum theory and the relativity theories. All physical and chemical properties of materials have since been understood within the framework of these theories. However, it was only in the latter half of the century that understanding of a huge field of materials science, i.e. science of biomaterials, underwent a breakthrough comparable to the advent of the quantum theory and the relativity theories. It was the discovery of the biological significance of the sequence of monomer in biopolymers. These achievements opened the door to the present development of biology and biotechnologies.

Materials science dealing with outer-nuclear phenomena has concentrated its efforts to understand the properties of widely diverse but individually distinguishable compounds formed by combinations of a comfortable number of some 100 elements in the language of quantum mechanics. However, the understanding obtained by this approach has not been sufficient to allow description of the characteristics of individual molecules according to a general principle. This becomes more difficult as the size of the molecule increases; it is particularly so in macromolecules. For example, there is as yet no satisfactory theory to estimate the high order structure of copolymers of different kinds of monomer with analogous molecular structures from its primary structure. Therefore, relating the primary structure of macromolecules with its properties is difficult even with the help of experiments.

The reactivity of macromolecules depends largely on the stage in which they are present and the high order structure as well as the primary structure and functional groups in the structure. In designing macromolecules, therefore, not only the position of functional groups but also the cooperative effects of the entire molecule is important. In this connection, it is to be noted that achievements have been made in the studies of solid phase polymerization of crystals of monomers, and those of cooperative phenomena in macromolecular solutions.

The method to design molecules with a function resembling the selective molecular differentiation of biopolymers, for which 1987 Nobel Prize for Chemistry was awarded to D. J. Cram, C. J. Pedersen, and J.-M. P. Lehn, was an advanced technique that, using cyclic polyether as the host molecules, allowed selective uptake of ions by the host molecules, selective induction of reactions in the host molecules, optical fractionation, and even designing of the transition state of a chemical reaction according to the solvation energy of metal ions and the ion radius based on differences in the gap diameter, the number of oxygen atoms and basicity, and the three-dimensional effect of the cyclic structure. Studies of methods of molecular design based primarily on the three-dimensional steric effects developed into those of the active transport and selective permeability of metal ions by the synthetic macromolecular membrane.

The exquisite sophistication of the three-dimensional architecture of biopolymers is typically demonstrated by the relationship between their physiological activity and optical isomerism and the molecular processes involved in the evolution of organisms over a long period. Especially, the replication of nucleic acid molecules and the synthesis of protein molecules in the living body are considered to be the supreme forms of designing of highly selective chemical reactions. Following these natural examples, studies of template synthesis utilizing the complementary interactions among bases of nucleic acid analogues have been carried out. How delicate and sophisticated the selectivity of bioorganic macromolecular reactions was demonstrated by the studies of M. Eigen, the winner of the 1967 Nobel Prize for Chemistry, and others. They treated the mixture of monomeric ribonucleotides with RNA replicase and succeeded in synthesizing RNA with a sequence of a fixed order of the monomers in vitro without using appreciable amounts of RNA templates. More amazingly, some specific RNA molecules were found to have an enzyme-like action in certain cellular reactions. The news that the discoverers S. Altman and T. Cech won this year's Nobel Prize for Chemistry still lingers in our ears. These studies had great implications in the synthesis of macromolecules with a fixed monomer sequence and exploration of the possibility of natural occurrence of biologically significant macromolecules.

The behavior of electrons in macromolecules has aroused vigorous interest in connection to the development of materials science, particularly, electronics materials. Electro-conductive macromolecules, especially superconductive macromolecules, were the topic of the time. Superconductivity was first discovered in low-temperature mercury in 1911 by H. Kamerlingh-Onnes and explained in 1957 by the BCS theory (J. Bardeen, L. N. Cooper, J. R. Schrieffer). The discovery of superconductive oxides by G. Bednorz and K. A. Müller in 1986 threw light on the feasibility of macromolecular "high-temperature superconducting materials", the possibility of which was once theoretically predicted.

DESIRABLE FUTURE OF SCIENTIFIC TECHNOLOGY

As the resources and energy reserve of the earth are becoming scarce, Science/Technology must first minimize their consumption. Science/Technology must be employed to recover resources already consumed and released into the ocean and atmosphere, to recycle resources, and further to create new resources and energy supply. Also, technology to repair the damaged natural environment or to prevent its further devastation must be developed.

The most difficult to cope with is the problem of global regional differences based on maldistribution of Science/Technology. In the future society, this inequity must be corrected so that every region and each individual can enjoy the benefits of scientific civilization, which is a common asset of mankind. For example, technologies to convert saltwater into fresh water and supply areas lacking in fresh water or to irrigate and cultivate deserts and provide them with energy are anticipated. This, however, requires extremely high intellect.

To study the complex relation between Science/Technology and humanity, we must consider the effects of technology on man and further venture to tackle the problem of how can man be adjusted to nature as his environment. When man makes an object, for example, physicochemical properties such as strength and durability has conventionally been the first matters to be considered. That is, the emphasis has been primarily on the "hard" technology until today, but the "soft" side is considered to bear increasing weight in technology. However, this in general is a very difficult field. Investigation, for example, to relate the human sensation of touch to the molecular properties of macromolecular materials will require tremendous elaboration.

Research and development of soft technology will entail fusion and integration of Science/Technology with art or various disciplines of cultural sciences. Human culture may be said to have originated by "integration". In the eras of ancient Greece and Rome, art, technology, and science were all regarded as creative activities characteristic of man and were not strictly distinguished by words. The Greek "$\tau \acute{\epsilon} \chi \nu \eta$" and the Latin "ars" both generally denoted all intellectual activities, skills, and proficiency. Although science differentiated into numerous specialized divisions as it developed, it is beginning to see an era of grand "re-integration". However, the significance as well as the difficulty of such re-integration and re-organization of science lied in the fact that it must proceed simultaneously with increasing specialization of scientific disciplines.

The above is Science/Technology of the future society as predicted from our review of its past history. These are what may be regarded as external factors given from outside Science/Technology as a framework of its development. At its opposite are so to speak internal factors, namely desire inherent in man to know what is not known and more practical enchantment of seeing fulfillment of one dream after another, which together have driven Science/Technology forward. These internal factors are still powerful and will continue to exert great influence on the future development of science and technology.

METHODOLOGY OF SCIENCE IN FUTURE

Next, let us consider the methodology of future science or how science and technology of the future society should be established and managed. At the onset, to review the methodology of conventional science and technology will be meaningful. The old Chinese proverb of "He that would know what shall be must consider what has been" is valid here.

Natural science began as natural history, which observed nature as it is and described it. As knowledge gradually increased, and as man became able to view things from various perspectives, man began to try to understand natural phenomena by the two conventional approaches of scientific thinking of analysis and synthesis. Science of life as well as science of materials has been built by

first dividing the objects of investigation and then re-assembling
them together. The Roman proverb of "divide et impera (divide and
rule)" was cited also in the well-known book "Die Physik im Kampf um
die Weltanschauung", 1937, by M. Planck.

Thus, we conceptually divided chemical compounds into
molecules and atoms, then to electrons and atomic nuclei, and then
to elementary particles. Next, we built them up again by the
methods of statistical mechanics and quantum mechanics to understand
the properties of the substances. In organisms also, we divided the
body into organs and cells, then cells into organelles and component
molecules, and attempted to re-assemble them to explain phenomena of
life. As new materials, new particles, and new constitutional elements were thus discovered one after another, the door to the mys-
tery of nature was opened. This methodology, by which concepts of
elements obtained by dividing the objects of investigation were organized according to more general concepts and theories, allowed understanding and explaining of the materials or events.

One important point in this method is that, by re-organizing
the information obtained by dividing the matter, we began to make
things that were not originally present in nature. This was a great
step ahead. This process of creating entirely new things during re-
assembling of the elements obtained by dissection will be employed
more and more in the future society.

A major advantage of the methodology of division and integration is that it enabled more systematic and generalized understanding of the peculiarity of each material and life phenomenon by means
of the commonness of the elements resulting from division. In this
way, natural science, being classified into physics, chemistry, and
biology according to the differences in the target of investigation,
has progressed to date by generalizing the uniqueness of chemistry
to physics, and the uniqueness of biology to chemistry. My frontier
orbital theory, for example, is nothing but a result of generalization of the specific facts of chemical reactions by the method of
physics.

The well-known conventional classification of natural science,
namely physics, chemistry, and biology, is just a tentative partition that developed as a consequence of science history. The
character of the discipline becomes less universal and more specific
and individual from physics to chemistry, and chemistry to biology.
In other words, the generality is the greatest in physics and small-

est in biology. Such stratification of the partitioning of science has favored the progress of science till today.

However, whether we may still resort to conventional approaches based on this stratified partitioning of science is one important problem in considering the methodology of future science. There are not essentially clear lines of partition in contemporary science any longer.

As for the tendency of generalization, disadvantages of lingering with this seem to have appeared in various areas. Generalization is made not only to systematically understand diverse phenomena but also to facilitate prediction of the outcome of an event or a series of events and use it for practical planning. However, generalization has been shown not to benefit the understanding of several phenomena. Very unique high-temperature superconductive ceramics and organic superconductive materials are typical examples of such phenomena. The findings in these materials have so far not been generalized, and if generalization is possible, whether it is of any use for planning the next step to take is quite dubious. What endows nature its subtlety is essentially the ability of molecules to recognize one another according to their uniqueness. I believe that the idea of generalization of this great mass of uniqueness toward complete integration is far from practical.

CHALLENGE TO SPECIALIZATION

Thus, it does not seem wise to be forever dependent on the conventional stratification of partitioning of science, namely physics, chemistry and biology. The situation may be quite the opposite in the future. As typically exemplified by the discovery of high-temperature superconductivity, specific examples of a group of materials which possess a certain physical attribute in common may be found, while their general requirements to be so are obscure, in the field of chemistry which locates in the lower order of the conventional hierarchy of sciences. In such cases, the specificity of chemical phenomena is not generalized by physical theories, but rather the universal physical concept is realized in individual chemical phenomena. This challenge of specialization will be a major theme of science of the future.

The same applies to the relationship between biology and chemistry. For example, studies to select compounds with specific activities in the body from a multitude of chemicals are an attempt of biological specialization of general chemical concepts, a reversion of the conventional methodology of specific to general. Only certain specific sequences of organic molecules such as nucleotides and amino acids are meaningful to organisms. Since such sequences have chemically no particular significance, they are considered to have distinguished specificity. Is such biological specialization of a general chemical concept possible? Maybe, it is almost eternally impossible. However, life, which is specificity selected from generality, is undoubtedly present. Therefore, life may be regarded as the supreme consequence of biological specialization of chemical generality.

Nature, in which we live, is extremely unique. The universe is considered to be the most peculiar of all that exists. Since the universe is one, no other matter is more special than this. It is literally incomparable. At the extreme opposite is the universality of mathematics. Nothing is more general than mathematics. Natural science, then, may be regarded as specialization of this mathematical generality in nature and universe. How to reproduce the inherent specificity of nature by physically, chemically, or biologically specifying the mathematical generality will be a very important problem and a challenging task of science of the future.

A word about why and how this task is difficult may be relevant. Science is logically constructed, and once laws of nature are formulated, future can be predicted by the logical process. This is the assumption of conventional science. However, with regard to specialization of mathematical generality, mathematical theories are present completely apart from actual nature, and they become laws of nature only after it is specified. Without specialization, it is simply a mathematical theory. How or with what experiments, then, can we specify this theory? There is no theory at present that allows logical simulation of this process. Can we predict whether there can exist a result of specialization of mathematical theories really in this universe? Probably, the answer is we cannot. We have only to actually materialize such things. The fact that the possibility of solving this problem cannot be logically estimated is considered to give it its novelty as well as difficulty.

For creation of new intellectual values, we will have to be able to manipulate this "specialization of the general" and conventional "generalization of the specific" with both hands and freely shuttle from one to the other. I call this "generalization/specialization paradigm", and expect that the chal-

lenge to achieve specialization corresponding to the specialty of
nature proves to be one of the greatest problems in operating in
this paradigm.

MACROMOLECULAR SCIENCE OF THE FUTURE

Let us apply the generalized discussion above again to macromolecular science. The exponential progress in Science/Technology
in the latter half of this century resulted in rapid exhaustion of
resources above and under the ground, changed natural environment,
and driving mankind to insatiable desires and competitions, disturbed their peace of mind. As a consequence, the industry that has
been dependent primarily on the potential value of resources has
begun to demand efforts of man toward creation of new industrial
values. In synthetic macromolecular industries, quantitative
priority was emphasized at first, but more quality-oriented values
were later introduced, and the industry has developed along these
two directions. One was the above-mentioned "hard" macromolecular
science aimed to produce state-of-the-art materials, and the other
was "soft" macromolecular science involved in the relationship and
harmony between man and nature. The latter is considered to contribute to solution of problems such as what psychological effects
nature has on man, how can man be adjusted to nature and environment, and how to reduce the consumption of energy and resources.
For these purposes, advanced scientific methods must be further
sophisticated.

Before the great leap made in generalization of natural
phenomena at the beginning of this century, a number of problems ac-
cumulated and awaited solution just as advance signs of a giant
earthquake. No such signs seem to be observed at present. Rather,
the inherent uniqueness of nature appears to restrict the methods of
generalization. To me, materials science seems to be entering the
era of specialization rather than generalization. Since macromolecular science seeks significance in application of general
physical principles to special systems, the dawning era may be
regarded as the era of macromolecular science. Promising fields of
macromolecular science include functional macromolecules, electro-
conductive macromolecules, electronics materials, macromolecular
materials with extreme properties, macromolecular liquid crystals,
macromolecules for composite materials, macromolecular complexes;
the list is thus endless. Experimental verification for natural

formation of specially sequenced polynucleotides or polypeptides will be one of the most fascinating problems in relation to the "origin-of-life" argument.

Macromolecules now play important roles in high-technology industries. A major academic breakthrough is likely to occur when a field of science, not necessarily macromolecular science, is linked to another field whose logical relation with the first field is not immediately clear. If the two fields were readily related logically, everyone would immediately connect them. It is difficult to link two scientific fields with obscure relationship as can be readily understood by seeing the apparent remoteness between the general theory of relativity and Riemannian geometry or between quantum spectra and eigenvalues of second order partial differential equations, that is, of the Schrödinger equation. These unexpected couplings become possible only by untiring exploration of other fields of basic science in advance. This appears even more true when we look back to the achievements of great pioneers of macromolecular science such as H. Staudinger, K. Ziegler, G. Natta, P. J. Flory, and many others. The importance of basic investigation lies in this very fact.

DESIREBLE DIRECTION OF SCIENCE AND TECHNOLOGY

Science and technology of the future must represent the wisdom that all human beings of the world work out together with the very survival of the earth and mankind at stake. It must help with man's sincere efforts to attain affluence rather than harshness and warmth rather than coldness or lessen the difficulty of this endeavor.

The 21st century will witness urgent and unremitting dilemma between the limitations of man on the earth and his unlimited ambitions toward science and technology. Man will have to strive and struggle against nature in which he is a part himself on the exhaustible earth. It is the fate of man to have to keep advancing while managing all these problems. It is as if balancing along a path with deep cliffs on both sides where a slip means death. I think that the reward of this strenuous struggle is nothing other than the survival of the earth and mankind.

Macromolecular science is considered to play no small role in the future struggle of mankind to properly guide the progress of science and technology.

Polymer Science and Technology:
An Industrial Perspective

Alexander MacLachlan

E. I. du Pont de Nemours & Company
Du Pont Building
Wilmington, DE 19898 U.S.A.

Throughout my more than 40 years' association and fascination with chemistry, I have been amazed at the value for society this science continues to provide. The promise for the future is truly spectacular. Clothing, shelter, food, transportation, electronics, superconductivity, information storage, the environment, medical progress, i.e. just about everything that is needed to advance the planet's and its inhabitants' well being, is inexorably dependent on and enabled by evermore elegant chemistry. And, more often than not, that chemistry is aimed at new or modified polymeric structures.

As the title of my talk indicates, I would like to give you an industrial perspective as to where I think human needs, which for the most part drive the marketplace, will require polymer science and engineering to go in the next few decades. I hope to illustrate that polymers are a never ending frontier, to more or less borrow a phrase, for the betterment of man, limited only by our imagination. I suspect most of you already believe this, but you also know many within and outside our field are raising doubts, especially in the environmental areas. So my theme today will encompass both the opportunity and the responsibilities we have as we develop new technology, especially polymers technology.

Let me start by going back about twelve thousand years to Ten Thousand B.C. [Fig. 1]. In this slide published two years ago by Ashby, we see the change in the materials that man has used to meet his needs. Initially man used naturally available materials. Our ceramics were stone, flint, and later, pottery. Our composites were bone, wood, and then bricks. Our polymers were skin, vegetable fibers, and gums. But cost, weight, and durability limited the utility of these materials and man soon learned that metals could be reduced from ores or purified from nuggets and nodules. By One Thousand A.D., gold, copper, bronze and iron were being used, and cast iron was the advanced material of the time.

B. C. Anderson · Y. Imanishi (Eds.)
Progress in Pacific Polymer Science
© Springer-Verlag Berlin Heidelberg 1991

30

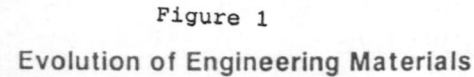

Figure 1

Evolution of Engineering Materials

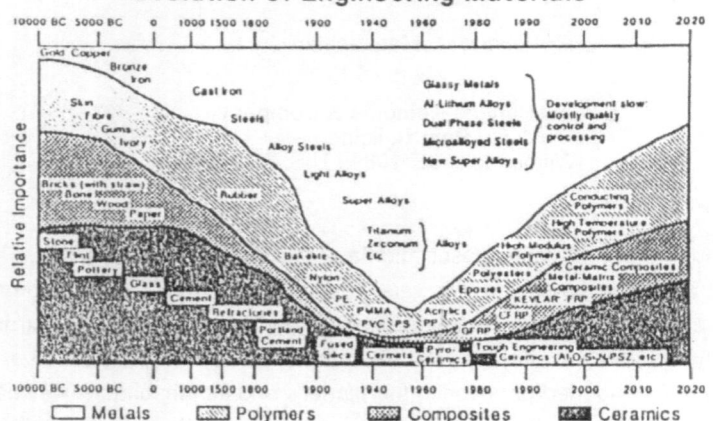

By 1900, metals had won the battle over natural polymers and composites and primitive ceramics. Basically, metals replaced wood and fired ceramics in most structural applications.

In the first third of this century, the petrochemical industry, supplying fuel and lubricants for internal combustion engines made many organic chemicals available at much lower cost than they had been. This led to early synthetic polymers like Bakelite, rayon and cellulose acetate. Then in 1937 nylon was introduced.

What characterized these early synthetic polymers was that they were made from a limited repertoire of simple, usually symmetric monomers.

These early polymers, not many in number, like the metals they complemented, were manufactured on a massive scale. For the most part a given polymer was utilized as a homogeneous material to be formed into many products: combs, radio cases, cups, bottle caps, etc.

But, by the 60's and 70's, we were rapidly gaining understanding at the molecular level. Analytical tools were developed that allowed us to understand molecular structure and how molecules are oriented or bonded to the molecules around them. This was a bit of a turning point. Polymers started to get more complex, by design. A product that better matched a group of market needs could be created by making copolymers such as spandex, modified acrylics, ABS, and other styrene copolymers. They also increased in complexity by becoming heterogeneous. Luster and opacity were controlled by the addition of titanium dioxide and other pigments. We were embarking on the era of "specialties": lower volume but higher value.

But let me stay for the moment with some statistics about polymeric materials.

The production of polymeric materials is a large and growing industry [Fig. 2]. In 1987, U.S. production was about 60 billion pounds, growing at about 8%/yr with the worldwide number about 200 billion pounds and also growing at 7%/yr. Numbers are less certain for the countries of the Pacific Rim, but, again we estimate at least a 7% growth rate. These trends seem likely to continue as industrial development spreads throughout the world.

Figure 2
U.S. Polymer Production

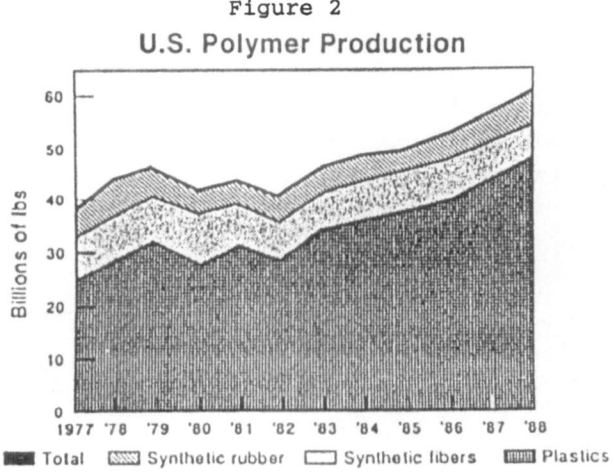

Our study of the literature shows there have been no new general purpose polymers developed in the last twenty years. The needs for designed polymers, for the heterogeneous materials and systems I just mentioned have been met instead by new product configurations based on existing polymers and development of new specialty polymers when needs cannot be met by older materials. The latter activity is striking.

A recent study [Fig. 3] showed that over 2,500 individual grades of thermoplastics were introduced between January, 1986 and June, 1987, with over 400 introduced in June, 1987, alone. Producers of existing volume thermoplastics are scrambling to meet the specialized needs of their customers with modifications of existing products.

This trend will continue and grow. This is clear when you realize that when the modified work horses are not adequate, a stable of new specialty polymers is waiting in the wings (Fig. 4). While volume of these specialties is small today, it is growing rapidly and even more rapid growth in the '90's is expected.

32

Figure 3

Thermoplastic Product Introductions
January 1986 - June 1987

Commodity	1,100
Performance	980
Blend/ alloy	150
High performance	290
Total thermoplastic introductions	2,675
Total grades available (U.S.)	9,700
Grades added in June 1987	> 400

Figure 4

High Performance Polymers
Trend Toward Specialties

	Sales, MM Lbs.	
	1985	1995
Polyarylate	3	17
PEEK, PEK	0.5	4
Polyetherimide	2	36
Liquid crystal polymer	2	15
Polyether sulfone	3	31
Sulfones	18	44
Polyphenylene sulfide	22	82
Total	50.5	229

The trend to replacement of metals by engineering plastics continues (**Fig. 5**) and shows no signs of slowing. Extrapolation to the year 2000 suggests that by that time 50% of metal parts will be replaced by polymers.

Figure 5

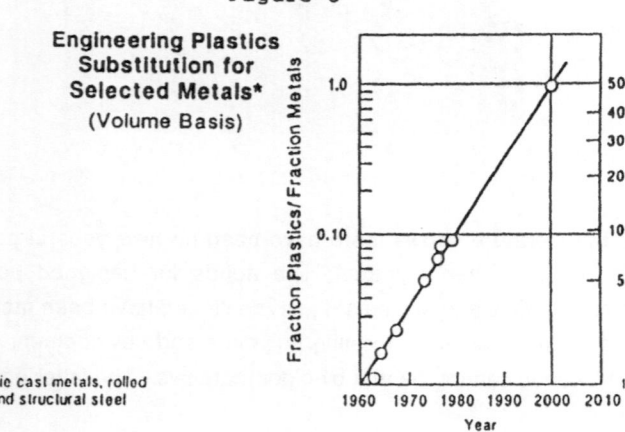

As polymeric materials fill every more specialized end uses, there is a strong trend toward heterogeneous or polyphasic materials. A part that was once molded from a simple polymer may now contain a reinforcing filler, a pigment, a toughener, or perhaps even fiber reinforcement to achieve optimum properties. The technology which must be developed to allow these advances is one of the most active in polymer science.

Well, as end uses of polymeric materials become more sophisticated, the trend to understanding and optimizing systems from synthesis through processing to components to optimum properties for the use at the lowest cost will continue and accelerate. The potential for progress here is great.

Manipulation of individual polymers can result in forms which meet a remarkable range of end uses in a highly functional and economic way. Polyethylene terephthalate for example, a general-purpose polymer in many fiber, film and plastic uses, can be engineered to have very special properties. As fibers, they can be hollow, sheath core or have a variety of cross-sections and surfaces. Experimenting with this almost infinite set of variables has already led to:

- Fabrics with feel and "hand" indistinguishable from silk.

- Bottles functionally superior to glass through special control of molecular parameters and processing systems.

- Clothing and filled products with properties superior to down-filled counterparts and relatively insensitive to moisture.

New ways to exploit the potential of polyester and the other general-purpose polymers will continue to develop as systems approaches to specific needs result in polymers and polymer shapes ingeniously tailored to give the desired properties. The challenge to bring these improvements to all parts of the world in an affordable form is continuously before us.

Fiber reinforced polymeric composite systems have the potential to do more than simply replace existing materials. They offer superior performance in many applications and will have a major impact on manufacturing technology. One study suggests that the body of an automobile, which currently consists of about 300 metal parts, could be reduced to less than 10 parts by using composites. The resulting structure would be just as strong but much lighter. Similar benefits are expected in aerospace manufacturing. Aircraft performance is enhanced by the lighter weight, maintenance is reduced, and we see the possibility of easier repair.

But we are only at the beginning for composites. To fully exploit their potential, we must gain a much deeper understanding of surface chemistry, rheology of polymer systems, thermal and oxidative stabilization, microstructure and failure analysis and so on. The development of this knowledge and its use in all parts of the world are major challenges. And certainly we have only to look at nature to see how far we have yet to go with composites.

Progress in polymer chemistry is also vital to the biological sciences. New polymers as aids for purifying proteins, nucleic acids, and associated sugars, as well as for their chemical modification, hold the future to much of the general usefulness of biotechnology. It's clear that biotechnology is fast approaching an exact useful science and much of it is polymer based.

To accomplish the challenges for more knowledge and understanding critical to advanced polymer chemistry many disciplines must collaborate. Those include synthetic polymer chemists, physical and analytical chemists, engineers, mathematicians and computer scientists. Let me just dwell for a moment or two on the role of mathematicians and computers in our future.

A growing trend in the chemicals and materials industry is the use of computers as a key experimental tool. In the early days of computers, theoreticians rejoiced at the expectation one would be able to calculate the course of chemical reactions and thus eliminate costly experimentation. That dream was not realized early on, but is now within grasp.

Today's high end computers, including supercomputers, are making impressive inroads into experimentation. They can often zero in on a narrow area and thus reduce total experimental effort, or in some cases eliminate experiments entirely. This trend will continue and accelerate. My own company has invested in supercomputers and benefitted significantly in making useful predictions of physical properties of polymers and monomers.

Computers of more modest capabilities are also growing in importance to technology development and are vital to competitiveness. Research personnel can now be linked to their company and university colleagues around the globe. Du Pont has established such a worldwide network and has seen the benefits of bringing together our total global research facilities and personnel as we respond to technology breakthroughs and customer needs worldwide. The efficient yet imaginative use of computers is a major challenge and promise for the future.

Let me turn finally to some of the societal challenges we face. First and foremost is the image of chemistry.

Richard Heckert, a recent past Chairman of Du Pont, has said that our biggest challenge as chemists is to convince the public that we are their friends and not their enemy. A strange state of affairs when you think of it. But true! And polymer science, as we all know it, is not an exception. In polymer science and technology the only way we can regain public confidence is through better stewardship of our products and the processes by which we make them to protect the users and the environment.

While it has been estimated that only 8% of the waste generated in the U.S. is from synthetic polymers, its durability and visibility magnifies it in the eyes of the public. Attitudes toward plastic products are becoming more negative worldwide. This problem must be addressed for continued growth and health of the polymer related businesses. New technologies to minimize waste and to recycle it, the most desirable solutions to the waste problem, must be developed.

Polymers which degrade after use by biological or photochemical procedures offer promise of environmentally acceptable solutions in specific instances. Incineration is lower on the list with entombment in landfills at the bottom.

Incineration of waste is, however, perhaps the best short-term solution and it's well along in Japan where 70% of waste is burned and Europe where 50% is burned, but lags in the U.S. where only 10% is burned.

In the future, emphasis on waste minimization, recycling and bio- and photodegradability must be integral to our plans for enveloping new and improved polymer-based products. If we do not have this mindset, we may find more and more environmental roadblocks appearing. In other words, we have to move from being problem creators to problem solvers in a modern world where concern for the present and future environment is required. To those of us in the scientific community these are intriguing challenges.

Well, enough on the science and critical environmental issues, although these deserve much more discussion. What about the ways in which we practice our science and technology? Are there trends that demand change here? Clearly, the answer is "yes." Three of the most important are competition, collaboration and conservation.

Competition: Competition, which results in the fruits of science and technology being widely available economically, means far more than companies, nations or groups of nations vying for the same local or global markets. Competition also means gaining access to the best technology and people. Since no nation has a corner on these assets, globalization in outlook and performance for industry is and will be a vital component of future success. Truly global companies must develop management where this behavior is as natural as today's export mentality.

Continuous quality improvement, flexibility, and fast product changeover, market focus, short product life cycles, total system optimization which includes just-in-time raw material and product delivery, computer-integrated manufacturing, and partnering with customers will all be second nature to successful companies. Technology will play a critical role in all this and no nation or part of the world will have a monopoly. Execution of these skills will often be a key differentiator among competitors.

Competition in some cases leads to underline. Relationships among companies, consortia, universities and government laboratories will become key components of business strategies. Those companies that master effective collaborations with these constituents will have significant competitive advantage in world markets.

Japan has been the most visible with effective collaborations among companies and their government through MITI. Supportive relationships between governments and companies have also long existed in most European countries and this continues. In the U.S. government has been neutral to somewhat antagonistic with little sign of real supportive involvement. Government-assisted collaborations like Semetech are now being tried in the U.S.A. because of the enormous technical expense associated with the development of many new technologies. Whether the U.S. companies will find this type of collaboration fruitful remains to be proven, but with initial success this could be a strengthening trend.

Collaborations between companies are becoming commonplace. Du Pont, for example, is involved in 22 joint ventures with Pacific Basin companies, 11 of which involve polymer technology.

Europe and the U.S.A. also have long histories of effective collaborations with universities. This trend is strengthening, especially in the U.S.A. In the U.S.A. such relationships are generally available to all companies, regardless of national origin, and have been increasingly attractive to foreign companies. The enormous basic research base of the U.S.A. is a rich source of new technology. At present it is being broadly shared. On balance, one would predict this will continue, but pressure to limit access will also continue, especially as the U.S. government and industry perceive unfair trade barriers in other nations.

Also in the U.S.A., government research laboratories are being encouraged to form relationships with industry to develop and transfer technology. Theoretically, this can be a rich source of new advances for American industry if the barriers to such collaboration can be removed. The early track record is spotty because of suspicions that such involvements might give a single company unfair competitive advantage, but the direction is encouraging.

And finally, conservation. Conservation of natural resources, the environment and human resources will become increasingly important elements of successful strategies for companies and nations. As natural resources like petroleum and precious metals become more scarce, process yields and recovery efficiencies become more critical. This need will spur major industrial thrusts to search for new processes based on quantitative reactions. In addition, biotechnological approaches to recovery and manufacture from abundant raw materials like plant matter or organic waste will grow in importance, a clearly daunting, but nevertheless exciting, challenge for all of us.

As I have already discussed, environmental legislation worldwide promises to accelerate and will have major impacts on competitiveness and conservation of all process materials is a key part of performing in this area. Companies that respond effectively with new technology to meet or exceed environmental limits will have increasing opportunities to have the best cost positions. New processes under development today will have to anticipate tomorrow's new limits. Major opportunities for new technologies to clean up exhaust gases, concentrate waste streams, recover valuable waste products and clean up raw materials before processing to final products to avoid pollutants, will occur. For many industrial products, the competitive position of companies could be strengthened or weakened markedly by performance in these areas.

Finally, we must conserve our most valuable resource, people. One major road block to a truly global company is the inability to fully utilize the capabilities of its human resources. Successful companies of the future must master this skill. Some are making good progress, but most have a long way to go. Full utilization of human resources goes much further than having nationals operate facilities in their own countries. Nationals must progress along professional and managerial lines without discrimination. Companies who achieve this will be difficult to beat because they will have the advantage of clear cultural understanding at the highest levels.

Similarly, females and minorities in virtually every country have not had the opportunity to contribute. Nations and companies that overcome this will gain a further advantage. This will occur not only in their home country, but all over the world, as the offices and laboratories of such progressive companies become favored places to work by talented individuals seeking to contribute.

38

SUMMARY

In summary, progress in polymer technology is clearly accelerating and promises great rewards to those countries and companies that can respond most effectively. The chemical industry and and the polymer-based industries have critical roles to play in bringing the promise of continued development to mankind in a timely and environmentally responsible and safe manner. Mastering of many new skills will be needed to succeed and these include accessing human talent as well as technology and markets on a truly global basis. All these trends are inevitable as we approach a globe where talent and technology are uniformly spread among nations, races, and gender. Companies and countries that effectively orchestrate all these forces will bring the most benefit to mankind and, after all, that is the most important bottom line.

Macromolecular Design and Architecture

Otto Vogl, Fu Xi, Gary D. Jaycox(a), William
Simonsick Jr.(b) and Koichi Hatada(c)

Polytechnic University,
333 Jay Street, Brooklyn, NY 11201

I. INTRODUCTION

In the last few decades, polymer science and the use of polymeric materials has grown tremendously. In the early phase of development, emphasis was placed on commodity plastics; large volume, relatively low cost materials. During this period, polymer production was
perfected, monomer syntheses were optimized and the availability of
other monomers was secured. Since the middle 1970's, a substantial
change of direction has occurred. This was first noticed in polymer
research. Emphasis shifted from polymers of high volume and relatively low return to materials possessing a high degree of sophistication,
i.e. tailor-made materials that are sometimes termed "specialty polymers". These changes required and still require scientists that are
knowledgeable in the broad subject of macromolecular science.

Progress in science often occurs in distinct stages, when people
in a developing field and their ideas reach a critical mass. In macromolecular science, we can identify specific areas that have coalesced
over the past 10 years. One area of research was that of anionic polymerization, i.e. living anionic polymerization, initially exemplified
by styrene and more recently by methyl methacrylate polymerizations.
Controlled anionic polymerization of methyl methacrylate was achieved
by the so-called group transfer process. This technique has allowed
for the preparation of living methacrylate polymers, block polymers
with controlled block lengths, macromonomers and their copolymers,
star polymers, specially designed dispersing agents, and microgels.
All of these developments flourished, in part, because methacrylates

(a) Department of Chemistry, Columbia University, New York, NY 10027;
(b) Marshall Laboratories, E.I.du Pont de Nemours & Co., Philadelphia,
PA 19146; (c) Department of Chemistry, Faculty of Engineering Science,
Osaka University, Toyonaka, Osaka 560, Japan.

B. C. Anderson · Y. Imanishi (Eds.)
Progress in Pacific Polymer Science
© Springer-Verlag Berlin Heidelberg 1991

and their polymers are a pivotal part of modern paint technology
which requires high solid contents and/or low solvent use at low
viscosity.

Cationic polymerization has also progressed considerably in
the direction of controlled living polymerization. For example, advances have been made in the polymerization of isobutylene and vinyl
ethers. These include controlled initiation processes, the use of ini-
fers, binifers, trinifers, the preparation of bifunctional monomers,
macromonomers, functional monomers and telechelic polymers based on
isobutylene and the vinyl ethers.

Functional olefins and epoxides, where a functional group is separated from the polymerizable moiety by a spacing arm, have been polymerized via coordinative-anionic and coordination polymerization processes. Metathesis polymerization has recently attracted a considerable
amount of interest as well.

II. FUNCTIONAL POLYMERS

A number of efforts in our laboratory have focused on functional
polymers. We have been active in this arena for about 20 years. Originally, the term "functional polymers" was reserved for those macromolecules that possessed specific or unique functional groups in their
structures. More recently, a second definition has been used for this
term, namely, polymers that perform definitive or definable functions,
e.g. polymeric membranes, polymeric drugs, etc.

A. Polymer-Bound Stabilizers

For the last two decades, we have been studying the concept of
polymerizable, polymeric or polymer-bound stabilizers. We have focused our efforts on non-leachable ultraviolet (UV) stabilizers with
a major emphasis on 2(2-hydroxyphenyl)2H-benzotriazole derivatives
and have reported extensively on this subject [1].

2(2,4-dihydroxyphenyl)2H-benzotriazole (BDH) has been introduced
into polymer chains, for example, in poly(methyl methacrylate), by:
(1) first preparing the polymerizable functional monomer and then
copolymerizing it, or (2) directly linking the triazole moiety to
a reactive group present in the polymer chain [2]. Equation 1 details
the reaction of BDH with glycidyl methacrylate giving an aliphatic

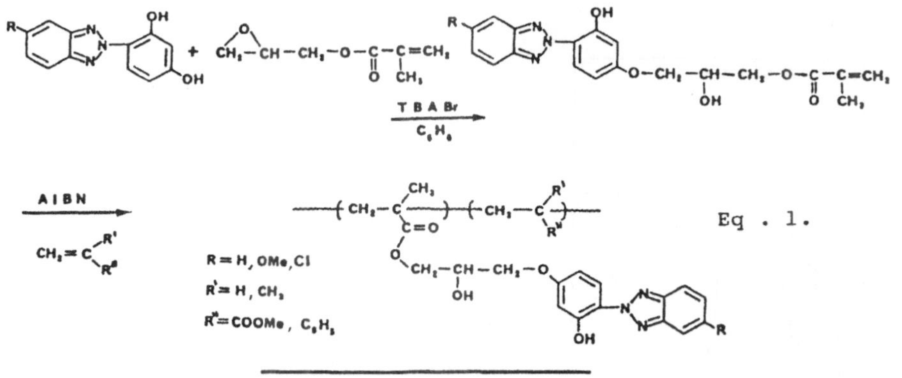

R—[benzotriazole]—N / OH + CH$_2$—CH—CH$_2$—O—C—C=CH$_2$ → R—[benzotriazole]—N / OH—O—CH$_2$—CH—CH$_2$—O—C—C=CH$_2$

$$\text{Eq . 1.}$$

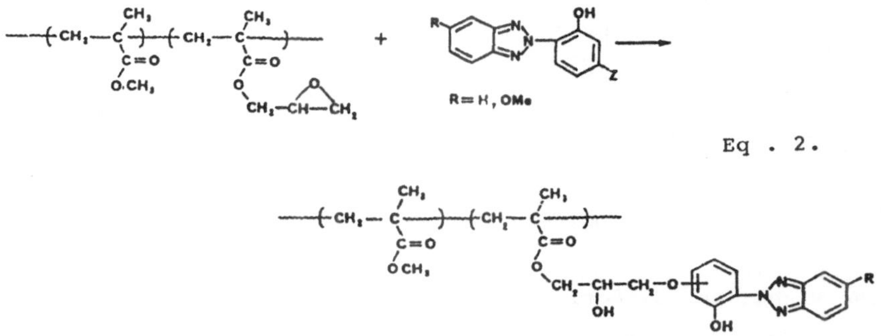

AIBN

CH$_2$=C R' R''

R = H, OMe, Cl
R' = H, CH$_3$
R'' = COOMe, C$_6$H$_5$

—(CH$_2$—C(CH$_3$)(C=O))—(CH$_2$—C R' R'')—

—(CH$_2$—C(CH$_3$)(C=O)(O·CH$_3$))—(CH$_2$—C(CH$_3$)(C=O)(O·CH$_2$—CH—CH$_2$)))— + R—[benzotriazole] OH, Z →

R = H, OMe

$$\text{Eq . 2.}$$

—(CH$_2$—C(CH$_3$)(C=O)(ÖCH$_3$))—(CH$_2$—C(CH$_3$)(C=O))—

ester of BDH which is then readily copolymerized. Equation 2 illustrates the second approach where BDH is allowed to react with a functional group already present on the formed polymer, in this case, an
acrylic polymer possessing glycidyl groups. BDH can also be introduced as an "end group" by endcapping a methacrylate polymer chain
prepared by group transfer polymerization with a BDH moiety. It
should be noted that under carefully controlled conditions, the
less-hindered hydroxyl group at the 4-position of the phenyl ring
reacts, allowing the hydrogen-bonded 2-hydroxyl group to play a
direct role in the stabilization process. Photophysical studies of
polymers containing the 2(2-hydroxyphenyl)2H-benzotriazole stabilizing group have afforded a number of interesting results.[3]

B. Photografting of UV Stabilizers

Grafting reactions on a polymer substrate, i.e. on the surface
of a polymer film, can be carried out via radical generation processes. For example, hydrophobic hydrocarbon surfaces can be made more
hydrophilic by a photografting reaction with acrylic acid. The photografting of a UV screen-type stabilizer would be highly desirable,

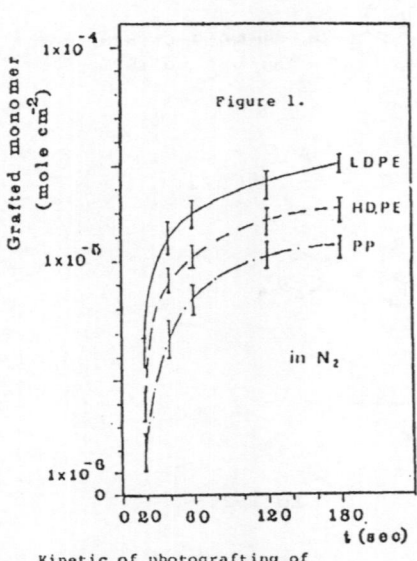

Kinetic of photografting of
monomeric photostabilizers onto polyolefins
under UV irradiation (254 nm)

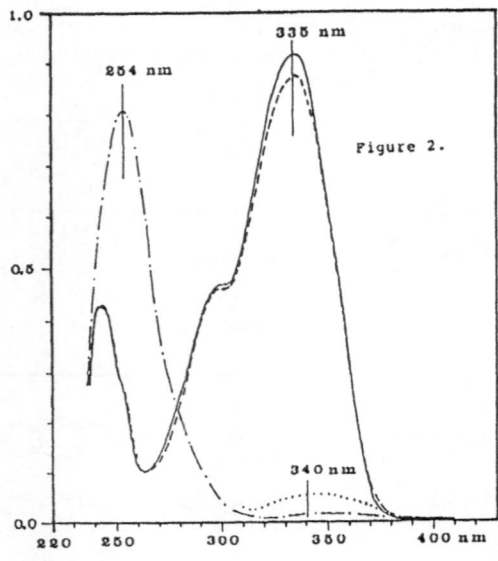

Absorption spectra of monomeric photostabilizers
$(3 \times 10^{-5}M$ in $CHCl_3)$: (——) BDHA; (--) BDIIM and
benzophenone $(3 \times 10^{-5}M$ in $CHCl_3)$ (—··—).

placing the stabilizer directly where needed at the polymer surface.
We have now shown that 2(2-hydroxy-5-vinyl(or isopropenyl)phenyl)2H-
benzotriazole can be efficiently photografted onto polyethylene (both
high and low density) and polypropylene surfaces (Figure 1) [4] providing that the photosensitizer employed in the reaction, i.e. benzophenone, exhibits an absorption maximum at the absorption minimum of
the photostabilizer (Figure 2).

C. Spacer Groups in Polymer Structures

In the last few years, it has been shown that the reactivity
of pendent groups present in functional polymers can often be enhanced through the use of properly designed spacing-arms. With sufficiently long linking arms, we have found that the reactivity of a
functional group is not perturbed by the presence of the polymer
chain. In naturally occurring polymers like proteins, pendent functionality, e.g. amino or carboxyl moieties, is often well separated
from the polymer backbone. This structural arrangement is essential
for the proper function of many biopolymers.

We have examined the behavior of functional monomers and polymers having spacer groups--specifically, functional olefins and their
polymerization. We have also studied functional epoxides and the corresponding polyoxyethylenes and their copolymers. Functional polyolefins
with terminal carboxylate groups at the end of polymethylene spacer

groups have been evaluated, particularly those having three or eight
methylene groups as spacing elements. Functional epoxides with terminal carboxylate groups were polymerized via a modified Vandenburg
initiating system. In these polymerizations, monomeric epoxides,
again with three or eight methylene spacer groups, were readily
polymerized and copolymerized as long as the spacing arm present
contained three or more methylene units. These studies are significant because they allow for the synthesis of a new class of ionomers
based on functional olefins and epoxides. An examination of the
solid-state properties of these new ionomers showed that cluster
formation, and consequently, full ionomeric properties, are already
present at the level of 1.5 mole percent of ionic groups. In contrast,
about 10 to 12 mole percent of carboxylate groups are required in polyolefins where the carboxylate moieties are directly attached to the
polyethylene backbone.

At present, we are examining reactions on functional polymers
having three different backbone structures: polyolefins, polyoxyethylenes and polymethacrylates. These structures provide amorphous
polymers with different basic Tg's. We have developed an effective
method to introduce a variety of groups via the free carboxylic
acids using various imidazolyl derivatives. Our objectives include
the incorporation of mesogenic groups, electrically conductive
groups and groups with nonlinear optical properties into polymer
structures.

III. NEW MONOMERS AND POLYMER INTERMEDIATES

An interesting development in monomer synthesis and subsequent
polymerization was disclosed earlier this year [5]. A synthetic
route for preparing giant rings containing about 20 to 40 monomer
units was devised. These macrocycles could be polymerized directly
to high molecular weight polymers (Equation 3).

It has been recognized for decades that medium or large sized
rings can only be prepared by the so-called "high dilution" technique.
Macrocycles of cyclic bisphenol A-carbonates can now be prepared from
bischloroformates by manipulating reaction conditions. The reaction
can be run at 20%, 30% and even higher concentrations with yields
as high as 90% of cyclic material. These macrocycles can be readily

Equ. 3.

X = $-(CH_2)_{\overline{m}}$ or $-\bigcirc-C(CH_3)_2-\bigcirc-$
m = 2, 4, 6, 10

Y = $-(CH_2)_{\overline{6}}$, $-\bigcirc-CH_2-\bigcirc-$, $-\bigcirc-$ or $-\bigcirc\hspace{-1em}\bigcirc$

Equ. 4.

polymerized with basic initiators to yield high molecular weight polymers. Macrocycles other than aromatic carbonates, e.g. arylates, etherimides, etherketones and ethersulfones have been prepared and polymerized.

In another development, the use of diketenes as polymer inter-mediates that are similar in reactivity to the aromatic diisocyanates has been described. A particularly interesting diketene is anthracene-9,10-diketene, shown in Equation 4. Polymers, polyamides and poly-esters, especially those from polyetherglycols and polyesterglycols, have been prepared. These diketenes, as reactive polymer intermed-iates, are also of potential interest for chain extension reactions on hydroxyl-terminated polyesters.

IV. HELICAL POLYMERS

Polymers that possess ordered or well defined helical structures are a subject of considerable experimental and theoretical interest [6]. Helical order is encountered among a wide variety of naturally occurring and synthetically derived macromolecules and is often responsible for the physical and biological properties of these systems. Isotactic polypropylene, for example, crystallizes in a 3/1 helical array with a repeat unit of 6.65 Å. This structure is comprised of a racemic mixture of left- and right-handed helical forms. Consequently, polypropylene lacks macromolecular asymmetry and is optically inactive in the solid state.

For the past three decades, we have been interested in preparing polymers having a so called "perfect helix", i.e. a single helical screw direction along the polymer backbone [7]. We have utilized polyaldehydes for our studies as these macromolecules have a strong propensity to form helical structures in the solid state. The simplest polyaldehyde, polyoxymethylene, is known to crystallize in a 2/1 helix. Higher polyaldehydes are isotactic, and typically display 3/1 to 4/1 helical order--depending on the size of their pendent side groups. Here, helical geometry, i.e. helix pitch, can be conveniently fine-tuned by altering the steric bulk of the side group in these macromolecules.

A. Helical Oligomers

Studies in our laboratory have shown that trifluoroacetaldehyde (fluoral) and trichloroacetaldehyde (chloral) are readily polymerized with a host of alkoxide initiators. Efforts to probe the stereochemistry associated with the early stages of helix development in these polymers have led to the preparation of a series of t-butoxy-initiated acetate-terminated fluoral and chloral oligomers [8]. Analysis of the fluoral oligomer mixtures by potassium ionization of desorbed species (K+IDS) mass spectroscopy has clearly shown the presence of a range of oligomers up to the decamer species (Figure 3). Analysis of the mixture by gas chromatography (GC) (Figure 4) has indicated that a number of diastereomers are present for each oligomer species, suggesting that the growth of the polyfluoral chain occurs with a low level of stereoselectivity for each step. Similar results have been obtained with the chiral initiator, lithium (-)borneoxide.

Analysis of chloral oligomers prepared under similar conditions by K+IDS mass spectroscopy [9] has again shown the presence of a range of oligomeric species, including the heptamer adduct at 1187 daltons (Figure 5). Here, assignments in the chromatogram are consistent with chlorine isotope distribution patterns obtained by the mass spectral technique. Interestingly, GC analysis of the chloral oligomer mixture has indicated that for the higher oligomeric adducts, e.g. trimer, tetramer, pentamer, only one major diastereomeric product is formed (Figure 6). These data suggest that, in stark contrast to the fluoral case, the growth of a polychloral chain becomes increasingly stereoselective after the initiation event, limiting the number of isomeric products formed during the reaction. This behavior, in agreement with theoretical studies [10], is probably a result of the larger steric bulk associated with the trichloromethyl side group in this system.

Fractional distillation of the chloral oligomer mixture under reduced pressure has allowed for the isolation of the unimer, dimer, trimer and higher oligomer fractions. Single crystals of the major isomer from each fraction, grown slowly from solutions of methanol, were analyzed by X-ray crystallography and 2D NMR spectroscopy [11]. These studies have helped to shed additional light on the structure of the oligomeric products isolated from the oligomerization reactions. We can now show that the individual chloral moieties that comprise the major isomers (for trimer to pentamer) are all linked together in meso fashion. Indeed, the tetramer exists as the first turn in the polymer's 4/1 helical screw, with a 4.8 Å repeat. This value compares favorably with the 5.2 Å repeat distance determined for the polymer. Thus, for the chloral case, our studies suggest that the embryonic stages of helix development seem to occur very early in the growth of the polymer chain.

Similar studies are presently underway with oligomers of triphenylmethyl methacrylate [12]. Like polychloral, poly(triphenylmethyl methacrylate) is known to possess a helical structure, although its exact geometry remains to be fully characterized. These studies are likely to compliment the work done on the chloral system.

B. Optically Active Helical Polymers with Macromolecular Asymmetry

Helical polymers with optical activity arising entirely from macromolecular asymmetry are a relatively new development and represent an exciting area for future study. From our oligomer studies,

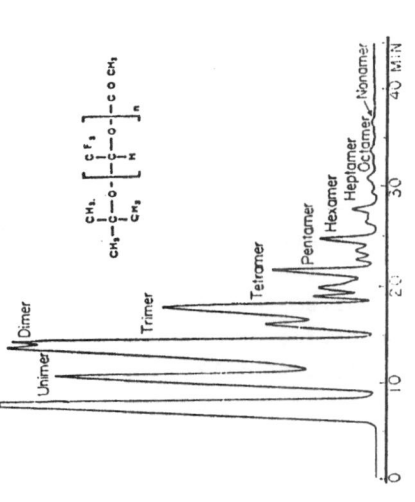

Figure 3. K+IDS Mass Spectrum of Tertiary Butoxide Initiated Acetate Endcapped Fluoral Oligomers

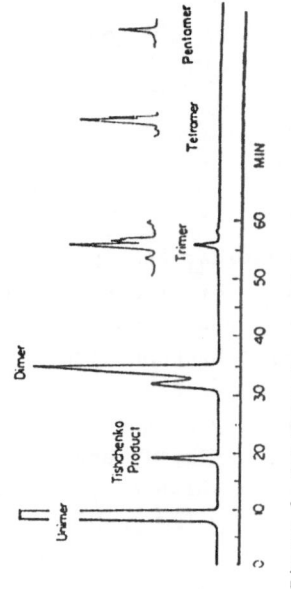

Figure 4. Gas Chromatogram of Tertiary Butoxide Initiated Acetate Endcapped Fluoral Oligomers

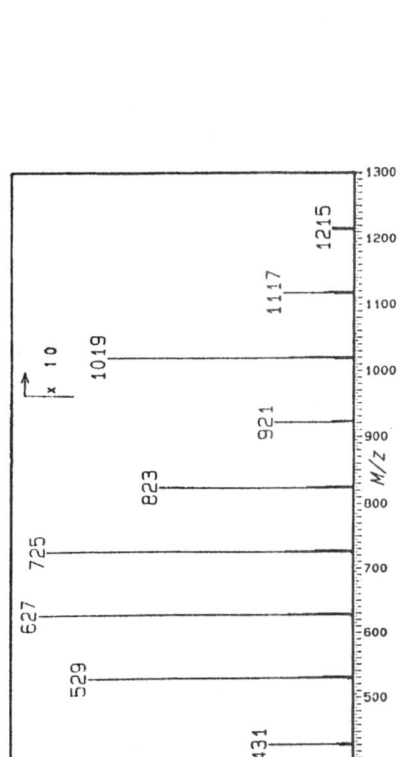

Figure 5. K+IDS Mass Spectrum of Tertiary Butoxide Initiated Acetate Endcapped Chloral Oligomers

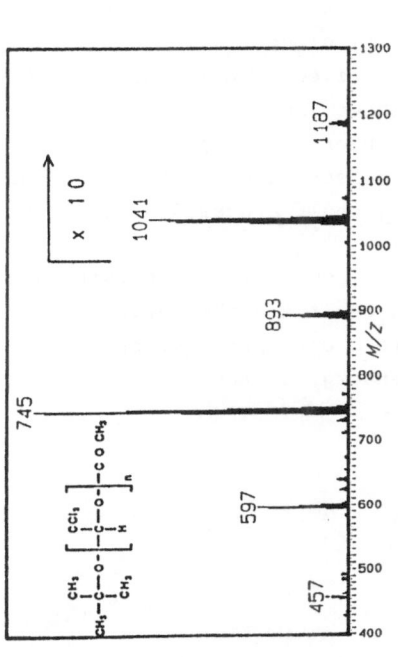

Figure 6. Gas Chromatogram of Tertiary Butoxide Initiated Acetate Endcapped Chloral Oligomers

$$\text{*I}^- \underset{\text{Cl}_3\text{C}}{\overset{\text{H}\cdots}{\diagdown}} \text{C}=\text{O} \quad \begin{cases} \longrightarrow \text{IM}_1^- \xrightarrow[(m)]{+\text{M}_2} \text{IM}_1\text{M}_2^- \xrightarrow[(m\ldots)]{+\text{M}\ldots} \text{OOOO} \\ \longrightarrow \text{IM}_1^- \xrightarrow[(m)]{+\text{M}_2} \text{IM}_1\text{M}_2^- \xrightarrow[(m\ldots)]{+\text{M}\ldots} \text{OOOO} \end{cases}$$

Equ. 5.

it is clear that polychloral, and not its fluoro analogue, is best suited for efforts of this type. We have used chiral initiators to induce the formation of a single helical screw sense along the poly-chloral backbone (Equation 5). Polymers prepared in this way are optically active, and their chiroptical properties are a result of asymmetry at the macromolecular, i.e. helical, level [13]. In this re-gard, optically active polychloral is somewhat similar to helicene ol-igomers which also display optical rotations due to molecular asymmetry.

The asymmetric polymerization of chloral can be effected with chiral initiators, or with achiral initiators that are coupled to chiral counterions. In the former case, it appears that the formation of a single screw sense may be limited by a low level of stereoselect-ivity in the initiation step. We are again probing the stereochemistry of the initiation and early propagation steps through the analysis of chloral oligomers, this time, prepared with the chiral lithium alkox-ide derived from (-)borneol. Oligomers up to the hexamer adduct have now been isolated and examined. We believe that the preparation of polychloral having an exclusive helical screw direction will require the selection of a chiral anionic initiator with a proper level of steric bulk centered near its anionic site.

Acknowledgements. This work was supported in part by the Herman F. Mark Chair and the National Science Foundation. We would like to express our appreciation to L. Kiliman, B. Ranby, B. Philipp, A. Eisenberg, J. Lucki and J. Rabek for their cooperation.

REFERENCES

1. O. Vogl and A. Sustic, Makromol. Chem. Makromol. Symp., 12, 351 (1987)

2. A. Sustic, C.L. Zhang and O. Vogl, Polymer Preprints, ACS Division of Polymer Chemistry, 28(2), 226 (1987)

3. a.) K.P. Ghiggino, A.D. Scully, S.W. Bigger, M.D. Yandell and O. Vogl, J. Polym. Sci., Polymer Lett. Ed., 26, 505 (1988) b.) A.D. Scully, S.W. Bigger, K.P. Ghiggino and O. Vogl, J. Photochem. Photobiol., in press

4. J. Lucki, J. Rabek, B. Ranby, B.J. Qu, A. Sustic and O. Vogl, Polymer, in press

5. J.W. Verbicky Jr., Pacific Polymer Preprints, 1, 19 (1989)

6. a.) O. Vogl and G.D. Jaycox, Polymer, 28, 2179 (1987); b.) R.A. Hegstrom and D. K. Kondepudi, Scientific American, 262, 108 (1990)

7. O. Vogl, The Chemist, 62, 16 (1985)

8. a.) J. Zhang, G.D. Jaycox and O. Vogl, Polymer J., 19, 603 (1987); b.) J. Zhang, G.D. Jaycox and O. Vogl, Polymer, 29, 707 (1988) c.) G.D. Jaycox, K. Hatada, F. Xi and O. Vogl, Pacific Polymer Preprints, 1, 267 (1989); d.) G.D. Jaycox, F. Xi, O. Vogl, K. Hatada, K. Ute and T. Nishimura, Polymer Preprints, ACS Division of Polymer Chemistry, 30(2), 167 (1989)

9. W.J. Simonsick Jr., M. Fulginiti, F. Xi and O. Vogl, Pacific Polymer Preprints, 1, 269 (1989)

10. A. Abe, K. Tasaki, K. Inomata and O. Vogl, Macromolecules, 19, 2707 (1986)

11. a.) O. Vogl, F. Xi, F. Vass, K. Ute, T. Nishimura and K. Hatada, Macromolecules, 22, 4658 (1989); b.) K. Hatada, K. Ute, T. Nakano, F. Vass and O. Vogl, Makromol. Chem., 190, 2217 (1989); c.) K. Ute, T. Nishimura, K. Hatada, F. Xi and O. Vogl, Makromol. Chem., in press; d.) K. Ute, M. Kashiyama, K. Oka, K. Hatada and O. Vogl, Makromol. Chem., Rapid Comm., in press.

12. Y. Okamoto, E. Yoshima, T. Nakano and K. Hatada, Chem. Lett., 1987, 759; b.) K. Hatada, K. Ute, K. Tanaka and T. Kitayama, Polymer J., 19, 1325 (1987); c.) K. Ute, T. Nishimura, Y. Matsuura and K. Hatada, Polymer J., 21, 231 (1989)

13. L.S. Corley, G.D. Jaycox and O. Vogl, J. Macromol. Sci., Chem., A25, 519 (1988) and references cited therein

New Frontiers of Polyolefins and Advanced Olefin-Based Alloys

P. Galli and J.C. Haylock

Himont Incorporated

2801 Centerville Road

Wilmington, DE 19850

Since the discovery of Ziegler-Natta catalysts and the first commercial exploitation and development of polypropylene, research has continued to understand and exploit the fundamental, molecular basis of this unique catalyst system. This understanding has expanded the potential of the polymerization process to allow development and production of olefinic polymers with unique combinations of properties never envisaged by its original discoverers. The initial titanium chloride/aluminum alkyl catalyst for the polymerization of propylene gave poor stereoregularity, resulting in a low percentage of the commercially viable isotactic polymer. The activities of the Natta Research School were directed towards a more stereospecific system and led to the first industrial process for the production of polypropylene in 1957. Although the stereoregularity of these polymers was sufficient, the efficiency of the $TiCl_3$ catalysts was low (3000-5000 g. polymer/g. Ti), and extracting of the polymer was required in order to remove catalyst residues and provide a practical, useful resin.

Improvements were later made in productivity and stereo-regularity of the $TiCl_3$/aluminum alkyl based catalysts with the addition of Lewis-base donors. This second generation catalyst system gave high stereospecificity and activity three to five times higher than the first generation $TiCl_3$ based catalysts. This increase in activity, however, was still not sufficient to avoid the need for polymer purification and such second generation, "self-supported" catalyst polymerization still requires washing in order to make a commercially viable polymer.

B. C. Anderson · Y. Imanishi (Eds.)
Progress in Pacific Polymer Science
© Springer-Verlag Berlin Heidelberg 1991

A real innovation in polypropylene production occurred with the development of a third generation catalyst system which introduced the technology of a catalyst support. In such a system, the Ti active centers are dispersed throughout the support and are all essentially available for the polymerization reaction. In the second generation, self-supported catalyst processes, less than 1% of the titanium active centers are available to take part in the polymerization reaction and elimination of the 99% of titanium that was present but inactive in the catalyst system reduced the residual catalyst level in the polymer to such a low level that the need for removal before melt processing and product fabrication was eliminated. Elimination of the need for catalyst removal led to significant process simplification, resulting in a 20% reduction in investment cost, as well as significant reductions in both maintenance and labor, raw materials and utilities consumption costs. Also eliminated was the need for treatment of aqueous wastes from the removal of catalyst residues, with a consequent reduction in potential pollution problems. (Figure I).

Figure I

COMPARISON OF DIFFERENT CATALYSTS FOR PROPYLENE POLYMERIZATION

Catalyst System	Catalyst Performance Polymerization In Hexane, 70° C, 7 Bar, 4 Hours				
	Activity		Isotactic Index (% Wt)	Polymer Morphology	Process Requirements
	(Kg PP/g Cat.	Kg PP/g.Ti)			
1st Generation TiCL$_3$AL ET$_2$CL	0.8 - 1.2	3 - 5	88 - 91	Irregular powder	Need of purification and atactic removal
2nd Generation TiCL$_3$AL ET$_2$CL	3 - 5	12 - 20	95	Regular Powder	Need of purification No atactic removal
3rd Generation TiCL$_4$,ED,MgCL$_2$/AL R$_3$	5	300	92	Irregular Powder	No purification, need of atactic removal
Super active 3rd gen. TiCL$_4$,ED,MgCL$_2$/AL R$_3$	20	800	96 - 98	Particles with regular shape & adjustable size and PSD.	No purification, no atactic removal, no pelletization.
	Bulk Polymerization, 70°C, 4 Hours				
	60	2400	90 to > 99		

Ziegler-Natta catalyst development did not stop with the introduction of the high yield, high isotactic third generation catalyst process. The superactive catalyst has provided the basis for a versatility in product and process development never previously thought possible. This system meets the four major objectives vital for the control of polyolefin properties.

(i)   A catalytic activity so high that removal of catalyst residues is not required without sacrificing polymer and product properties.

(ii) The ability to produce a wide range of molecular weight distributions, from very narrow to very broad.

(iii) The ability to control the degree of stereo-regularity in the polymer without the need for subsequent costly operations to remove unwanted polymeric fractions.

(iv) The ability to obtain, in polymerization, a granular polymer particle with a controlled morphology, particle size, distribution, density and form suitable for product fabrication without the need for melting, extrusion and pelletization.

It is the development of this granular polymer particle that has allowed production of unique, reactor made polyolefin alloys and blends without melt compounding.

In this unique, third generation, superactive catalyst, the
physical and chemical structure of the active sites on the support
provide a three-dimensional "catalyst architecture that is duplicated
by the growing polymer particle. As polymerization continues, the
active sites spread out from the original small catalyst particle in
a kind of "controlled explosion", each site becoming accessible to
additional monomers. Magnesium chloride, which constitutes the best
type of activating support, is unusual in that it can be shaped into
uniform spherical or globular particles having different sizes.
Replication of the catalyst particle occurs when the mechanical
strength of the catalyst granule is in balance with the catalyst
polymerization activity. If reactivity is too high, an uncontrolled
"explosion" occurs, generating a fine polymer powder. If the
mechanical strength of the catalyst particle is too great, a low
level of reactivity occurs because the internal active sites cannot
generate polymer for lack of space. Good replication only occurs
when the mechanical strength and the polymerization reactivity of
the supported, superactive third generation catalyst are well
balanced. (Figure II).

Figure II a

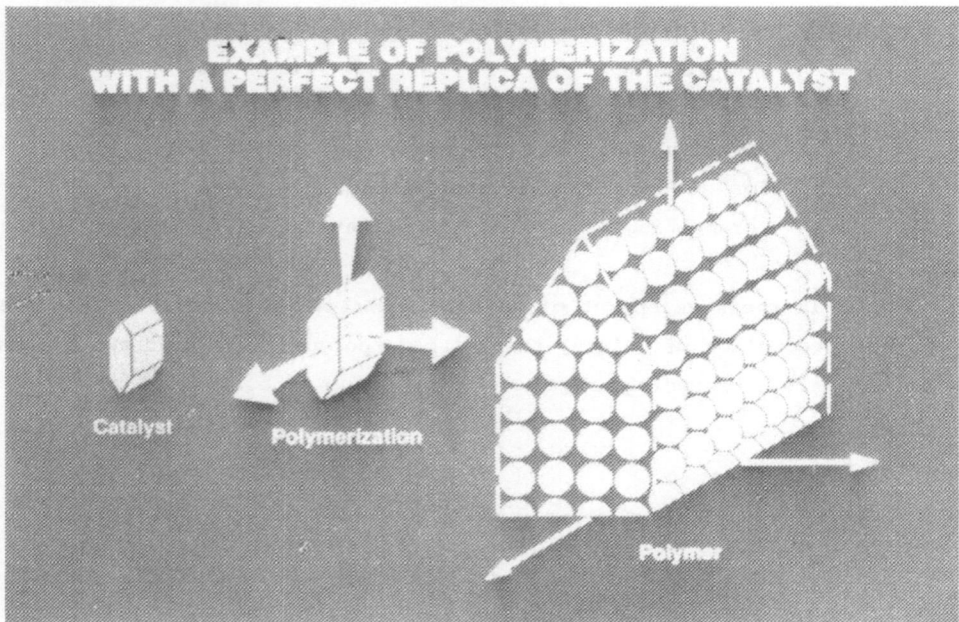

Figure II b

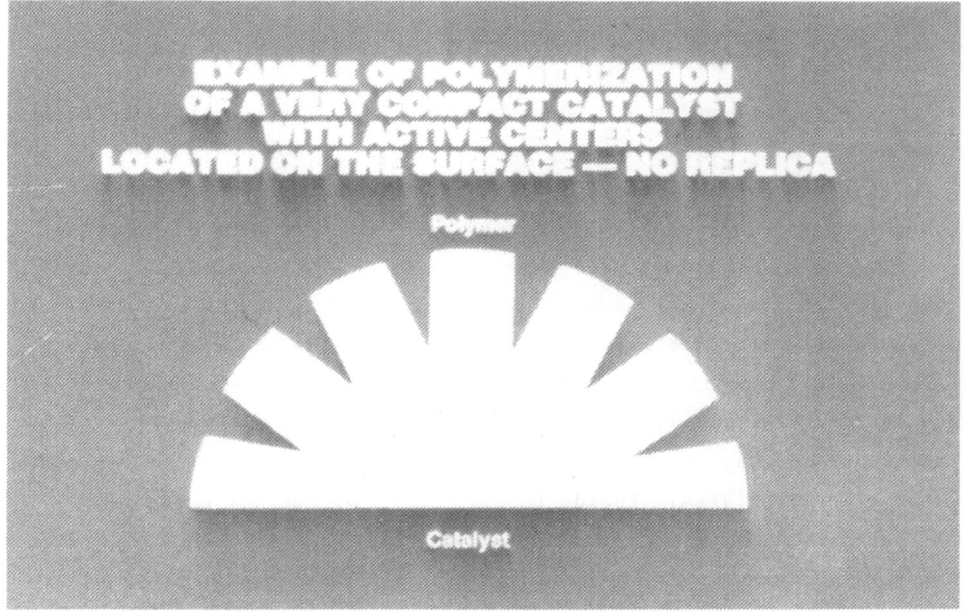

The effectiveness of the third generation, superactive Ziegler-Natta catalyst is dependent on reproducible dispersion of the active sites on the support as well as the chemical and physical properties of the support itself. The five basic indispensable requirements for this superactive catalyst are:

(i)   High surface area.

(ii)   High porosity, with a large number of cracks evenly distributed throughout the mass of the granule.

(iii)   High enough mechanical strength to withstand mechanical processing, but low enough to allow the force developed by the growing polymer to break down the catalyst into the microscopical particles that remain entrapped and dispersed in the polymer.

(iv)   Homogenous distribution of the active centers.

(v)   Free access of the monomers to the innermost regions of the catalyst.

When such conditions are met, it is possible for the polymer to grow uniformly and progressively on each active center, replicating the shape of the catalyst.

It is the understanding and control of all these interdependent phenomena that has been the basis of the continued research in Ziegler-Natta catalysis. A sufficient degree of understanding and control has now been achieved, permitting the production of the "ideal polyolefin catalyst" and the development of a polymerization process that can engineer the molecular structure of the growing olefinic polymer chain to build into the final resin a wide range and unique balance of properties. Himont's Spheripol process takes full advantage of this catalyst technology. As the polymerization takes place, a growing skin of polymer is formed from the expanding catalyst active sites. Inside this solid skin, further polymerization can take place, controlled by the monomer diffusion rate. The most favored active centers, from a diffusion point of view, are located on the external layer of the catalyst particle, and generate polymer at the maximum polymerization rate. The polymerization rate will progressively decrease in the various catalyst granule layers from the external surfaces to the core. The thus-generated first polymer granules assume the shape of an empty shell; and, as the surface layer polymerization rate declines, the second catalyst layer's active centers became the most-favored polymerization sites, giving rise to a layered, lattice structure for the growing polymer particle, somewhat like an onion. (Figure III).

Figure III

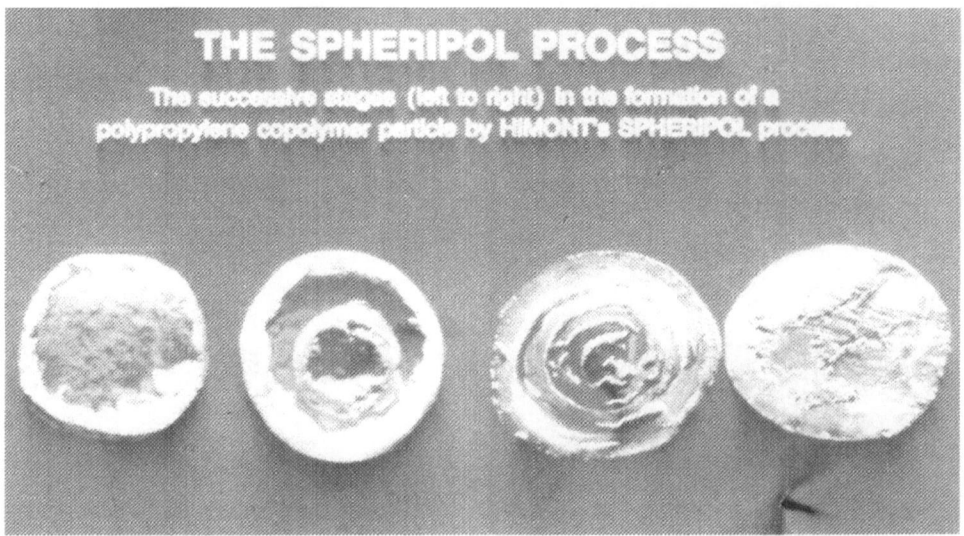

The fundamental, morphological importance of this catalyst
and polymerization mechanism is that the formation of an outer,
continually growing polymer skin provides a reaction bed within
which copolymer, terpolymer or multimonomer olefinic polymer systems
can be produced through the introduction of other monomers during
the polymerization process. It is thus possible to generate in the
reactor in a spherical form, a polymer alloy suitable for subsequent
processing and fabrication, with an extremely wide range of
properties, dependent on the morphology and chemical composition of
the monomers introduced at different stages in the polymerization
process. Figure (IV) gives examples, in the case of homopolymer
production, of perfect replication of the original catalyst particle,
magnified 40-50 times. Figures (V) & (VI) diagramatically illustrate
how the introduction of other monomers, such as ethylene, can
produce an ethylene-propylene rubber phase finely dispersed inside

a polypropylene matrix. Figure (VII) illustrates diagramatically, the truly remarkable virtuosity of this "ideal catalyst" where a third polymeric phase such as polyethylene, is produced and finely dispersed inside an ethylene-propylene rubber phase which is, in turn, finely dispersed in the spherical polypropylene matrix.

Figure IV

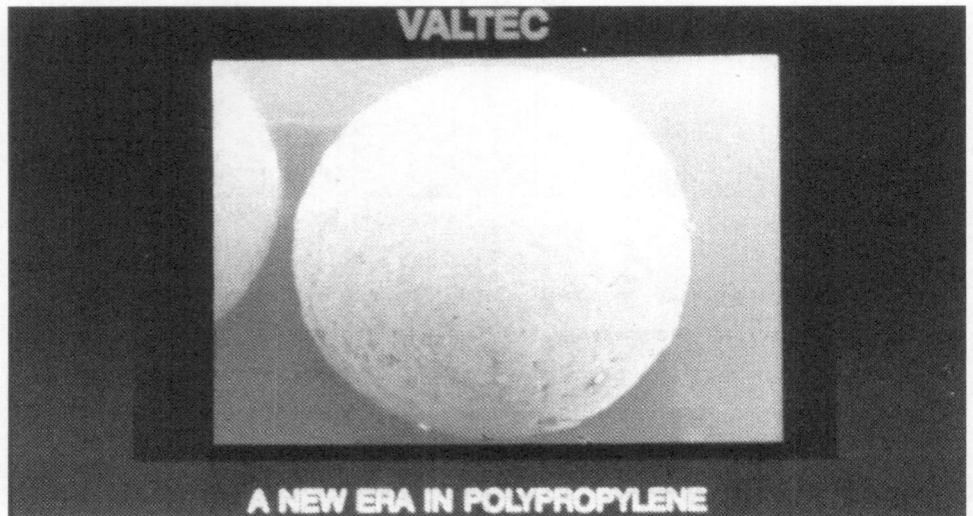

Figure V

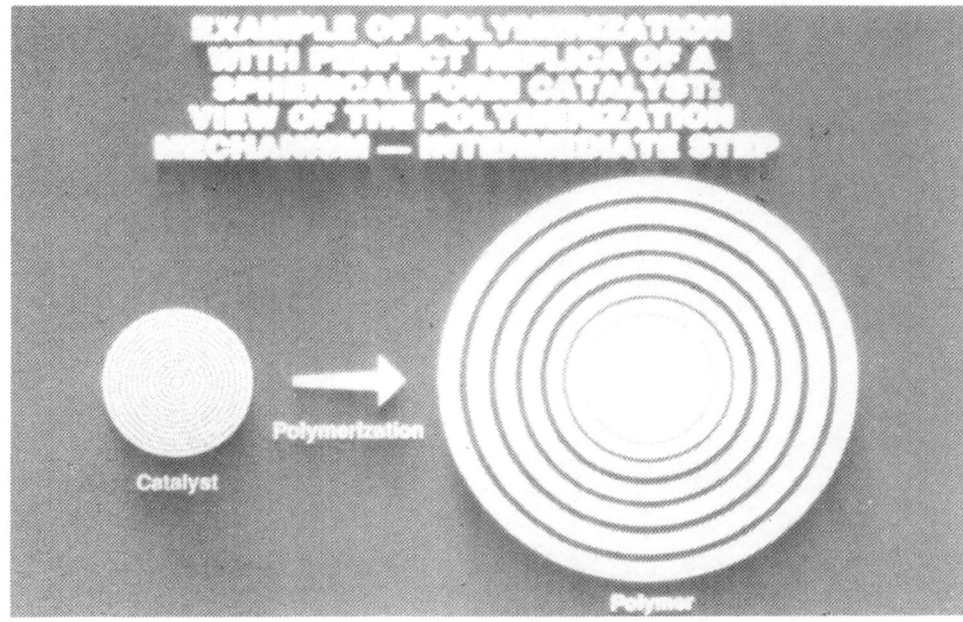

Figure VI

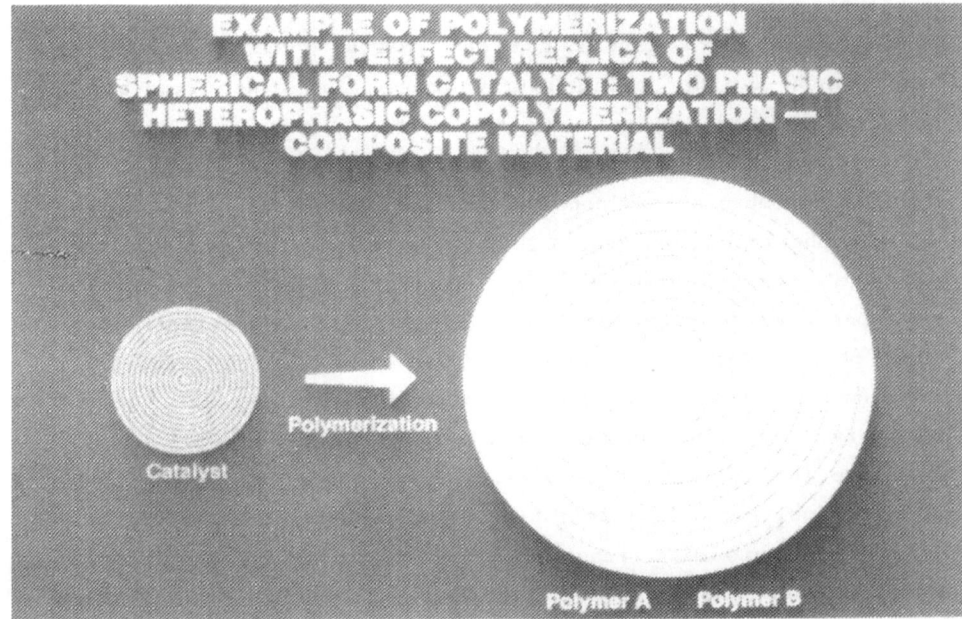

Figure VII

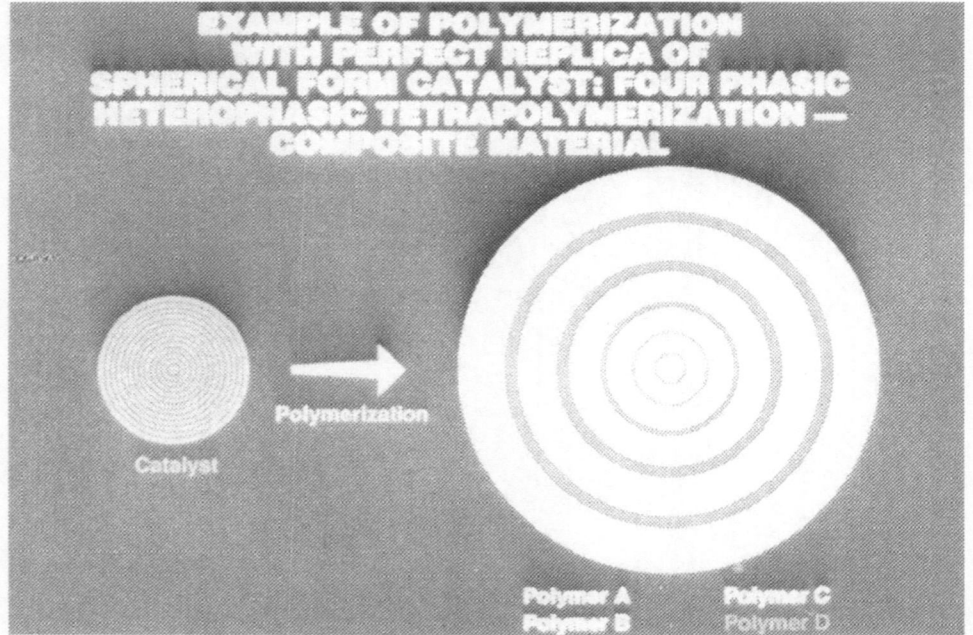

In order to fully exploit this catalyst system, a unique
commercial polymerization process was *developed*. Himont's
Spheripol process is, in practice, a two-stage hybrid process
consisting of both liquid and gas phases. Homopolymerization
takes place in the liquid slurry phase, and after removal of
unreacted monomer and solvent, the solid, porous particles pass
into the gas phase part of the process in which the introduction
of other monomers allows copolymerization to take place within the
solid, spherical particle. This growing polymer particle has
become a "reactor granule" and represents a revolution in the
development of Ziegler-Natta catalysis. Since polymerization can
take place within a solid polymer shell, the mechanical
containment aspects of the polymerization process became
secondary. Bulk, gas phase, and slurry processes are all equally
adaptable for use with this catalyst system and are chosen based
on economics, mechanical reliability and reaction control criteria
in order to maximize reactivity and productivity. It removes
almost all of the previous process constraints on Ziegler-Natta
catalysis, allowing reactor-made resins to be produced with

properties no longer limited by mechanical considerations of the process. For instance, polypropylene containing a very high content of amorphous rubber cannot be produced with conventional catalysts in slurry or gas phase reactors because the low melting, tacky nature of the rubber produced after flashing off the monomer would cause coalescence of the particles and prevent subsequent processing. In contrast, the hybrid Spheripol process allows production of a solid homopolymer skin, enclosing the tacky, amorphous copolymer rubber phase. The polypropylene skin allows subsequent processing without difficulty. Figure (VIII) gives a general comparison between product characteristics obtainable from the Himont catalyst and Spheripol process and conventional catalyst and processes used by other polypropylene producers. It clearly illustrates the dramatic differences in versatility between the two families of third generation polyolefin catalysts and processes.

Figure VIII

PRODUCT CAPABILITY COMPARISON

Type of Reactor Grades	HIMONT Spheripol	BASF Novolen	UCC/Shell Unipol
Homopolypropylene	15	5	16
Random Copolymers	8	2	4
Terpolymers	1	0	0
Impact Copolymer	15	4	4
Polyolefinic Alloys	9	0	0
Spheripol Products	(1)	-	-
Total Reactor Grades	48	11	24

(1) In commercial development stage for most grades

The capability of obtaining directly by synthesis in the reactor a polyolefin resin with a wide range of properties and with a spherical shape comparable to that of pelletized products has required the development of a new technology for adding stabilizers to the resin since they would no longer be added during extrusion and compounding into pellets. The resulting polypropylene completely formulated with the stabilizers and other additives needed for processing into a fabricated part, is marketed by Himont under the trade name "Valtec".

"Valtec" is the culmination of many years of Ziegler-Natta catalysis development and shows many processing advantages compared to conventional, pelletized polypropylene. Since the polymer resin, exiting the reactor, possesses a less organized crystalline structure compared to a melt extruded, pelletized resin, it exhibits an absence of spherulites compared to conventional polypropylene pellets. This lower degree of crystallinity requires less energy to melt the resin during a fabrication step, resulting in energy savings of up to 10%. Its packing characteristics in processing equipment have been demonstrated in commercial trials to provide higher output and better mixing homogeneity. This is illustrated in Figures (IX). Additionally lower temperature, faster processing of the polypropylene granule gives a significant reduction in degradation on melt processing compared to the more crystalline pellets.

Figure IX

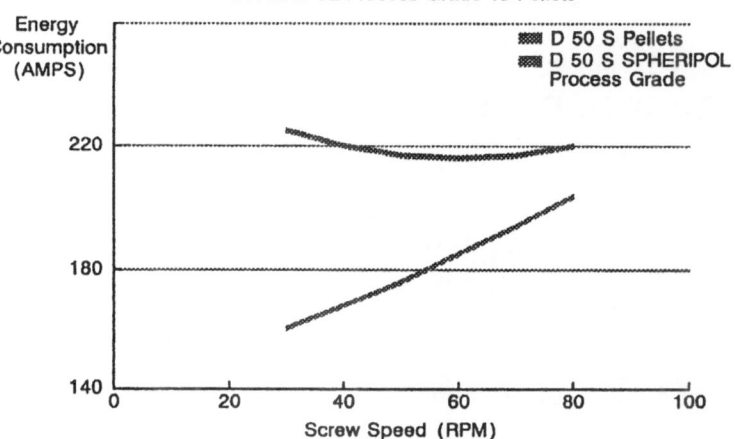

These processing advantages are a consequence of the spherical form, reactor-made product characteristics compared to a conventional pelletized product. In addition to these processing advantages, however, this unique catalyst and polymerization process technology has enabled the development of completely new grades of polypropylene with unique balance of properties, significantly enlarging the property envelope for polypropylene resins. This is illustrated in Figure (X).

64

Figure X

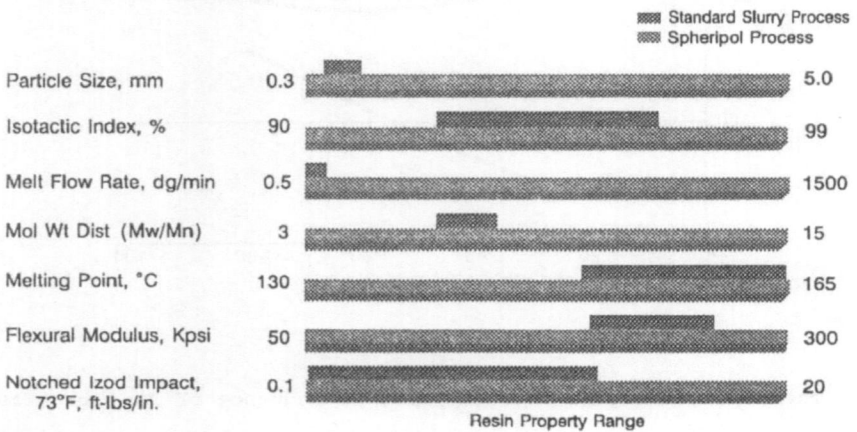

VERSATILITY OF SPHERIPOL PROCESS

Standard Slurry Process
Spheripol Process

Particle Size, mm	0.3	5.0
Isotactic Index, %	90	99
Melt Flow Rate, dg/min	0.5	1500
Mol Wt Dist (Mw/Mn)	3	15
Melting Point, °C	130	165
Flexural Modulus, Kpsi	50	300
Notched Izod Impact, 73°F, ft-lbs/in.	0.1	20

Resin Property Range

In summary, therefore, continued investment in the understanding of Ziegler-Natta catalysis has developed an "ideal catalyst" for
polyolefins. Polypropylene now stands at a crossroads with many more
new opportunities possible. The production of the final polymer
resin in the reactor without the need for subsequent remelting and
pelletizing has expanded the potential for modifying this most
versatile polymer to develop unique combinations of properties that
enable it to compete economically with a wide range of existing

polymers. The ability to use the growing polymer particle as a reaction bed in which to introduce and polymerize other monomers will enable "Valtec" olefinic resins to demonstrate properties ranging from low modulus, flexible, tough, chemically resistant polymers to tough, high modulus, high HDT engineering plastics. Heterophasic copolymer resins have been developed to compete with PVC in applications requiring high impact and clarity. Random copolymers with very low seal-initiation temperatures have been developed for film applications, and combination with other Himont proprietary technology has developed homopolymer and copolymer resins that allow polypropylene to compete economically in foam sheet and bead applications.

Finally, the vastly expanded ability to control so many aspects of the polymer composition, morphology, properties, and subsequent processability, has led to a whole new level of potential developments in the use of these materials in ways that were previously impossible. Truly, we have reached a new frontier in polyolefinic chemistry.

Recent Progress in Fluoropolymers

M. Yamabe

Research Center, Asahi Glass Co., Ltd.

1150 Hazawa, Kanagawa-ku, Yokohama 221, Japan

INTRODUCTION

The characteristics of fluoropolymers are summarized in Table 1. Thermal and chemical resistance is in general with most of plastics, elastomers and perfluorinated membranes. Weather resistance with the outdoor durability for more than 20 years is specific for fluorinated paint resins. Surface properties such as water and oil repellency are provided by acrylic polymer-based textile finishes and coatings with long-chain per-fluoroalkyl groups. Electrical properties as well as a low refractive index are important for optoelectronics applications like optical fibers.

Table 1   Characteristics of Fluoropolymers

A chronological survey of fluoropolymers is shown in Table 2. Since the discovery of polytetrafluoroethylene by Dr. Plunkett of du Pont in 1938, a great number of fluoro-polymers has been studied and has come into market. These fifty years will be roughly divided into three periods.

The first period until the end of '40s might be the era of homopolymers. Not only PTFE but also PCTFE, PVdF and PVF were commercialized.

B. C. Anderson · Y. Imanishi (Eds.)
Progress in Pacific Polymer Science
© Springer-Verlag Berlin Heidelberg 1991

Table 2   Chronology of Fluoropolymers

1930 ⟨ 1940s	PTFE(C_2F_4), PCTFE(C_2F_3Cl), PVdF(CH_2CF_2), PVF(CH_2CHF)

⟨ THE ERA OF HOMOPOLYMERS ⟩

1950 ⟨	FEP(C_2F_4-C_3F_6), ETFE(C_2F_4-C_2H_4), ECTFE(C_2F_3Cl-C_2H_4), PFA(C_2F_4-CF_2CFORf),
1970s	VdF-HFP(CH_2CF_2-C_3F_6-(C_2F_4)) TFE-P(C_2F_4-C_3H_6), TFE-PMVE(C_2F_4-CF_2CFOCF_3),

⟨ THE ERA OF COPOLYMERS ⟩

1980s	MEMBRANES THERMOPLASTIC ELASTOMER PAINT (C_2F_3Cl-CH_2CHOR) CYCLOPOLYMERS

⟨ THE ERA OF FUNCTIONAL POLYMERS ⟩

The second period from '50 to mid '70s might be the era of copolymers. Major improvements of processabilities or porperties of homopolymers like PTFE were attained in this period. The copolymers include FEP, ETFE and PFA which is a copolymer of TFE and perfluoropropyl vinyl ether. Another progress in this era is the commercialization of fluoroelastomers such the most popular elastomer as Viton[R], that is the co-polymer of vinylidene fluoride and hexafluoropropylene or sometimes with TFE as a termonomer, which was followed by tetrafluoroethylene-based elastomers with propylene or perfluoromethyl vinyl ether comonomers.

The third period since mid '70s might be called as the era of functional polymers. The commercialized examples are membranes, a thermoplastic elastomer, weather resistant paints and some cyclopolymers. The thermoplastic fluoroelastomer developed by Daikin involves a quite interesting technology. This is a very unique block copolymer with hard and soft segments based on the specially designed iodine-containing living fluoropolymer (1).

Some alternating copolymers and functional polymers mainly developed by Asahi Glass will be described in this paper.

ALTERNATING COPOLYMERS

Table 3 summarizes the performance and applications of three alternating copolymers commercialized by Asahi Glass, that is, ETFE as Aflon Cop[R] in 1971, P-TFE as Aflas[R] in 1980 and fluoroolefin and vinyl ethers as Lumiflon[R] in 1983.

Table 3 Applications of Alternating Copolymers

Polymer	Performance	Application
$-(CF_2CF_2-CH_2CH_2)-$ AFLON COP®	melt-processable cut-through resistance	Wire coating Film applications
$-(CF_2CF_2-CH_2CH)-$ $\quad\quad\quad\quad CH_3$ AFLAS®	peroxide curable elastomer resistance to harsh chemicals electrical resistivity $(10^{15}-10^{16}\Omega cm)$	Wire coating Oil well Packer Automobile Food Processing
$-(CF_2CFX-CH_2CH)-$ OR LUMIFLON®	amorphous (gloss, transparent) ambient cure weather resistant paint	Roof, Siding walls Outdoor constructions (plants, bridges etc.) Vehicles

The ETFE copolymer is a melt processable resin with a crystalline melting point of around 270°C. Its excellent cut-through resistance has made the copolymer advantageously applicable to wire coating.

The change of the comonomer from ethylene to poropylene resulted in an amorphous elastomeric copolymer with highly alternating sequences. The P-TFE elastomer is peroxide curable. With the resistance to harsh chemicals and the excellent electrical resistivity, the application field of Aflas covers flexible wire coating, oil well packers, automobile parts and food processing.

The alternating copolymer of fluoroolefin and hydrocarbon vinyl ether provides an amorphous, solvent-soluble and ambient cure type paint resin with the excellent weather resistance. Lumiflon is rapidly expanding the market of protective coating in the field of roofs, siding walls of buildings, out door constructions and vehicles.

All these copolymerization proceeds either in a bulk, solution or emulsion system with a proper radical initiator.

Ethylene-Tetrafluoroethylene Copolymer

The copolymerization behavior of E-TFE is shown in Fig. 1 (2). The curve shows rather random character with some alternation tendencies and one of the key technologies for commercialization was to isolate the ethylene unit as much as possible to reduce the content of thermally and chemically weak sequences by controlling the monomer feed. The other important technology was to overcome the thermal stress crack in the highly crystalline ETFE coplymer by incorporating small amounts of termonomer such as perfluoroalkyl ethylene (3).

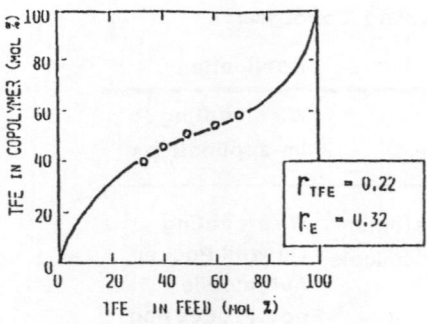

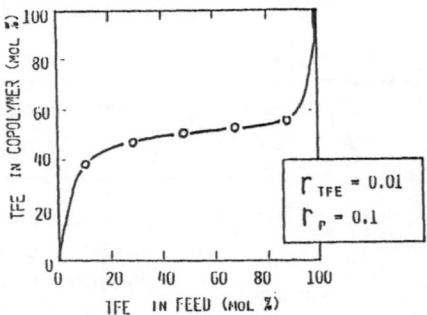

Fig. 1 E-TFE Copolymerization Curve   Fig. 2 P-TFE Copolymerization Curve

Propylene-Tetrafluoroethylene Copolymer

Figure 2 shows the P-TFE copolymerization curve. The higher tendency of alternating sequences than in the ETFE system can be seen in this figure. The key technology for commercialization of this elastomer was to establish the peroxide vulcanization system with dialkyl peroxide in the presence of polyallyl compounds like triallyl isocyanurate (4). In order to meet the requirement for automobile applications, a new grade of Aflas with the improved low temperature performance, has been developed by incorporating a suitable termonomer (5).

Fluoroolefin-Vinyl Ether Copolymer

Figure 3 shows the almost perfect alternating behavior in the copolymerization of various vinyl ethers and chlorotrifluoroethylene or tetrafluoroethylene.

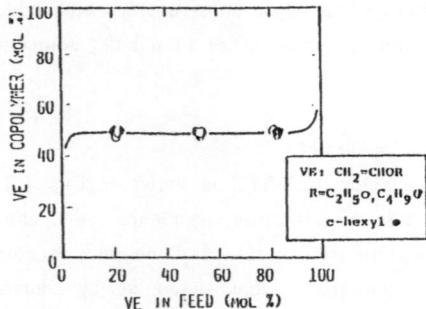

Fig. 3 (C)TFE-VE Copolymerization Curve

The copolymer has been designed and modified to meet a wide variety of applications as a highly weather resistant paint. The ambient cure system as well as the quick cure system at high temperature have been established by introducing suitable cure sites in combination with the selection of curing agents (6).

Figure 4 shows the concept of the molecular design of Lumiflon® with multiple vinyl ether comonomers, which function fo afford solubility, flexibility, curability and compatibility to pigments.

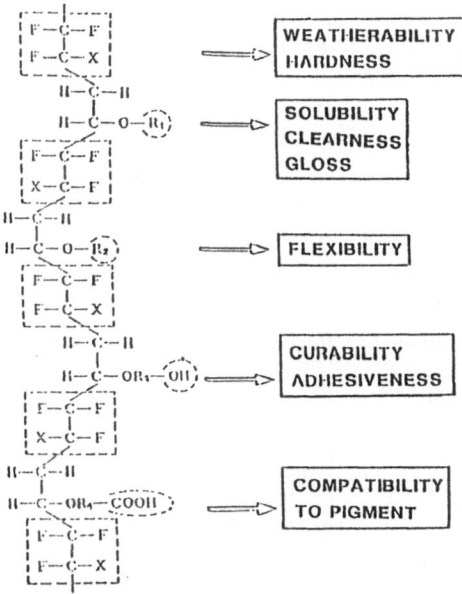

Fig. 4   Structural Features of Lumiflon® Polymer

Fluoroolefin-Vinyl Ester Copolymer

The copolymerization of fluoroolefin and vinyl ester has also been studied. one example is CTFE and vinyl pyvalate, which afford a random copolymer, as expected from these monomer reactivities shown in Fig. 5. The copolymer is featured by ultra thin film formation by casting and the resultant film shows a relatively high oxygen permeability. One practical application of thie ultra thin film supported by a porous sheet is an oxygen enricher for medical use, supplying the air with 40 wt% oxygen concentration (7).

$$-(CF_2CFCl)_x-(CH_2CH)_y-$$
$$\underset{\underset{O}{\overset{\parallel}{|}}}{O-C-C(CH_3)_3}$$

CTFE - Vinyl Pyvalate (HI-SEP®:1985)

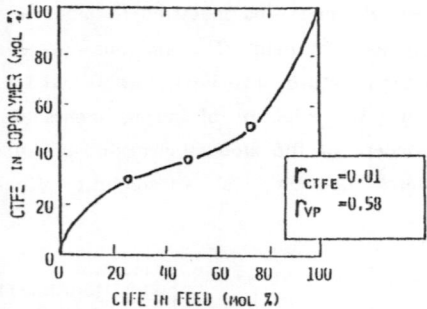

Fig. 5   CTFE-VP Copolymerization Curve

FUNCTIONAL POLYMERS

The hexafluoropropylene oxide (HFPO) chemistry developed by du Pont in 1960s has played a crucial role in the progress of functional fluoropolymers.

HFPO is prepared by the simple oxidation of hexafluoropropylene in a considerable yield (8). Furthermore, HFPO was found to rect with perfluoroacyl fluoride in the presence of fluoride ion to afford the corresponding perfluorovinyl ether (9).

$$CF_2=CFCF_3 \xrightarrow[150\text{-}160°C]{O_2/R\text{-}113} \underset{O}{CF_2\text{-}CF\text{-}CF_3}$$

(HFPO)

$$C_3F_7O\underset{CF_3}{C}FCOF \xleftarrow{HFPO} \left[CF_3CF_2CF_2O^{\ominus}\right] \xleftarrow{F^{\ominus}} $$

$$\xrightarrow{\Delta} \boxed{C_3F_7OCF=CF_2}$$

$$\xrightarrow{COF_2/F^{\ominus}} CF_3O\underset{CF_3}{C}FCOF \xrightarrow{\Delta} \boxed{CF_3OCF=CF_2}$$

$$\xrightarrow{R_FCOF/F^{\ominus}} R_F'O\underset{CF_3}{C}FCOF \xrightarrow{\Delta} \boxed{R_F'OCF=CF_2}$$

$$R_F'=CF_2CF(CF_3)OCF_2CF_2SO_2F$$

$$:(CF_2)_3CO_2CH_3$$

For example, perfluoromethyl vinyl ether can be obtained by the reaction with carbonyl fluoride. The dimerization of HFPO followed by pyrolysis, leads to the formation of perfluoropropyl vinyl ether. Moreover, functionally substituted perfluorovinyl ether can be synthesized if corresponding acyl fluoride with a functional group are used.

$$CF_2=CF_2 \begin{cases} CF_2CFOCF_3 & \longrightarrow \text{Kalrez}^{®} \text{(du Pont)} \\ CF_2=CFOC_3F_7 & \longrightarrow \text{Teflon PFA}^{®} \text{(du Pont)} \\ CF_2=CFOCF_2CF(CF_3)OCF_2CF_2SO_2F & \longrightarrow \text{Nafion}^{®} \text{(du Pont)} \\ CF_2=CFO(CF_2)_3CO_2R & \longrightarrow \text{Flemion}^{®} \text{(Asahi Glass)} \end{cases}$$

The combinations of these perfluorovinyl ethers with tetrafluoroethylene have prompted the commercialization of various fluoropolymers of industrial importance, such as Kalrez; perfluoroelastomer, Teflon PFA; thermoprocessable perfluororesin without deteriorating the excellent properties of PTFE. Nafion and Flemion are sulfonated and carboxylated perfluorinated ion exchange membranes, respectively.

Ion Exchange Membrane

The copolymerization curve of TFE and carboxylated perfluorovinyl ether is shown in Fig. 6. This is a typical behavior of copolymerization of TFE and perfluorovinyl ether. The copolymerization proceeds either in solution or emulsion system with a radical initiator similar to the alternating copolymer systems mentioned above.

$$-(CF_2CF_2)_x-(CF_2CF)_Y-$$
$$O(CF_2)_3CO_2CH_3$$

TFE-carboxylated perfluoro(vinyl ether)(M)
(Ion-Echange Membrane : FLEMION[®]:1977)

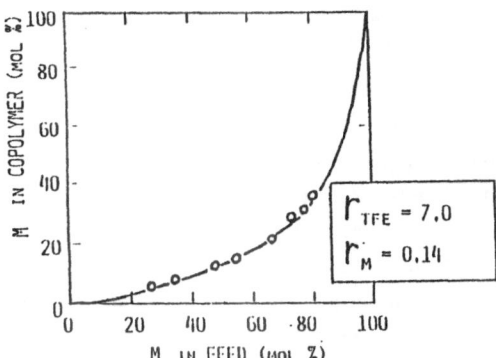

Fig. 6   TFE-M Copolymerization Curve

The carboxylated ion exchange membrane has been advantageously used to obtain up to 35 wt% of caustic soda with the current efficiency of more than 96% in the industrial ion exchange membrane process for chlor-alkali production by electrolysis of sodium chloride. (10)

A specially designed composite membrane has made the direct production of 50 wt% of caustic soda possible and the durability test of the membrane is under way in the pilot plant.

Cyclopolymerization

There were only a few reports on the attempted cyclopolymerizations so far. The

α,ω-perfluorodiolefin polymerized at an extremely high pressure under the irradiation of γ-ray only to afford a low molecular weight grease with a cyclic structure (11).

On the other hand, the α,ω-perfluorodivinyl ether polymerized smoothly, but the gel formation was observed when the yield exceeded 12%. Some cyclic structures with six- and seven-membered ring were detected in the resultant polymer mixture (12).

Previous Attempts :

$$CF_2=CF(CF_2)_NCF=CF_2 \xrightarrow[\text{13,000 ATM.}]{Co^{60}\ \gamma\text{-RAY}} -(-CF_2-CF\overset{CF_2}{\underset{(CF_2)_N}{\diagup}}CF-)-$$

GREASE

$$CF_2=CFO(CF_2)_2OCF=CF_2 \xrightarrow[\text{SOLUTION POLYM.}]{} -(CF_2-CF-CF-CF_2)- + -(CF_2-CF\overset{CF_2}{\underset{O}{\diagup}}CF-)-$$

The Present Approach :

$$CF_2=CFO(CF_2)_3CO_2CH_3 \xrightarrow{NaOH} CF_2=CFO(CF_2)_3CO_2Na \xrightarrow{\Delta} CF_2=CFOCF_2CF=CF_2$$

$$CF_2=CFOCF_2CF=CF_2 \xrightarrow[\text{IPP OR } (C_3F_7COO)_2]{25-40°C} -(CF_2 -\cdot CF\overset{CF_2}{\underset{O-CF_2}{\diagup}}CF-)_N-$$

Recently, an extremely selective cyclopolymerization has been developed by using specially designed difunctional perfluoromonomers such as perfluoroallyl vinyl ether (13). This monomer is readily derived from the carboxylated perfluorovinyl ether according to the scheme shown above, and characterized by the different reactivities of the two double bonds, that is, a vinylether type and an allyl type.

The cyclopolymerization proceeds smoothly in a bulk or solution system with a radical initiator. The cyclic structure in the resulting polymer was confirmed by NMR (14).

Based upon these findings, Asahi Glass commercialized a novel transparent perfluoropolymer, Cytop[e], with a cyclic back bone structure. Interestingly, du Pont also announced the development of cyclopolymer, Teflon AF[k]. This polymer is reportedly prepared by the copolymerization of TFE and a perfluoro dioxole (15).

The characteristics of those unique polymers are the exceptional transparency with more than 95% of UV and visible light transmission, solubility in a specific solvent, chemical and electrical properties similar to PTFE and the very low refractive index (16).

Now the market development of these new polymers has been carried out intensively and optical lens, protective coatings, substitutes for quartz glass and insulation materials would be expected as a part of practical applications.

⟨Novel Fluoropolymers with Cyclic Structures⟩

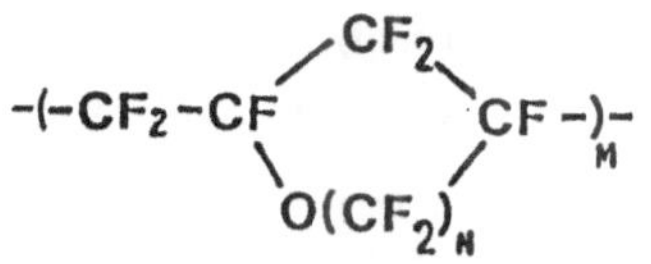

$$-(-CF_2-CF \overset{\displaystyle CF_2}{\underset{\displaystyle O(CF_2)_N}{\diagup \diagdown}} CF -)_M-$$

(CYTOP®
Asahi Glass :1989)

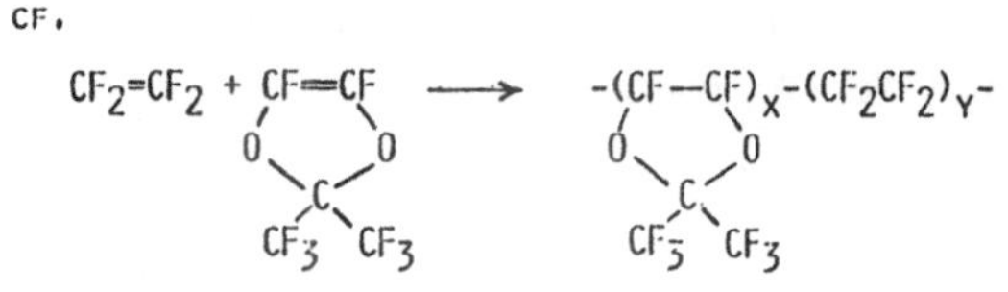

(Teflon AF®du Pont :1989)

CONCLUSION

As described herein, most of the industrially available fluoropolymers are produced by radicalic polymerization. However, many attempts have been reported as approaches to novel fluoropolymers, a part of which are listed in Table 4.

Table 4   Chronology of Fluoropolymers (2)

1930 ↕ 1940s	THE ERA OF HOMOPOLYMERS
1950 ↕ 1970s	THE ERA OF COPOLYMERS
1980s	THE ERA OF FUNCTIONAL POLYMERS

• Cyclopolymerization
• Anionic polymerization

$$CH_2=CCO_2R \ (or \ R_F)$$
with CF_3

• Condensation polymerization

Polyarylates :

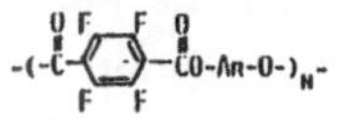

Aramides :

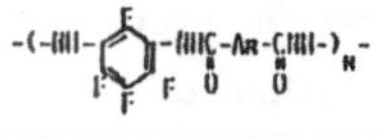

| 1990s | WHAT WILL COME NEXT? |
| | THE ERA OF MOLECULAR DESIGN |

These include anionic polymerizations of some conjugated monomers such as -tri-fluoromethyl acrylates, fluorostyrene derivatives and 1,4-perfluoro-butadiene (17), and condensation polymerization to afford such aromatic fluoropolymers as polyarylates and aramides (18). In addition, fluorinated polyimides are on a way of evaluation for practical applications in the electronics field.

Finally rapid diversification of market needs in fluoropolymers would necessarily require the era of molecular design in combination with existing technologies and emerging technologies toward the 21st century.

ACKNOWLEDGEMENT

The author would like to dedicate this paper to the late Dr. Hiroshi Ukihashi who directed Asahi Glass' R&D in fluorochemicals for a long time with his strong leadership.

References:

1 Daikin Co. (1983) Japanese Pat 58-4728
2 Nishimura H, Yamabe M (1974) Rept Res Lab, Asahi Glass Co. 24: 57
3 Asahi Glass Co. (1978) US Pat 4,123,602
4 Kojima G (1988) In: Bhowmick A, Stephens H (eds) Handbook of Elastomers. Marcel Dekker Inc NY and Basel, Chapter 14
5
6 Yamabe M et al. (1984) In: Ranfitt G, Patsis A (eds) Organic Coatings, Science and Technologies. Marcel Dekker Inc NY and Basel, 7: 25
7 Yamabe M (1988) Symposium of Fluoropolymers, Third Chemical Congress of North America, Tronto
8 Du Pont (1962) Brit Pat 904, 877
9 Tarrant P (ed) (1971) Fluorine Chemistry Reviews. Marcel Dekker Inc NY 5:77
10 Ukihashi H et al. (1986) Progress in Polymer Science 12: 229
11 Brown DW, WAll LA (1969) J Polymer Sci A2 7: 601
12 Du Pont (1968) US Pat 3,418,302
13 Asahi Glass Co. (1988) Laid Open Japanese Pat 63-23811
14 Nakamura M et al. (1989) Pacific Polymer Preprints 1; 559
15 Resnick PR (1989) Polymer Preprints Japan 38(5): 1254
16 Kawasaki T et al. (1989) Pacific Polymer Preprints 1: 555
17 Narita T et al. (1989) Symposium of Fluoropolymers, Third Chemical Congress of North America, Tronto
18 Imai Y (1989) Symposium of Fluoropolymers, Third Chemical Congress of North America, Tronto

Macromonomers with Activated Allyl End Groups: Synthesis and Copolymerization

E. Rizzardo*, D. S. Harrison, R. L. Laslett, G. F. Meijs, T. C. Morton and S. H. Thang

CSIRO Division of Chemicals and Polymers
Private Bag 10, Clayton, Victoria 3168, Australia

Abstract: Macromonomers of structure **4**, prepared by the free radical polymerization of MMA using cobalt complexes as chain transfer agents, become incorporated into polymer chains by copolymerization but also act as efficient chain transfer agents by an addition-fragmentation reaction. Macromonomers of general structure **1** have been prepared by utilizing appropriately substituted allylic sulfides as chain transfer agents in free radical polymerizations. Preliminary experiments have shown that a polystyrene macromonomer (**1**, R = Ph) copolymerizes efficiently with ethyl acrylate.

Introduction

The synthesis and copolymerization of macromonomers have received a great deal of attention worldwide in recent years. This is because the copolymerization of macromonomers with simple monomers provides a means of generating well defined graft copolymers which, in turn, are of interest for their unique properties (1). This paper is concerned with the preparation and free radical copolymerization of olefin-terminated oligomers and polymers of structure **1**.

1   R = COOR, Ph, CN, $CONH_2$

It should be noted that these differ from the more usual macromonomers in that they are derivatives of methacrylates, α-methylstyrene, methacrylonitrile or methacrylamides in which the polymer chain is attached to the α-methyl substituent. In the more common macromonomers terminated with an acrylate or methacrylate function, the polymer chain is part of the ester group, as shown in **2**, and in those that contain a styryl end group, the chain is usually attached to the aromatic ring, e.g. **3**.

B. C. Anderson · Y. Imanishi (Eds.)
Progress in Pacific Polymer Science
© Springer-Verlag Berlin Heidelberg 1991

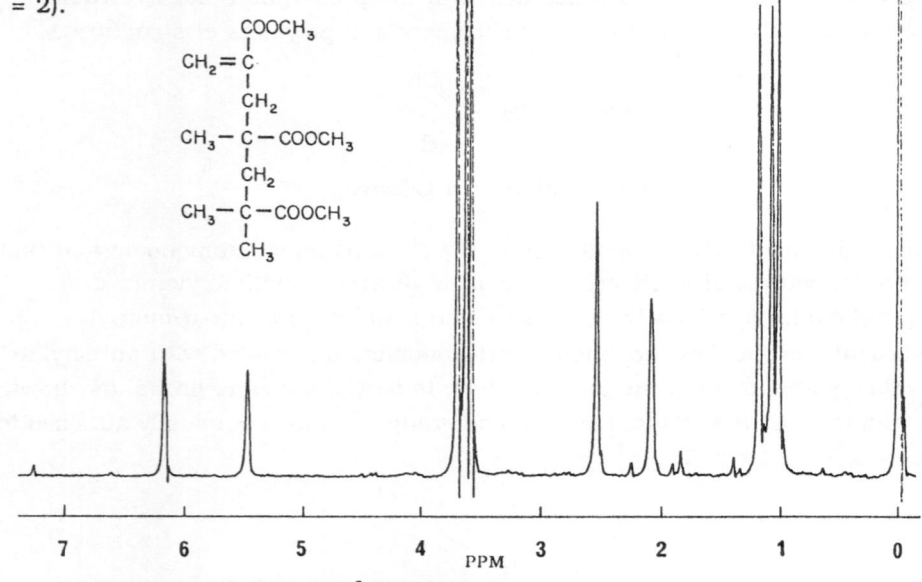

2 R = H, Me   **3** R = H, Me

Existing knowledge of free radical polymerizations would suggest that macromonomers of type **1** should be less reactive than those of type **2** or **3** owing to greater steric hindrance in the former. It is well known, for example, that α-substituents larger than methyl in acrylates retard both homopolymerization and copolymerization (2).

Preparation of PMMA Macromonomers **4**

Olefin-terminated poly(methyl methacrylate) (PMMA) **4** can be prepared by free radical polymerization of MMA using cobalt complexes as catalytic chain transfer agents (3).

4

By appropriate choice of conditions, oligomers and polymers **4** with different number average degrees of polymerization can be readily prepared. The lower oligomers, n = 1, 2 or 3, can be isolated in pure form by either chromatography or fractional distillation (4). Figure 1, shows the ^{1}H NMR spectrum of the trimer (n = 2).

Figure 1. 250 MHz ^{1}H NMR Spectrum of **4** (n = 2) in CDCl$_3$

Copolymerizations of PMMA Macromonomers **4**

Table 1 shows the results of copolymerizations of ω-unsaturated PMMA **4** of number average molecular weight ($\overline{M}_n$) = 1000, 540 and 680 with ethyl acrylate (EA), styrene (St) and methyl methacrylate (MMA), respectively (4). It will be noted that copolymerization of the macromonomer occurred in each case and that the efficiency decreased in the order EA > St > MMA. Most noteworthy, however, was the much lower $\overline{M}_n$ of the copolymers when compared to the homopolymers of EA, St and MMA produced under the same conditions. This suggested that the macromonomer **4** was acting as a chain terminator or a chain

Table 1. Copolymerization of Macromonomers **4** with Various Monomers[a]

Monomer	Macromonomer $\overline{M}_n$	Mole % Macromonomer in Feed	% Conversion	Product $\overline{M}_n$ x 10^{-3}	Mole % Macromonomer in Product
EA	—	—	95	270.0	—
	1,000	12.0	95	36.0	11.0
St	—	—	92	21.0	—
	540	7.7	51	4.8	5.3
MMA	—	—	99	26.0	—
	680	5.9	71	9.3	1.7

[a]10 mmol monomer, 2 ml benzene, 7 mg AIBN, 60°C, 7 days

transfer agent, or both. At this point it was felt that a better understanding could be achieved by investigating the free radical chemistry of pure single oligomers of **4**, namely, dimer (n = 1), trimer (n = 2) and tetramer (n = 3). The principal questions to be answered were: (a) how reactive is the double bond of **4** towards radical addition? and (b) what is the fate of the new radical so formed?

Rate of Radical Addition to Macromonomers **4**

The rate of t-butoxy radical addition to the double bond of **4** relative to MMA was determined by the radical trapping technique devised in our laboratories some years ago (5). In this, t-butoxy radicals are generated (by thermal decomposition of di-t-butyl peroxyoxalate) in the presence of the olefin and a stable nitroxide radical. The t-butoxy radical adds to the olefin and the carbon-centered radical so formed is trapped by the nitroxide to give a stable alkoxyamine (Scheme 1). The method functions effectively because nitroxides do not react with olefins nor couple with alkoxy radicals (e.g. t-butoxy), but combine with carbon-centered radicals at near-diffusion controlled rates. When two different olefins are present in the reaction mixture the relative yields of the two alkoxyamine products (determined by HPLC) reflect the relative reactivities of the olefins towards the alkoxy radical. This reaction on an equimolar mixture of MMA and dimer **4** (n = 1) gave, amongst other products, the alkoxyamines **5** and **6** in the ratio of 1.6 : 1.

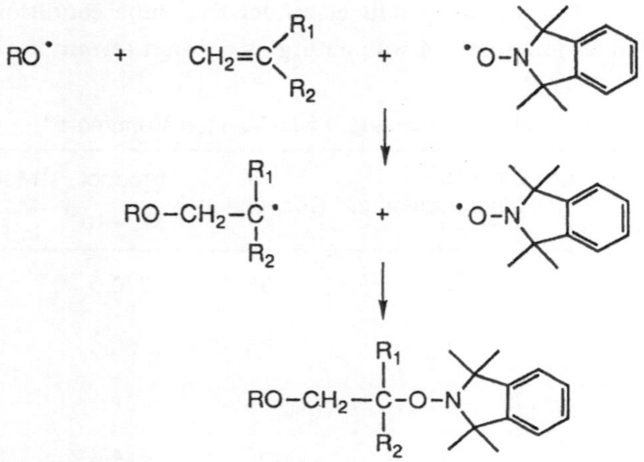

Scheme 1: Trapping of Radicals with Nitroxide

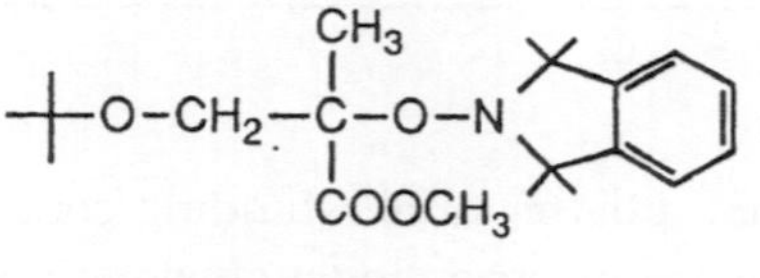

Similar competition experiments were conducted on mixtures of MMA with trimer **4** (n = 2), and dimer **4** (n = 1) with trimer **4** (n = 2). These established the relative reactivities of MMA : dimer : trimer to be 1.0 : 0.63 : 0.56 towards the t-butoxy radical. It is evident that steric effects retard radical addition to **4**, when compared with MMA, but the oligomers remain, nevertheless, highly reactive.

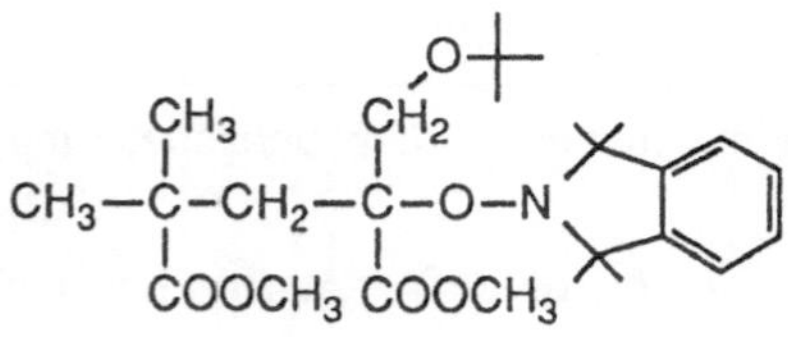

Reaction of Trimer **4** (n = 2) with Cyanoisopropyl Radicals

Some important information was obtained by examining the reaction of trimer **4** (n = 2) with AIBN in degassed benzene solution at $60^{\circ}C$. The trimer did not homopolymerize; instead, it gave rise to several low molecular weight products most of which could be rationalized by the addition-fragmentation reactions shown in Scheme 2 (4).

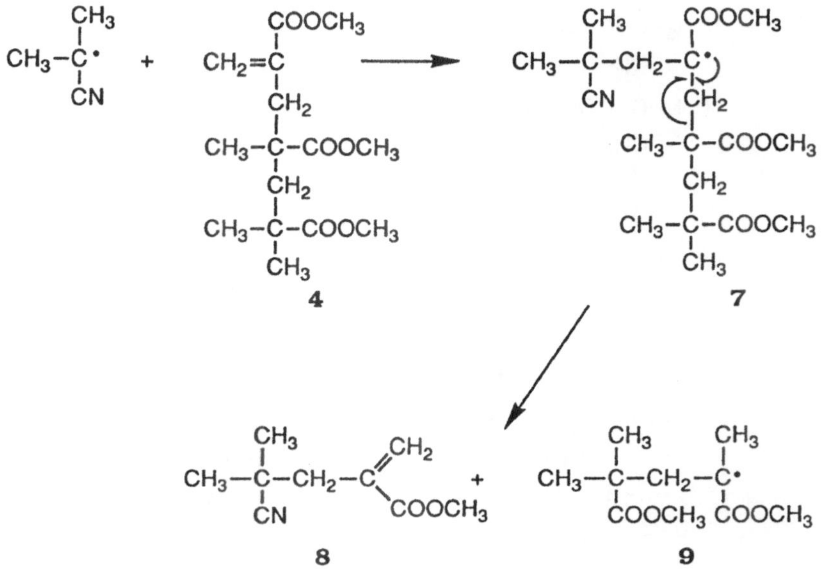

Scheme 2

This experiment demonstrates that the cyanoisopropyl radical is capable of addition to the double bond of trimer **4** but the radical **7** so formed is incapable of sustaining propagation and fragments to olefin **8** (a major product) and radical **9**. Radical **9** proceeds to stable products by combination or disproportionation with itself or other radicals (4). The mechanism in Scheme 2, therefore, is likely to be the process responsible for the chain transfer observed in the copolymerization experiments summarized in Table 1. An additional but minor reaction pathway for radical **7** was observed to be termination by hydrogen abstraction. Other radicals in the mixture (e.g. **9** or cyanoisopropyl) would be the most likely hydrogen donors (i.e. disproportionation).

Copolymerization Scheme for Macromonomers 4

From the evidence presented thus far and other experiments conducted to estimate chain transfer constants and number of macromonomer units incorporated per copolymer chain, it is possible to formulate the generalized copolymerization mechanism shown in Scheme 3.

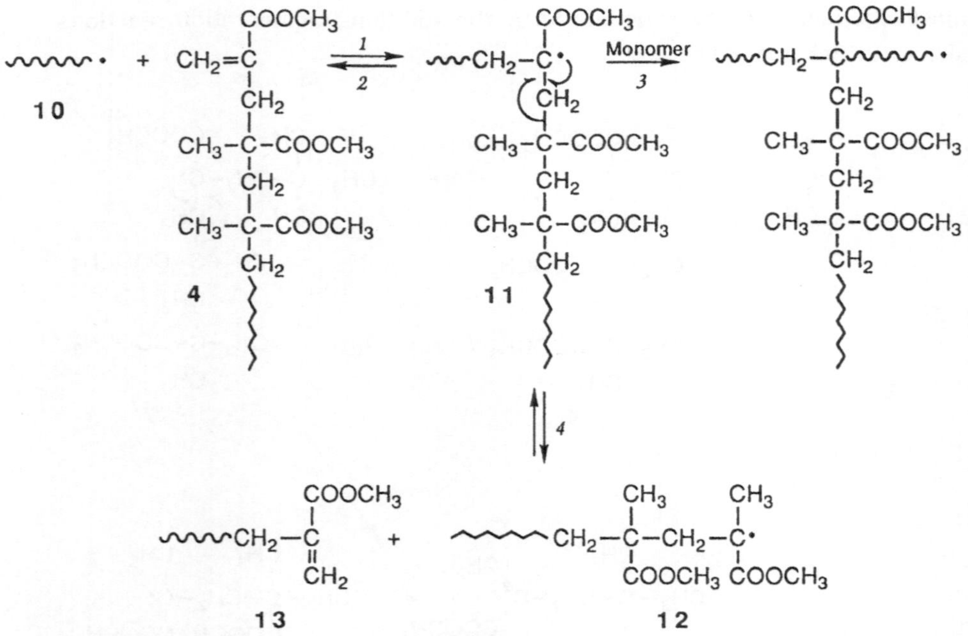

COOCH₃ structures — Scheme 3 (chemical reaction diagram)

$$
\begin{array}{c}
\text{COOCH}_3 \\
| \\
\text{wwww} \cdot + \text{CH}_2=\text{C} \underset{2}{\overset{1}{\rightleftharpoons}} \text{wwwCH}_2-\text{C}\cdot \xrightarrow[3]{\text{Monomer}} \text{wwwCH}_2-\text{Cwww} \cdot
\end{array}
$$

10 **4** **11** (fragment/propagation) **4** → **13** + **12**

Scheme 3

Addition of a propagating radical **10** to macromonomer **4** (reaction *1*) is expected to occur readily to give radical **11**. This can then undergo one of three reactions. It can: (a) add to monomer, in which case the macromonomer will become incorporated into the copolymer chain (reaction *3*), (b) fragment, and in doing so generate a polymeric re-initiating radical **12** and a new olefin-terminated polymer **13** (reaction *4*) or (c) revert to starting species (reaction *2*) in which case the propagating radical **10** can continue to grow by further addition of monomer. The relative rates of these pathways depend on the comonomer used. When the comonomer is an alkyl acrylate, fragmentation by reaction *4* is expected to be more favorable than fragmentation in the reverse direction (reaction *2*) since the former reaction produces a tertiary radical and the latter a less stable secondary radical. Indeed the chain transfer constant for trimer **4** (n = 2) was estimated to be 0.9 in the copolymerization of **4** with EA. Propagation (reaction *3*) was also relatively favorable and up to twelve trimer units on average have been incorporated into EA polymer chains. This can be attributed to the EA being relatively unhindered (monosubstituted olefin) and can be attacked by the bulky radical **11**. When the comonomer is ethyl methacrylate on the other hand, only one macromonomer unit per chain could be introduced into the copolymer, irrespective of the reaction conditions chosen. This can be best explained by a slow rate of addition of the bulky radical **11** to a hindered disubstituted olefin (reaction *3*) when compared with fragmentation (reactions *2* and *4*). All forward reaction would occur by reaction *4* and the one macromonomer unit per chain would consist of the fragments in **12** and **13**. With styrene as the comonomer the situation is somewhere in between the acrylate and methacrylate cases. Up to four macromonomer units have been incorporated into polystyrene chains. It should be noted that polymer chains which propagate by reaction *3* will also eventually terminate by reaction *4* and that the products **13** are themselves macromonomers which can take part in further polymerization. From this it may be predicted that copolymerizations of macromonomer **4** with acrylates or styrene will result in highly branched and segmented copolymers when taken to high conversion.

The ^{1}H NMR spectrum of a copolymer of trimer **4** (n = 2) and EA from a copolymerization in benzene solution taken to low conversion (<10%) is shown in Figure 2. The copolymer was purified by several precipitations from light petroleum and the absence of unreacted trimer confirmed by GPC. The spectrum exhibits signals at δ ~5.5 and ~6.2 ppm for the terminal olefinic protons and at δ 3.9-4.2 ppm and 3.5-3.8 ppm for the methyleneoxy of EA and methoxy of trimer respectively. The latter two signals provide an estimate of the ratio of trimer to EA in the copolymer while the small signal at δ 3.72 ppm confirms the presence of a carbomethoxy group attached to a double bond.

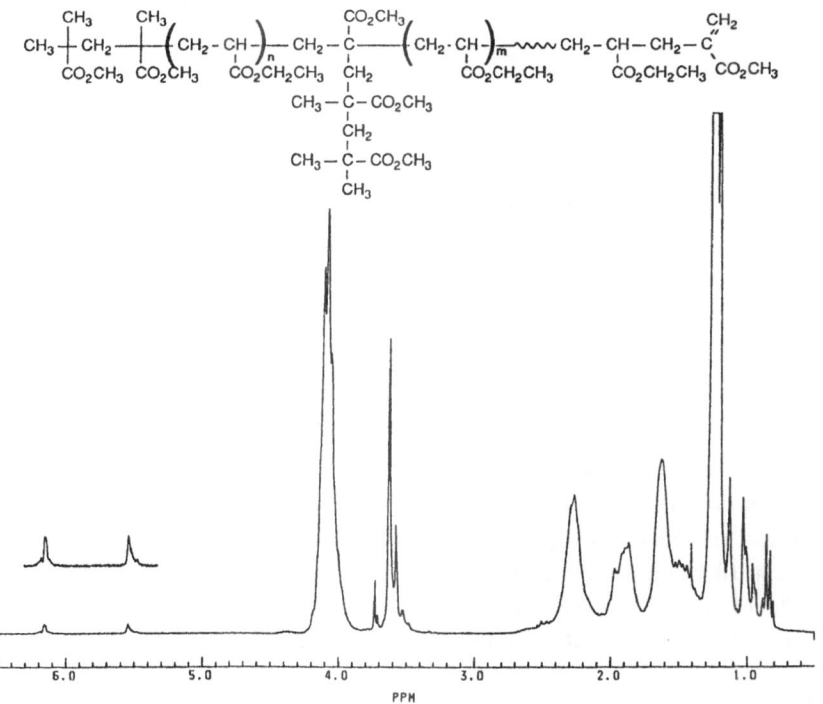

Figure 2. 250 MHz ^{1}H NMR Spectrum of Copolymer of **4** (n = 2) with EA in CDCl$_3$

Macromonomers by Chain Transfer with Allylic Sulfides

The results obtained in the work described above suggested that it should be possible to design simple and effective chain transfer agents which would give rise to polymers with activated allyl end groups by addition-fragmentation reactions. Appropriately substituted allylic sulfides of the general formula **14** have been shown to meet these criteria. The mechanism of chain transfer may be visualized (Scheme 4) as addition of the propagating radical to the olefinic center of **14** followed by fragmentation to give **15**. The process regenerates the double bond adjacent to its original position and a thiyl radical is expelled (6). The driving force for addition of propagating radicals to **14** is provided by suitable activating groups, R, and fragmentation to **15** is facilitated by the expulsion of a stabilized sulfur-centered radical and the formation of a double bond.

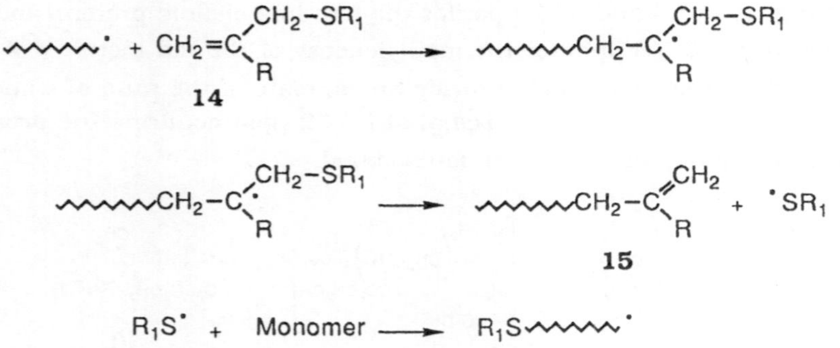

$$\text{wwww}^{\bullet} + CH_2=C\begin{smallmatrix}CH_2-SR_1\\R\end{smallmatrix} \longrightarrow \text{wwww}CH_2-C\begin{smallmatrix}CH_2-SR_1\\ \bullet\\R\end{smallmatrix}$$

14

$$\text{wwww}CH_2-C\begin{smallmatrix}CH_2-SR_1\\ \bullet\\R\end{smallmatrix} \longrightarrow \text{wwww}CH_2-C\begin{smallmatrix}CH_2\\R\end{smallmatrix} + {}^{\bullet}SR_1$$

15

$$R_1S^{\bullet} + \text{Monomer} \longrightarrow R_1S\text{wwww}^{\bullet}$$

$$R = Ph, COOCH_3, CN, CONH_2, OAc, \text{etc.}$$

$$R_1 = \text{alkyl}, CH_2COOH, CH_2CH_2OH, CH_2CH_2NH_2, \text{etc.}$$

Scheme 4

Unlike the cobalt catalytic chain transfer agents mentioned earlier, which are essentially restricted to the preparation of alkyl methacrylate macromonomers, allylic sulfides **14** can be used to generate macromonomers **15** from a wide range of common monomers. In addition, the olefin-activating group R (Scheme 4) can be varied and used to control the reactivity of both the chain transfer agent **14** and the resulting macromonomers **15**. Moreover, the technique can be used to prepare either mono-end-functional or di-end-functional macromonomers by introducing functionality into the groups R and R1 (7).

The control of molecular weight and end group functionality of polymers by chain transfer in free radical polymerization can be achieved most accurately and conveniently when the chain transfer constant has a value in the range 0.5 - 2; 1.0

being the ideal (8). In this respect, allylic sulfides have proved to be very useful as evidenced by the chain transfer constants of three differently substituted allylic sulfides shown in Table 2. The values with styrene (0.8-1.8) are particularly noteworthy as they contrast the very high chain transfer constants (10-20) of mercaptans with this monomer (9).

Table 2. Chain Transfer Constants of Allylic Sulfides

Chain Transfer Agent	Chain Transfer Constant, 60^0 C		
	MMA	St	MA
$CH_2=C\overset{CH_2-S-C(CH_3)_3}{\underset{Ph}{}}$	1.2	0.8	3.9
$CH_2=C\overset{CH_2-S-C(CH_3)_3}{\underset{COOEt}{}}$	0.7	1.0	2.2
$CH_2=C\overset{CH_2-S-C(CH_3)_3}{\underset{CN}{}}$	1.4	1.8	1.6

If the mechanism shown in Scheme 4 operates efficiently, a polystyrene macromonomer prepared by the use of allylic sulfide **16** as chain transfer agent should have structure **18**. That this is so, is confirmed by the ^{1}H NMR spectrum (Figure 3) in which the olefinic protons appear at δ ~6.1 and ~5.3 ppm and the methylene attached to oxygen at δ ~3.5 ppm in the ratio of 1 : 1 : 2 respectively. Careful integration of the NMR signals and knowledge of $\bar{M}_n$ of the polymer (GPC) indicated approximately one CH2OH and one C=CH2 per polymer chain on average. Infrared spectroscopy showed the presence of the carboxylic acid group (1710, 2500-3000 cm^{-1}) and titration confirmed the existence of approximately one such end group per chain.

The ^{1}H NMR spectrum of a polystyrene macromonomer, **19**, prepared by chain transfer with allylic sulfide **17**, is shown in Figure 4. The end group olefinic protons appear at δ ~5.1 and 4.7 ppm and the t-butyl end group at δ ~1.2 ppm. (The sharp signals at δ 2.0 and 2.2 ppm are due to d5-acetone and water respectively, present in the solvent).

16

17

18 $\overline{M}_n = 4300$

19 $\overline{M}_n = 3600$

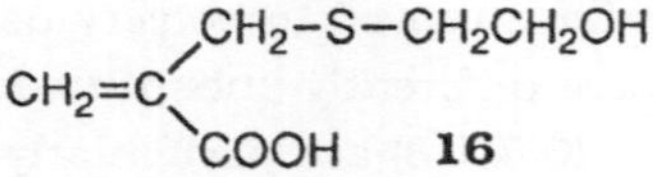

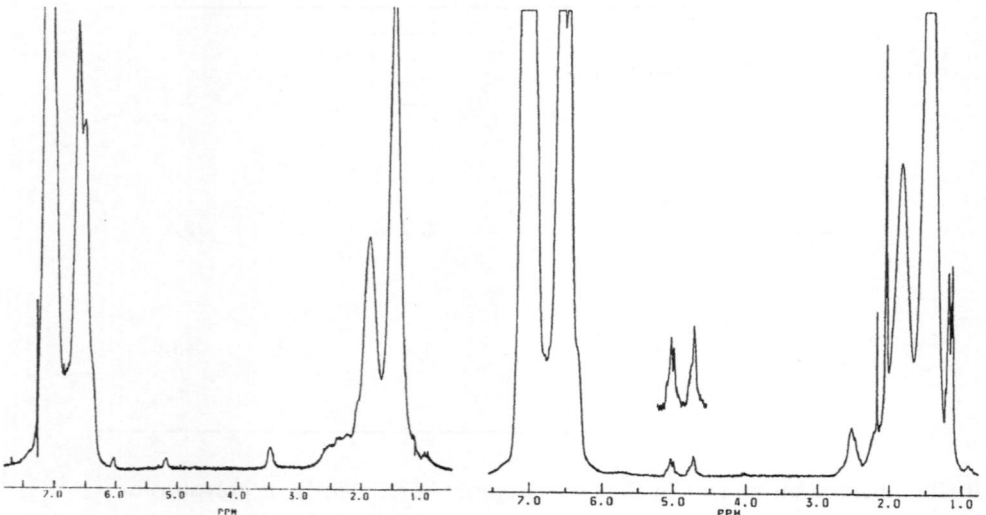

Figure 3. 250 MHz ^{1}H NMR Spectrum of Macromonomer **18** in $CDCl_3$

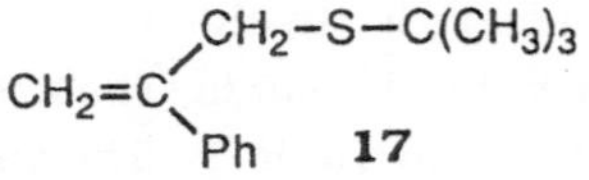

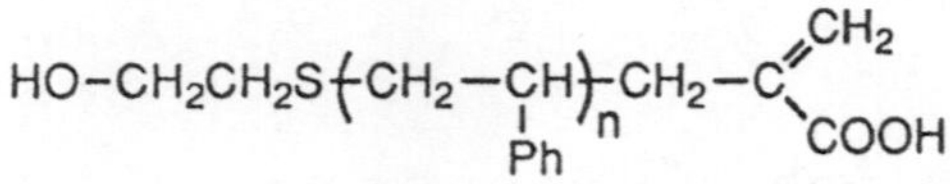

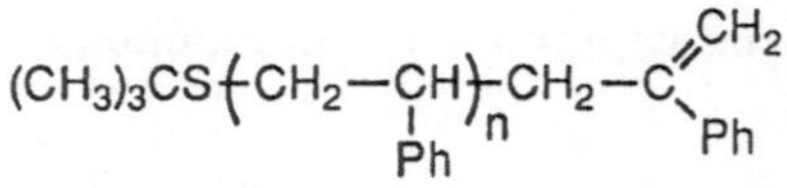

Figure 4. 250 MHz ^{1}H NMR Spectrum of Macromonomer **19** in CCl_4/d_6Acetone

Copolymerization of Macromonomer **19** with EA

The copolymerization of macromonomer **19** of $\overline{M}_n$ = 5100 with EA (1 : 50 molar ratio) in benzene solution at 60°C (AIBN initiation) taken to approx. 90% conversion gave a copolymer of $\overline{M}_n$ = 28000 as estimated by GPC. ^{1}H NMR analysis of the copolymer showed that, within the limits of detection, all the macromonomer had been consumed. Integration of the signals due to phenyl of the macromonomer (δ 6.2-7.4 ppm) with respect to those of methyleneoxy of EA (δ 3.7-4.2 ppm) indicated that the copolymer was slightly richer in styrene (macromonomer) than was the feed. This suggests that the EA propagating radical has a preference for addition to macromonomer over EA, as might be expected from consideration of reactivity ratios for the copolymerization of EA with α-methylstyrene.

The GPC traces of the starting macromonomer **19** and its copolymer with EA are shown in Figure 5. Fractionation of the copolymer by preparative GPC yielded fractions ($\bar{M}_n$ = 72400, 55000, 34700 and 19300) whose compositions were not too different from the bulk copolymer, indicating favorable reactivity ratios for the copolymerization.

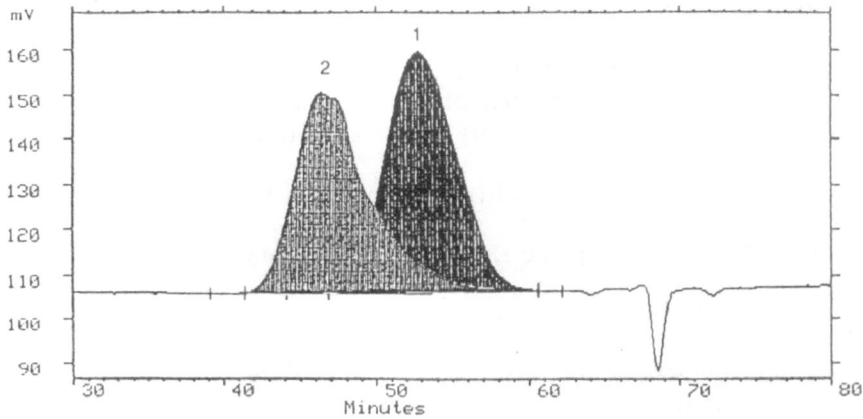

Figure 5. 1, GPC of Macromonomer **19** and 2, its Copolymer with EA

The preliminary results described in this paper suggest that macromonomers derived from chain transfer with cobalt complexes or allylic sulfides have potential for the convenient synthesis of unusual graft copolymers. Work in this area is continuing.

Acknowledgments: The technical assistance of Y. K. Chong, D. McIntosh and N. B. Dao is gratefully acknowledged. We are also grateful to M. N. Galbraith for the large scale preparation of allylic sulfide **17** and to R. I. Willing for NMR spectra.

References

1 Kawakami Y (1987) In: Kroschwitz JI (ed) Encycl Polym Sci Eng, Wiley-Interscience, New York 9:195
  Chujo Y, Yamashita Y (1989) In: Telechelic Polymers: Synthesis and Applications, CRC Press, Boca Raton, Florida
  Meijs GF, Rizzardo E (1990) J Macromol Sci, Rev Macromol Chem Phys, in press

2 Chikanishi K, Tsuruta T (1965) Makromol Chem 81:198, 211
  Mertwoy HE, Gisser H (1977) Macromolecules 10:794
  Brand E, Stickler M, Meyerhoff (1980) Makromol Chem 181:913

3 Enikolopyan NS, Smirnov BR, Ponomarev GV, Belgovskii IM (1981) J Polym Sci, Polym Chem Ed 19:879
   Burczyk AF, O'Driscoll KF, Rempel GJ (1984) J Polym Sci, Polym Chem Ed 22:3255

4 Cacioli P, Hawthorne DG, Laslett RL, Rizzardo E, Solomon DH (1986) J Macromol Sci Chem A23:839

5 Rizzardo E, Solomon DH (1979) Polym Bull 1:529
   Griffiths PG, Rizzardo E, Solomon DH (1982) J Macromol Sci Chem 17:45
   Grant RD, Rizzardo E, Solomon DH (1985) J Chem Soc Perkin Trans II 379

6 Meijs GF, Rizzardo E, Thang SH (1988) Macromolecules 21:3122

7 Meijs GF, Morton TC, Rizzardo E, Thang SH Macromolecules, to be submitted

8 Corner T (1984) Adv Polym Sci 62:95

9 Brandrup J, Immergut EH (eds) (1975) Polymer Handbook, Wiley-Interscience, New York

Condensation Macrocyclic Oligomers: Synthesis and Polymerization

J.W. VERBICKY, JR.

Chemical Research Center, GE Corporate Research and Development

P.O. Box 8, Schenectady, NY 12301, USA

The processing of engineering thermoplastics is frequently limited by the high melt viscosity of high molecular weight linear polymers. This limitation in the processing of engineering polymers could, in principle, be overcome by processing low viscosity macrocyclic oligomers which could be converted to high molecular weight polymer in a ring opening polymerization process. This approach has been largely unexplored due to the lack of a convenient and practical synthesis of such macrocyclic oligomers of condensation polymers without the use of high dilution techniques and prolonged reaction times.

We have recently developed an effective synthetic methodology for the preparation of macrocyclic oligomers of aromatic condensation polymers and have investigated their ring opening polymerization chemistry.

Macrocyclic aromatic carbonates, arylates, etherimides, ether-ketones and ethersulfones have all been prepared in yields of 50-90%. For example, macrocyclic aromatic carbonate oligomers of a variety of bisphenols have been prepared in 70-90% yields by a kinetically

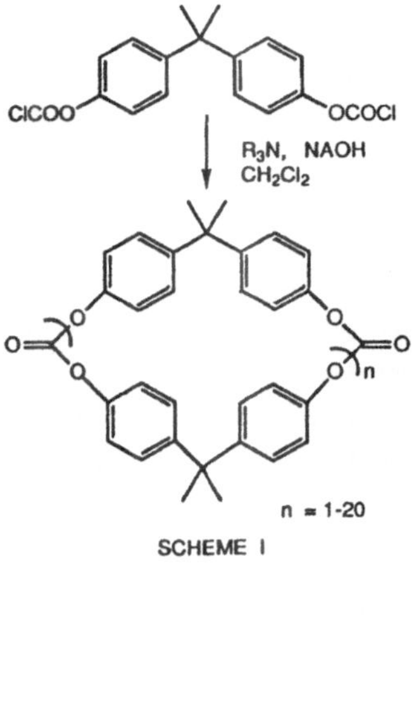

SCHEME I

B. C. Anderson · Y. Imanishi (Eds.)
Progress in Pacific Polymer Science
© Springer-Verlag Berlin Heidelberg 1991

controlled process involving the hydrolysis of bischloroformate oligomers in a two phase reaction mixture of methylene chloride and water in the presence of an alkali base and a tertiary amine catalyst[1] (Scheme I).

This process produces almost exclusively a mixture of cyclic oligomers and high molecular weight polymer. Small amounts, typically less than 0.1%, of low molecular weight linear oligomers are also formed in the process. The absence of low molecular weight linear oligomers is important in the subsequent ring opening polymerization since the presence of significant quantities of these species could limit the ultimate polymer molecular weight obtained.

Cyclic oligomeric carbonates are isolated as an amorphous white solid with a softening point of approximately 200°C.

The polycarbonate cyclic oligomers exhibit a melt viscosity of approximately 10 poise at 250°C versus the melt viscosity of the corresponding high molecular weight polymer of 50,000 to 100,000 poise at 250°C.

These cyclic carbonates can be converted to high molecular weight linear polymer in a ring opening polymerization in the presence of basic catalysts. (Scheme II). Polycarbonate polymers with

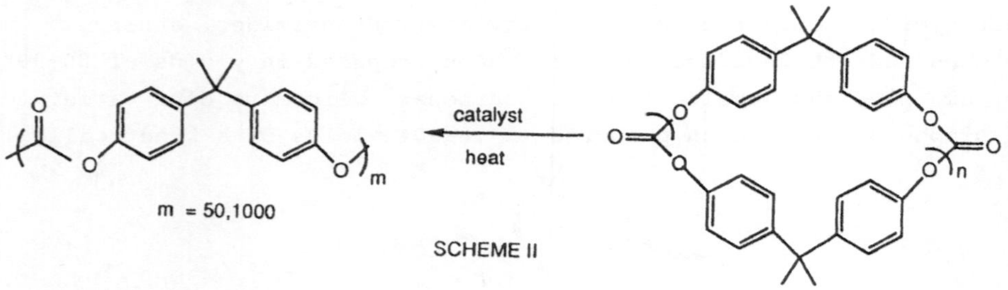

m = 50,1000

SCHEME II

molecular weights ($\bar{M}_w$) in excess of 1,000,000 have been achieved via this approach.

The molecular weight of the polycarbonate polymer produced in the ring opening polymerization can be controlled by the level of initiator used in the polymerization. Final polymer molecular weight increases linearly with the ratio of cyclic oligomer to initiator used.

Polymerization reactions can be carried out in the melt or in an appropriate solvent. One interesting feature of these polymeriza-

tions is their "psuedo living" characteristics. Increases in
polymer molecular weight are observed when additional cyclic
oligomer is added to a fully polymerized reaction mixture.

A variety of aromatic bisphenols can be condensed with isophthaloyl
dichloride in a phase transfer catalyzed reaction to produce cyclic
polyarylate oligomers moderate to excellent yields[2]. Most notably,
spirobiindane bisphenol 1 displays a unique propensity to form
high yields of cyclic arylate oligomers with isophthaloyl dichloride.

1

The propensity of this bisphenol to form high yields of cyclic
macromers exhibits itself in the condensation reaction of 1 with
4,4'-dihalo substituted diphenylketones and diphenylsulfones to
produce cyclic etherketones and ethersulfones. Polymerization of
the polyarylate, polyetherketone and polyethersulfone macromers
can also be accomplished in the presence of anionic initiators to
produce linear polymers.

References:

1. Brunelle, D.J., Evans, T.L., Shannon, T.G., U.S. Patent No.
   4,644,053 (1987)
2. Guggenheim, T.L., McCormick, S.J., Kelly, J.J., Brunelle, D.J.,
   Colley, A.M., Boden, E.P., Shannon, T.G., "Synthesis and
   Polymerization of Cyclic Oligomeric Arylates Based on Bisphenol
   A and Isophthaloyl Dichloride", paper presented at the Miami,
   FL ACS meeting Sept. 10-15, 1989; See Polymer Preprints 30(2)
   Sept. 1989

Recent Progress in Living Cationic Polymerization

T. Higashimura and M. Sawamoto

Department of Polymer Chemistry
Kyoto University, Kyoto 606, Japan

Abstract: This paper discusses the principle, initiator design, and synthetic application of living cationic polymerization of vinyl monomers. The principle is based on the nucleophilic stablization of the growing carbocation; some new initiating systems have been developed with use of phosphate esters; and end-functionalized and star-shaped polymers have been synthesized.

1. INTRODUCTION

1.1. Background

This decade is the age in which the use of polymers has been expanding dramatically from simple structural materials to more advanced and functional materials. Because of these changes, polymer chemists have been facing increasing demands for the synthesis of new polymers that have functional groups as well as controlled structures and molecular weights. In such controlled polymer synthesis, one must control a number of factors (formula $\underset{\sim}{1}$), including pendant functional group "R", terminal functionality "X", molecular weight and its distribution, and the steric structure of the backbone. Another new aspect of interest is the control of the spatial shape of macromolecules, directed towards not only block and graft polymers but other interesting architectures, such as star-shaped, tree-like, spherical, rod-like, two-dimensional sheet, and so on.

In general, polymerization of vinyl monomers involves chain reactions, where the structural control of the resulting polymers is inherently difficult. One of the approaches to overcome this difficulty is so-called "living" polymerization, which is particularly effective in regulating polymer molecular weight and molecular weight

B. C. Anderson · Y. Imanishi (Eds.)
Progress in Pacific Polymer Science

distribution as well as terminal functionality. A truly living process was first found in anionic polymerization in 1956 by Szwarc and his associates (1). Since then, a large number of living anionic systems have been developed, by which various block and graft polymers have been prepared. Even in anionic polymerization, however, living polymers with polar functional groups have been difficult to synthesize above room temperature.

Except for some heterocycles (2), living cationic polymerization has been considered almost impossible, particularly for vinyl monomers, which generate unstable carbocationic intermediates. Despite this pessimistic view, we have recently found that living cationic vinyl polymerization is indeed possible by stabilizing the growing carbocations with nucleophilic counteranions or with externally added bases (3).

This paper discusses (i) the principles of living cationic polymerization, (ii) new initiating systems for living polymerization of vinyl ethers, which are based on phosphate esters as initiators, and (iii) some selected topics in the controlled polymer syntheses by our living cationic processes, specifically focused on end-functionalized polymers and star-shaped or multi-armed macromolecules.

1.2. Principles of Living Cationic Polymerization

In general, cationic vinyl polymerization is initiated by addition of an initiator $A^{\oplus}B^{\ominus}$ to a monomer, and the resulting carbocationic intermediate (2) propagates through electrophilic addition onto the monomer (Eq. 1). Because of the neighboring cationic charge, the terminal β-hydrogen in 2 is acidic and then subject to abstraction by nucleophiles including the monomer and the counteranion. This is the reason for frequent chain transfer reactions that ultimately render living cationic polymerization very difficult to occur.

$$A^{\oplus}B^{\ominus} + CH_2=CH \atop R \longrightarrow ACH_2-\overset{\oplus}{CH}\cdots B^{\ominus} \atop R \xrightarrow{CH_2=CHR}$$

(1)

$$A\!\!\sim\!\!\sim\!\!CH-\overset{\oplus}{CH}\cdots B^{\ominus} \atop R \longrightarrow A\!\!\sim\!\!\sim\!\!CH=CH + H^{\oplus}B^{\ominus} \atop R$$

$$\xrightarrow{CH_2=CHR} A\!\!\sim\!\!\sim\!\!CH=CH + CH_3-\overset{\oplus}{CH}\cdots B^{\ominus} \atop R \quad R$$

2

$$\stackrel{\delta\oplus}{\mathrm{\sim\!\!\sim\!C}}\ \stackrel{\delta\ominus}{\mathrm{B}} \rightleftharpoons \mathrm{\sim\!\!\sim\!C}\!\cdots\!\overset{\oplus}{\mathrm{B}}{}^{\ominus} \rightleftharpoons \overset{\oplus}{\mathrm{\sim\!\!\sim\!C}} + \mathrm{B}^{\ominus}$$

$$(2)$$

$$\text{Slow} \xrightarrow{\text{Propagation}} \text{Fast}$$

$$\text{Slow} \xrightarrow{\text{Side reaction}} \text{Fast}$$

It is then expected that decreasing the cationic charge at the active site would stabilize the growing carbocation 2 and then permit the formation of living cationic polymers, because in a less cationic species, the β-proton is less acidic and accordingly more difficult to eliminate. Such a decrease in the cationic charge may indeed retard propagation, but it may also suppress chain transfer as well (Eq. 2). Recently, we have established two methods for such stabilization of the growing carbocation 2 through decreasing its cationic charge (3):

(i) carbocation stabilization by **nucleophilic counteranions**;
(ii) carbocation stabilization by **added bases.**

This paper discusses the recent progress in the first approach based on the use of nucleophilic counteranions; an overview on the second method is available elsewhere (4).

2. INITIATING SYSTEMS FOR LIVING CATIONIC POLYMERIZATION

2.1. Selection of Initiating Systems

As already reported (5), hydrogen iodide has been employed as an initiator that would give a nucleophilic iodide counteranion for polymerization of vinyl ethers (Eq. 3). This protonic acid, however, soon turned out to yield an adduct (3) only, without inducing any polymerization, because the polarization of the carbon-iodine bond of 3 is too small (6). In order to activate this linkage, we added a weak Lewis acid (Y), like iodine and zinc halides (ZnX_2), to a quiescent mixture of 3 and a vinyl ether, and then a cationic polymeriza-

$$CH_2{=}CH \xrightarrow{\ HI\ } CH_3{-}CH{-}I \xrightarrow{\ Y\ } CH_3{-}\overset{\delta\oplus}{CH}\!\cdots\!\overset{\delta\ominus}{I}\!\cdots\!Y \xrightarrow{\ CH_2{=}CHOR\ }$$

positions below: OR under first two, OR under third

$$\underset{3}{}$$

$$(3)$$

$$\mathrm{\sim\!\!\sim\!CH_2}{-}\overset{\delta\oplus}{CH}\!\cdots\!\overset{\delta\ominus}{I}\!\cdots\!Y \qquad \textbf{LIVING Polymer}$$

tion proceeded. Therefore, hydrogen iodide is called as "initiator" and the Lewis acid Y as "activator" (3,6). These hydrogen iodide-based initiating systems (HI/I_2, HI/ZnX$_2$, etc.) readily polymerize a variety of vinyl ethers, including those with polar functional substituents, to give narrowly distributed living polymers (3).

With zinc halides as the activators, living cationic polymerization is possible not only for vinyl ethers but also styrenic monomers. For example, in addition to the p-methoxy (7) and p-methyl (8) derivatives, p-t-butoxystyrene (9) can be polymerized by the HI/ ZnI$_2$ initiating system in toluene at +25 °C or below to form well-defined living polymers (Eq. 4). The number-average molecular weight ($\bar{M}_n$) of the polymers increases linearly as the reaction proceeds; such a controlled polymer growth can be continued by adding a second feed of monomer into a completely polymerized reaction mixture; and the observed $\bar{M}_n$ is very close to the calculated value assuming that each hydrogen iodide molecule forms one living polymer chain. In addition, the molecular weight distribution (MWD) of the polymers is almost monodisperse ($\bar{M}_w/\bar{M}_n \leq 1.1$). Polymers of p-t-butoxystyrene are also interesting as materials, because the pendant butoxy groups can readily be transformed into alcohols to give poly(p-hydroxystyrene) (Eq. 4).

$$CH_2=CH \quad \xrightarrow[\text{Living Polymn}]{HI/ZnI_2} \quad -(CH_2-CH)_{\overline{n}} \quad \xrightarrow{HBr} \quad -(CH_2-CH)_{\overline{n}} \qquad (4)$$

2.2. Living Polymerization by Phosphate Esters

The living cationic polymerizations discussed above are invariably based on the nucleophilic iodide counteranion (activation of the carbon-iodine terminal bond; Eq. 3). It is expected, however, that similar living processes are equally possible with other counteranions that can exert, as the iodide anion does, a suitably strong nucleophilic interaction with the growing carbocation. We have in fact found the phosphate anions to meet this requirement (10). Similarly to hydrogen iodide, monoacidic phosphate esters [HOP(O)R'$_2$; R': alkyl, alkoxyl, etc.] like diphenyl phosphate (4) form a stable adduct (5) with a vinyl ether (Eq. 5). Zinc chloride or iodide then activates the phosphate bond in 5 by increasing its polarization (as in 6), and living cationic polymerization proceeds via an intermediate (7) where the carbocationic site is stabilized by a phosphate anion coupled with the zinc halide activator.

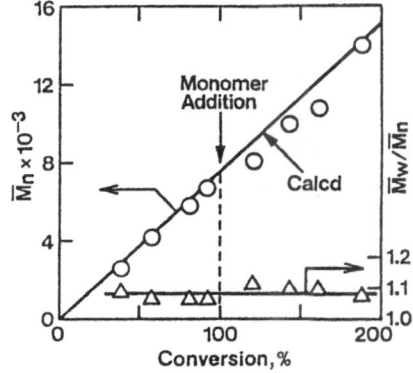

Fig. 1. Living polymerization of IBVE by the diphenyl phosphate/ $ZnCl_2$ initiating system in toluene at 0 °C: $[M]_0 = [M]_{add} = 0.38$ M; $[HOP(O)(OPh)_2]_0 = 5.0$ mM; $[ZnCl_2]_0 = 10$ mM.

$$\text{(5)}$$

In the presence of zinc chloride (as an activator), for example, diphenyl phosphate (as an initiator) rapidly and quantitatively polymerizes isobutyl vinyl ether (IBVE) in toluene at 0 °C. As shown in Figure 1, the $\bar{M}_n$ of the produced polymers is directly proportional to monomer conversion and in good agreement with the calculated value based on the feed molar ratio of IBVE to the initiator (one polymer chain per phosphate). On addition of a second feed of IBVE at the end of the polymerization, polymer molecular weight further increases with conversion, while the polymers invariably exhibit very narrow MWDs ($\bar{M}_w/\bar{M}_n \leq 1.1$). All these data show that the diphenyl phosphate/ $ZnCl_2$ initiating system induces living polymerization of IBVE.

In this way, we have demonstrated that careful selection of an initiatior ($A^{\oplus}B^{\ominus}$) and a Lewis acid activator (Y) permits living cationic polymerization, as generalized in Eq. 6. According to this approach, there may be a variety of candidates for the initiators and the activators that should lead to new initiating systems.

$$CH_2=\underset{R}{CH} \xrightarrow{\text{AB(Initiator)}} \overset{\delta\oplus \; \delta\ominus}{A CH_2-\underset{R}{CH}-B} \xrightarrow{\text{Y (Activator)}} \overset{\oplus \quad \ominus}{A CH_2-\underset{R}{CH}\cdots B\cdots Y} \quad \text{(6)}$$

3. POLYMER SYNTHESIS BY LIVING CATIONIC POLYMERIZATION

Living cationic polymerization, as those by other mechanisms, offers a variety of synthetic applications that lead to well-defined functional polymers. The following is a partial list of such possibilities, of which the synthesis of (ii) end-functionalized polymers and (iv) star-shaped macromolecules are discussed below:

    (i) Pendant-Functionalized Polymers
    (ii) End-Functionalized Polymers
    (iii) Sequential Block Polymers
    (iv) Star-Shaped Polymers
    (v) Polymers with Unique Spatial Shapes
    (vi) Polymers with Controlled Repeat-Unit Sequences

These applications are all based on an advantage of living cationic polymerization that one can readily prepare well-defined living polymers even from monomers carrying polar functional pendant substituents. Typical examples of functionalized vinyl ethers include:

$CH_2=CH$ X: (saturated ester) $OCOCH_3$, $OCOC_6H_5$, $CH(COOEt)_2$

(unsaturated ester) $OCOCMe=CH_2$, $OCOCH=CH_2$, $OCOCH=CHC_6H_5$;

(poly(oxyethylene)) $O(CH_2CH_2O)_n Et$ (n = 0-4);

(silyloxyl) $OSiMe_2tBu$, $OSiMe_3$;

8

(imide) $N \overset{CO}{\underset{CO}{\big\langle}} \bigcirc$

3.1. Synthesis of End-Functionalized Polymers

As already reported (11), various end-functionalized polymers of vinyl ethers have been prepared through the living cationic polymerization that is based on the carbocation stabilization by the nucleophilic iodide counteranion (Eqs. 7 and 8). The first method (Eq. 7) starts from the addition of hydrogen iodide to a vinyl ether (**8**) with

$$CH_2=CH \xrightarrow{HI} CH_3-CH-I \xrightarrow[CH_2=CH]{I_2} CH_3-CH \sim\sim CH_2-CH-I\cdots I_2 \qquad (7)$$

$$\underset{\substack{\text{O}\\(X)\\ \textbf{8}}}{} \qquad \underset{\substack{\text{O}\\(X)\\ \textbf{9}}}{} \qquad \underset{\text{OR}}{} \qquad \underset{\substack{\text{O}\\(X)}}{} \qquad \underset{\text{OR}}{} \qquad \text{LIVING POLYMER } \textbf{10}$$

$$\sim\sim\sim CH_2-CH-I\cdots I_2 + :Nu(X) \longrightarrow \sim\sim\sim CH_2-CH-Nu(X) \qquad (8)$$

$$\underset{\text{OR}}{} \qquad \qquad \underset{\text{OR}}{}$$

$$\text{LIVING POLYMER } \textbf{11} \qquad \textbf{12} \qquad \qquad \textbf{13}$$

a functional groups X (see above); the resulting adduct (9) is activated by iodine; and then another vinyl ether is polymerized in a
living fashion to give a polymer (10) that has the functionality X in
the initiation terminal. The second method (Eq. 8) involves endcapping of an HI/I_2-generated living polymer (11) with a functional
nucleophile (12), and the product (13) carrys a terminal functional
group X coming from the terminator. Combination of the two methods
then gives so-called telechelic polymers with, for instance, terminal
carboxyl groups (12): $HOCOCH_2-CH_2CH_2OCHMe+CH_2CH(OiBu)+_nCH_2COOH$.

When a counteranion is not so nucleophilic as the iodide anion,
the propagating carbocation may be stabilized instead by adding an
base (Z) so that living polymerization proceeds (Eq. 9; see also section 1.1) (4). This method is particularly effective for the polymerization initiated with ethylaluminum dichloride ($EtAlCl_2$) (13);
and typically, the bases may be 1,4-dioxane and related ethers that
form a "base-stabilized" carbocationic species like 14 where Z is an
ether oxygen. We have recently synthesized end-functionalized polymers via these base-stabilized living species (14).

$$\text{} C^{\oplus}\cdots B^{\ominus} + Z\diagup \longrightarrow \text{} C \overset{\oplus}{\cdots} Z\diagup \cdots B^{\ominus} \tag{9}$$

(not living)   14 (living)

Eq. 10 presents an example of the synthesis of hetero-telechelic
poly(vinyl ethers) via the base-stabilized living species 14. The
ester-type adduct 15 is the functional initiator of choice, which is
obtained from trifluoroacetic acid and a vinyl ether (8) with a pendant functionality X. In n-hexane solvent at temperatures from 0 to
+60 °C, the trifluoroacetate 15 can initiate living polymerization of

X: Functional Group

IBVE through the activation of the ester bond by $EtAlCl_2$; 10 vol% 1,4-dioxane is needed to generate the base-stabilized living species (16). The resulting polymers have very narrow MWDs ($\bar{M}_w/\bar{M}_n \leq 1.1$) and molecular weights controllable by the molar ratio of monomer to initiator 15.

The growing living species 16 can be quenched cleanly with sodiomalonic ester to form the target hetero-telechelic polymers (17). The terminal (head) function X (arising from 15) may be acetate or imide that in turn leads to a carboxylic acid or an amine, respectively. Another terminal carries a malonate (from the malonate terminator), which can be converted to a carboxylic acid by hydrolysis/ decarboxylation. One of the advantages of the polymer synthesis via the base-stabilized species is the rather high operational temperature up to +60 °C.

3.2. Synthesis of Multi-Armed Polymers

Multi-armed polymers are a new class of macromolecules with interesting spatial shapes. Particularly when endowed with pendant or terminal functional groups, they may show unique characteristics and functions that are thus far unknown and may differ from those of the linear counterparts. Such functions may include adsorption or recognition of molecules and regulation of chemical reactions, all based on the unique spatial shapes. We have recently started the synthesis of star-shaped, multi-armed polymers by the reactions of our living cationic polymers (18) with divinyl ethers (e.g., 19) (Eq. 11). The resulting polymers (20) are expected to consist of a microgel core of the divinyl ether to which linear arms are attached. This strategy is well-known in anionic polymerization (15) but, to our knowledge, none in the cationic counterpart so far.

(11)

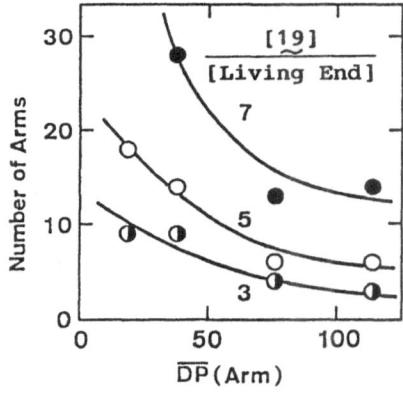

Fig. 2. Number of arms per molecule in star-shaped poly(IBVE) (**20**) as a function of the degree of polymerization [$\overline{DP}$(arm)] of the arm chain and the molar feed ratio of the divinyl ether (**19**) to the living end (**18**).

In a typical example (Eq. 11), IBVE is polymerized with the HI/ ZnI_2 system in toluene at -40 °C into a linear and near monodisperse living polymer **18**, which is then mixed with a small amount of the divinyl ether **19** (16). The products are totally soluble in common organic solvents (toluene, chloroform, etc.) and have 3 to 60 arms per molecule depending on the reaction conditions. The arm number is determined from the weight-average molecular weights of the star-shaped polymer (**20**) and of the starting living polymer (**18**), which are measured by laser light scattering in THF.

Figure 2 plots the number of arms of the star polymers as functions of the mole ratio of the divinyl ether to the living end, as well as on the length [degree of polymerization, $\overline{DP}$(arm)] of the arm chain. When the arm length is constant, the arm number sharply increases at higher concentrations of the divinyl ether **19**. This trend is due to the formation of a larger microgel core to which a larger number of the arm chains can be attached without steric crowding. When the feed ratio of the divinyl ether to the living end is constant, on the other hand, the arm number decreases for longer arms. The use of a longer arm chain may cause a greater steric hindrance on the formation of the microgel core, and thus the coupling reaction between the living end and the divinyl ether may be increasingly difficult.

As we have already reported, sequential living cationic polymerizations of functionalized vinyl ethers (**8**) readily give amphiphilic block copolymers (17). Such sequential living polymers are equally applicable to the star polymer synthesis. A typical example utilizes an AB living block copolymer (**21**) that consists of 2-(acetoxy)ethyl vinyl ether and IBVE (10 and 30 units per chain, respectively). The

102

living polymer 21 is then treated with a small amount of the divinyl
ether 19 in toluene at -15 °C to give star-shaped polymers. The products are initially insoluble in methanol, but on hydrolysis of the
acetate groups into alcohols, the polymer becomes soluble in methanol. The star-shaped polymer
has a double-layered amphiphilic structure in which the
inner and outer layers are
hydrophobic and hydrophilic,
respectively.

$$\text{H}(\text{CH}_2\text{-CH})_{10}(\text{CH}_2\text{-CH})_{30}\text{-I}\cdots\text{ZnI}_2 \qquad 21$$

By the same strategy, we have prepared a variety of star-shaped
multi-armed poly(vinyl ether)s by our living cationic polymerization.
As illustrated below, the nature of the arms can be varied by changing the structures of the pendant substituents (derived from functionalized vinyl ethers 8), including hydrophobic (22), hydrophilic
(23), amphiphilic (24), and terminally functionalized (25).

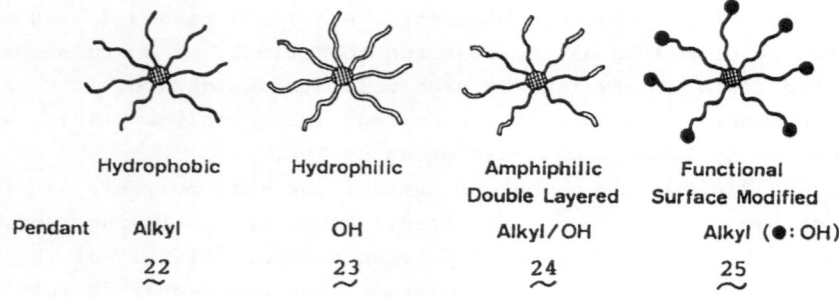

	Hydrophobic	Hydrophilic	Amphiphilic Double Layered	Functional Surface Modified
Pendant	Alkyl	OH	Alkyl/OH	Alkyl (●: OH)
	22	23	24	25

4. CONCLUSIONS

We have achieved a variety of living cationic polymerizations on
the basis of the two principles for stabilization of the growing carbocations, either by nucleophilic counternanions or by added bases.
These living processes allow us to prepare, by simple procedures and
in most cases at room temperature, various types of new polymers from
functionalized vinyl monomers, specifically vinyl ethers and styrene
derivatives.

Acknowledgment
    This study was supported in part by a grant for International
Joint Research from the New Energy and Industrial Technology Development (NEDO), Japan.

REFERENCES

1 (a) Szwarc M (1956) Nature 178:1169; Szwarc M, Levy M, Milkovich R (1956) J Am Chem Soc 78:2656
  (b) As a review, see: Szwarc M (1968) Carbanions, living polymers, and electron transfer processes, Interscience, New York
2 Dreyfuss MP, Dreyfuss P (1965) Polymer 6:93; Bawn CEH, Bell RM, Ledwith A (1965) Polymer 6:95
3 For recent reviews, see: Higashimura T, Aoshima S, Sawamoto M (1988) Makromol Chem Macromol Symp 13/14:457; Higashimura T, Sawamoto M (1989) In: Comprehensive polymer science, Pergamon, London, Vol 3, Part I, Chapter 42
4 Higashimura T, Sawamoto M, Aoshima S, Kishimoto Y, Takeuchi E (1989) In: Saegusa T, Higashimura T, Abe A (eds) Frontiers of macromolecular science, Blackwell, Oxford, p 67
5 Miyamoto M, Sawamoto M, Higashimura T (1984) Macromolecules 17:265
6 Higashimura T, Miyamoto M, Sawamoto T (1985) Macromolecules 18:611
7 Higashimura T, Kojima K, Sawamoto M (1988) Polym Bull 19:7
8 Kojima K, Sawamoto M, Higashimura T, J Polym Sci Part A Polym Chem, submitted
9 Higashimura T, Kojima K, Sawamoto M (1989) Makromol Chem Suppl 15:127
10 Sawamoto M, Kamigaito M, Higashimura T (1988) Polym Bull 20:407
11 Sawamoto M, Aoshima S, Higashimura T (1988) Makromol Chem Macromol Symp 13/14:513
12 Sawamoto M, Enoki T, Higashimura T (1987) Macromolecules 20:1
13 Higashimura T, Kishimoto Y, Aoshima S (1987) Polym Bull 18:111; Kishimoto Y, Aoshima S, Higashimura T (1989) Macromolecules 22:3877
14 Shohi H, Sawamoto M, Higashimura T (1989) Polym Bull 21:311
15 Decker D, Rempp P (1965) Compt Rend 261:1977; and subsequent publications
16 Kanaoka S, Sawamoto M, Higashimura T (1989) Polym Prepr Jpn 38:186, 1515
17 Minoda M, Sawamoto M, Higashimura T (1987) Macromolecules 20:2045

The Use of Free Radical Ring-Opening Polymerization to Produce Biodegradable Polymers

William J. Bailey
Department of Chemistry
University of Maryland
College Park, MD 20742

Although ionic ring-opening polymerization of heterocyclic monomers and the Ziegler-Natta ring-opening polymerization of cyclic olefins by metathesis are quite common, free radical ring-opening polymerizations are quite rare. On the other hand, free radical ring-closing polymerizations are common. Since a carbon-oxygen double bond is some 40-50 Kcal more stable than a carbon-carbon double bond, it was reasoned that the formation of a stable carbon-oxygen could be used to promote free radical ring-opening polymerization. Thus it was shown that cyclic ketene acetals could be used to promote such polymerizations with the introduction of an ester group into the backbone of an addition polymer. Thus if 2-methylene-1,3-dioxalane (1) was treated with a peroxide at high temperatures, a high molecular weight polyester resulted.

The mechanism of the polymerization involves first an addition of a radical to the cyclic ketene acetal 1 followed by ring opening with the driving force for the ring-opening being the formation of the stable ester group. At lower temperatures not all the rings open, but under high dilution ring opening is again favored. Copolymerization of these cyclic ketene acetals with a wide variety of common monomers, such as ethylene, styrene, and methyl methacrylate, produces copolymers containing ester groups in the backbone. These copolymers can be used to synthesize functionally terminated oligomers capped with hydroxyl groups, carboxylic acid groups, amine groups and thiol groups; to product thermally stable copolymers by stopping depolymerization; and to prepare biodegradable polymers.

B. C. Anderson · Y. Imanishi (Eds.)
Progress in Pacific Polymer Science
© Springer-Verlag Berlin Heidelberg 1991

Since Potts had shown that low molecular weight aliphatic polyesters were the only general class of synthetic polymers that were biodegradable, it appeared the copolymers with the cyclic ketene acetals with various monomers should lead to copolymers that were really polyesters and therefore should be biodegradable. Thus, the copolymerization of the cyclic ketene acetal **1** produced a series of copolymers that were in fact polyesters, but with physical properties very similar to polyethylene.

$$CH_2 = C \underset{O}{\overset{O}{\diagup\diagdown}} \overset{CH_2}{\underset{CH_2}{|}} \quad + \quad CH_2 = CH_2 \quad \xrightarrow[\text{peroxide}]{140°C, \text{ toluene}}$$

$$\underset{m}{-\!\!\!\left[\!CH_2\!-\!CH_2\!\right]\!\!\!-}CH_2-\overset{O}{\overset{\|}{C}}-O-CH_2-CH_2\underset{n}{-\!\!\!\left[\!CH_2\!-\!CH_2\!\right]\!\!\!-}CH_2-\overset{O}{\overset{\|}{C}}-O-CH_2CH_2-$$

If the ester monomer content was greater than 6%, the copolymer underwent biodegradation faster than polycaprolactone. In a similar fashion biodegradable polystyrene, polyacrylic acid, and poly-α-hydroxyacrylic acid could be prepared. In addition the reaction can be modified to produce a series of polymerized biodegradable vesicles.

The screening method for biodegradation involved complete conversion of the polymers to carbon dioxide and water by bacteria or fungi. A control for a baseline involved hydrolyzed casein and used polycaprolactone as a standard. On this basis a linear polyethylene copolymer containing 6% ester monomer units was metabolized slightly faster than polycaprolactone while a polystyrene copolymer containing 20% ester units degraded only half as fast as polycaprolactone. Copolymers containing both keto groups and ester groups in the backbone were shown to be both photodegradable and biodegradable. Furthermore, the residue from the photodegradation was shown to be biodegradable.

Effect of Dopants and Solvents
on Chronoamperometry of Polypyrrole

J. M. Ko, H. W. Rhee and C. Y. Kim*

Korea Institute of Science & Technology, Polymer Materials Lab.,
P. O. Box 131, Cheongryang, Seoul, Korea

ABSTRACT: PPy was galvanostatically formed on a platium plate in the system of TBADS-AN or NaDS-H_2O. DS^- embedded in the polymer on polymerization leaves on reducing in AN but sticks to it in H_2O. The aliphatic portion of the dopant contracts to become immobile in H_2O. Na^+ moves into polymer on reducing to counter the electrically freed anion. Diffusion coefficient of Na^+ deduced by chronoamperometry is 5.2×10^{-10} cm^2/sec in H_2O while that of TBA^+ is 2×10^{-11} cm^2/sec in H_2O. The coefficient of DS^- of the polymer formed with NaDS is 3.6×10^{-10} cm^2/sec in AN and that of the polymer formed with TBADS 4.7×10^{-10} cm^2/sec.

INTRODUCTION

Easy formation and good stability of polypyrrole(PPy) attracted many interests(1,2) and flexibility of the polymer was improved by using dopants with long aliphatic chains(3,4). It was found that hydrophobic nature of the aliphatic segment of dopants in an aqueous solution embedded the dopants tightly in PPy on polymerization(5). The dopants were hardly removed on reduction of the polymer in an aqueous system.

The rate and the degree of reaction of an electrically conductive polymer in repeated redox reaction are important factors in application of the polymer. The fast response of the polymer to an external stimulation may find uses in a sensor or a display. The reaction rate must depend on the mobility of ions in the polymer toward the reactive sites under an applied potential. The degree of the reaction in the cycled oxidation-reduction process predicts applicability of the electrically conductive materials in battery, sensor, transistor, solar cells, etc(6,7).

B. C. Anderson · Y. Imanishi (Eds.)
Progress in Pacific Polymer Science
© Springer-Verlag Berlin Heidelberg 1991

This work has a goal of improvement of PPy in stability. Energy dispersive X-ray spectroscopy and chronoamperometry were employed to chase movement of ions in the polymer in redox reaction. Mobility of ions involved on the redox reaction was correlated qualitatively with conformation of the dopants in the solvent used.

EXPERIMENTAL

Pyrrole and acetonitrile(AN) were dried in CaH_2 and distrilled before use. The electrolytes used are sodium dodecylsulfate(NaDS) and tetrabutylammonium dodecylsulfate(TBADS). NaDS was used as obtained but TBADS was synthesized in this laboratory(5). PPy films were prepared by applying a constant current of 5 mA/cm^2 to a platinum electrode with a working area of 1cm x 1cm in H_2O or AN with 0.36M of pyrrole and 0.036M of one of the electrolytes depending on the solvent. Scanning electron micrographs(SEM) were taken with a Hitachi SEM Model S-510. The search of ions in the PPy film was carried out by means of energy dispersive X-ray spectroscopy(EDS) using a device attached to the scanning electron microscope. Chronoamperograms(CAs) as well as cyclic voltammograms(CVs) were recorded employing a Hokuto Denko HA-301 potentiostat-galvanostat along with a Hokuto Denko HA-301 function generator and a Linseis Model LY1800. The reference electrode used was an Ag/AgCl, saturated KCl electrode.

RESULTS AND DISCUSSION

CAs of PPy reveal redox reactivity of the polymer as shown in Figure 1. The polymer with a thickness of 0.23μm was deposited galvanostatically on the platinum electrode. Chronoamperometry(CA) is performed by reducing the polymer at ·0.8V for 10 sec., switching the potential to 0.5V to hold for three seconds for oxidizing, and then switching back to ·0.8V for reducing. PPy formed in the aqueous solution of NaDS gives a sharp current change when CA is performed in the AN solution of TBADS as seen in Figure 1a. Figure 1b shows that the same polymer gives a broad current change when CA is performed in the aqueous solution of TBADS.

It is apparent that the solvent used in CA controls the rate of the redox reaction. The polymer formed in the system of NaDS-H_2O should have the aliphatic portion of the anion contracted. It is difficult to remove the anion from a thick PPy film on reducing in an aqueous solution(5). However, the polymer is swollen in AN and the anion leaves the polymer quick on reducing. The bulky dodecylsulfate ion (DS$^-$) moves easily in and out of the thin polymer film on redox reaction. It is interesting to note that the charges involved on redox reaction both in the systems of TBADS-AN and TBADS-H_2O are almost the same although the sharpness of peaks is distinctively different.

The different reaction rate was also noticed when CA of PPy formed in the TBADS-AN system was performed both in AN and H_2O as shown in Figure 2. The CA in the TBADS-AN system shows a sharp current change while the one in the TBADS-H_2O system gives a slow change in the current as shown in Figure 2a and 2b, respectively. The CA curves look the same as the ones in Figure 1, which indicates that the solvent in the redox reaction rather than the condition of the polymer formation controls the redox reaction rate. Since DS$^-$ stick to the

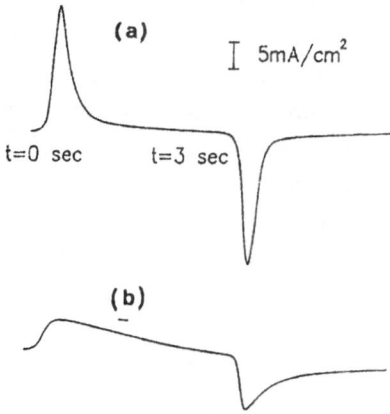

Fig. 1. CAs of PPy films formed with NaDS in systems of TBADS-AN(a) and TBADS-H_2O(b).

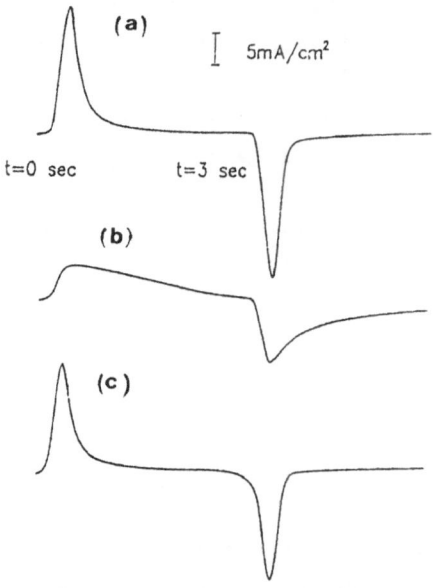

Fig. 2. CAs of PPy films formed with TBADS in systems of TBADS-AN(a), TBADS-H_2O(b) and NaDS-H_2O(c).

polymer on reducing as described previously, tetrabutylammonium cation (TBA+) should move but slow into the polymer on reducing and leave the polymer on oxidizing. The mobility of the cation is poor to show CAs broad. However, Figure 2c shows a sharp current change although the CA is obtained in the NaDS-H_2O system. The polymer as well as the anion are assumed to be contracted in H_2O and the anion with a long aliphatic chain hardly leaves the polymer on reducing. Na+, instead, moves in and out of the polymer to balance the ionic strength in the polymer.

Figure 3 is an EDS spectrum of PPy formed in the system of NaDS-H_2O and reduced for 37 hours in the same system. The strong sulfur peak proves that DS⁻ sticked to the polymer on reducing while Na+ moved into the polymer to counter the immobile anion freed from the neutralized polymer. The sharp CA curve of the polymer in the system of NaDS-H_2O suggests that Na+ moves freely in and out of the polymer in redox reaction while the bulky anions remain tight in the polymer.

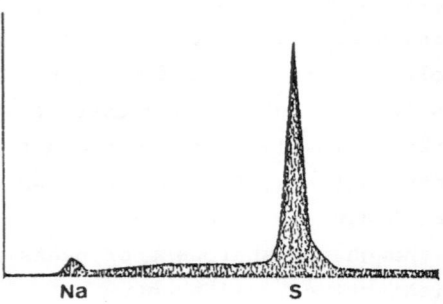

Fig. 3. EDS spectrum of PPy reduced in NaDS-H_2O for 37 hrs.

The current in CA changes with time as shown in equation(1)

$$i(t) = \frac{4nFAD_oC_o^*}{1} \exp \left(\frac{-\pi^2 D_o}{1^2} \right) t$$

$$= i_o \exp(-pt) \tag{1}$$

where n is the number of electrons transferred per molecules, F Faraday constant, A the surface area, D_o diffusion coefficient, C_o^* concentration of reaction sites, 1 sample thickness, respectively. The plot of of log {i(t)/i_o} vs. t may show a straight line and produce diffusion coefficients of the moving ions in the polymer from the slope.

Figure 4 shows the change of the normalized current in CAs with time. When TBADS is used as the electrolyte in the aqueous system, the slope of the CA curve is gentle to show a slow reaction. The slope of the curve in the case of the organic solvent is steep to prove a rather fast reaction. The steepest curve is obtained in the NaDS-H_2O system. It is significant to find that a fast redox reaction is feasible by employing a small cation as a balancing ion while fixing a long-chain dopant in the polymer for better mechanical properties.

The diffusion coefficients of the different ions are caculated from the slopes as shown in Table 1. The diffusion coefficient of Na$^+$ in the aqueous system is 5.2×10^{-10} cm^2/sec and that of TBA$^+$ 0.2×10^{-10} cm^2/sec, respectively. It is understandable that a small cation diffuses faster in the polymer than a large one does. DS$^-$ shows comparably high diffusion coefficient of 3.6×10^{-10} cm^2/sec with the polymer formed in the aqueous system and 4.7×10^{-10} cm^2/sec in the organic solvent, respectively. The difference of

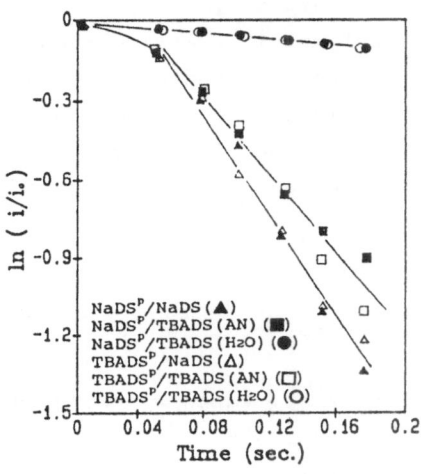

Fig. 4. Plot of ln(i/i$_o$) vs. time.

1.1×10^{-10} cm^2/sec in the diffusion coefficients may be originated from the conformational difference of the aliphatic segment in DS$^-$. The aliphatic segment in the anion may be stretched in the polymer on formation in the organic solvent while contracted in the aqous system.

Table 1. Diffusion Coefficients of Ions

Electrolyte used in polymerization	Electrolyte and solvent used in electrolysis	Diffusing ion	Diffusion coefficient ($\times 10^{-10}$ cm^2/sec)
NaDS	NaDS/H_2O	Na$^+$	5.2
TBADS	NaDS/H_2O	Na$^+$	5.2
NaDS	TBADS/AN	DS$^-$	3.6
TBADS	TBADS/AN	DS$^-$	4.7
NaDS	TBADS/H_2O	TBA$^+$	0.2
TBADS	TBADS/H_2O	TBA$^+$	0.2

CONCLUDING REMARKS

A dopant with a long aliphatic tail embedded in an electrically conductive polymer on electrochemical polymerization makes the polymer
mechanically superior to the one with an inorganic dopant. The
aliphatic nature of contracting in H_2O holds the dopant in the
polymer on reduction and brings a small cation into the polymer to
neutralize the freed anion. When a large cation is used instead of
small one, the redox reaction goes slow. It seems that redox reaction occurs only when the reactive site is electrically balanced.
The polymer is oxidized when dopant is nearby and is reduced when
the dopant is countered by an opposite ion. If a dopant is forced
to remain in the polymer on reduction, a countering ion must be
around. Otherwise, the reaction hardly goes. It may be possible to
make a mechanically strong conductive polymer with a lasting redox
reactivity if an electrolyte as well as a solvent are properly
chosen.

REFERENCES

1. Diaz AF, Kanazawa KK, Gardini GP (1979) J. Chem. Soc., Chem.
    Commun. 635
2. Tourillion G, Garnier F (1982) J. Electroanal. Chem. 135:173
3. Wernet W, Mokenbusch M, Wegner G (1984) Makromol. Chem. Rapid
    Commun. 5:157
4. Bates N, Cross M, Lines R, Walton D (1985) J. Chem. Soc., Chem.
    Commun. 871
5. Ko JM, Rhee HW, Kim CY (1990) Makromol. Chem., Macromol. Symp.
    33:353
6. Cowan DO, Wiygul FM (1986) C & EN July 21 28
7. Baughman RH (1984) Contemporary Topics in Polymer Science 5:321

ACKNOWLEDGEMENT

The work is financed by the Ministry of Science and Technology,
Korea. Thanks go to Dr. H. N. Cho and K. S. Min who helped to synthesize tetrabutylammonium dodecylsulfate.

Recent Developments in the Field of Engineering Plastics

Kunio Maeda

Plastics Technical Center, Asahi Chemical Industry Co.,Ltd.
3-1, Yako-1-Chome, Kawasaki-ku, Kawasaki City 210, JAPAN

Abstract: Many people point three directions in the developments of engineering plastics ---- high performance, high function and multifunction. When I look at the recent trends of the plastic-made products in Japanese market, I believe we must add to them the fourth direction --- so called "high touch". "High-touch" materials, which have been seen often in the field of fibers, have high sensuous characteristics which appeal directly to the five senses. On the contrary, there have been few examples developed for this target in the field of plastics. I considered what kind of plastics could be of "high touch", referring four examples recently developed by Asahi Chemical.

1. SITUATION OF PLASTICS IN THE CAR INDUSTRY

For the start of the discussion, let me think about the car-makers' wants as an example of those of the plastics users.

People of car manufacturers are always considering how well their products appeal to the consumers. Needless to say that it depends on the time and also on the country.

Thirty years ago in Japan, it was reliability, namely, to fulfill its three essential functions, to move, to turn and to halt surely.

Twenty years ago, when highway networks began to appear also in Japan the consumer required higher-performanced cars such as faster-driving ones, more powerful ones and bigger-sized ones.

Fifteen years ago, fuel economy, clean exhaust and long life became the targets. Later, various convenient functions including those for safety were added due to the progress of electronics. Of course comfortableness as well as good styling has been always considered.

And now, their recent advertising messages are, for example, "Feel the beat" or "Fun to drive". They believe that the consumers' motive

B. C. Anderson · Y. Imanishi (Eds.)
Progress in Pacific Polymer Science
© Springer-Verlag Berlin Heidelberg 1991

Table 1. REQUIREMENTS FOR CARS

Essential Functions:	Move, Turn, Halt
Higher Performances:	Speed, Power,Life
Higher Functions	: Convenience, Safety
Additional Items	: Amenity, Appearance
And More ---?	: To fit "KANSEI"

for buying is now dominated by "KANSEI", at least in the Japanese market.

The word "kansei" itself is usually translated to "sensibility" or "sensitivity", but recent Japanese people, especially in the field of marketing, favorably use it in a little bit different nuance. According to the literature cited as 1, it may be something like an integrated feeling received by five senses. A consumer decides to buy something when he or she feels it is nice, instead of logical consideration, even if the thing is a rather sophisticated mechanism like a car.

By the way, plastics used in an automobile have reached about 8% by weight in Japan, and are still increasing. Up to now plastics have been useful for upgrading performances and functions of cars, and also for saving production cost. However, people of car makers do not seem to think about plastics as the key-point material of cars for fitting "kansei".

Their ideal materials seem to be wood, leather, fur or even silk. These are believed to be the "real" materials, and consumers would, they believe, prefer such "real" articles. Plastics are, in this

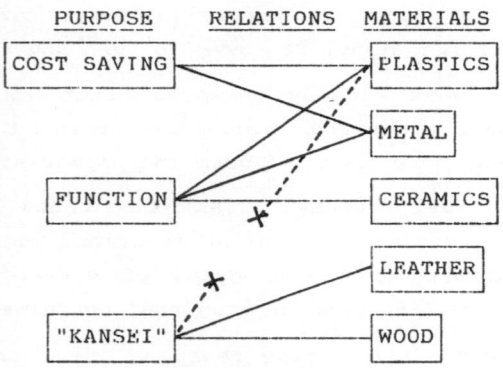

Fig.1 MATERIALS EXPECTED IN THE CAR INDUSTRY

point of view, regarded only as a "substitute" or even an "imitation" of the "real" materials!

What is the reason that plastics are given such poor status?

2. DEVELOPMENT OF "HIGH-TOUCH" MATERIALS

A wide variety of plastics have been developed so far, in order to have higher performances, higher functions, a better processability and a lower manufacturing cost. However, I am afraid that little effort have been made for developing plastic materials which meet "kansei". I have not heard of any trial to make the "real" plastics instead of an "imitation".

It is said that the major directions of developing plastics, especially engineering plastics, are "high performance", "high function" and "multi-function". Here I propose that we would add to our targets the development of plastics for fitting "kansei" as the fourth direction. In order to express this trend, I followed the literatures 2,3 and 4, using the word "high-touch".

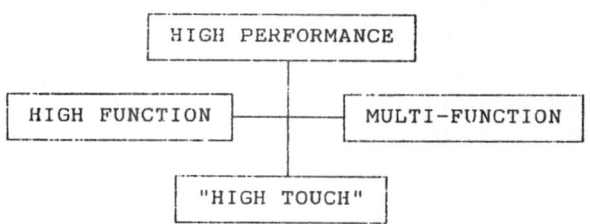

Fig.2 FOUR DIRECTIONS IN THE
DEVELOPMENT OF ENGINEERING PLASTICS

To get a hint for supposing what kind of plastics would be characterized as a "high-touch" materials, I will see how the existing "real" materials are. Marble feels massive, hard and cool. Wood smells mild and feels friendly. Crystal glass looks brilliant and sounds clear. Leather feels soft and flexible.

However, no matter how wonderful they are, we should not try to develop plastics of similar characteristics to these materials. That brings inevitably another Imitation. Plastics should find out its own characteristics, which only plastics can achieve.

There exist a few examples of plastic-made article which may be regarded as "real". Golf clubs made of carbon fiber composite are generally believed to be superior to ones made of metal. Skis made of glass fiber composite have won a higher status than those made of wood. ---But it is an irony that people call them "carbon shaft" and "glass ski" instead of "plastic shaft" and "plastic ski".

Anyhow, these examples suggest that the satisfication of performances and functions is prior to appearance or feeling. In that point of view, "high-touch" is an additional direction to high performance and high function, rather than being an opposite concept to those.

In the case of clothes, the essential functions had been already satisfied in the old days, and the major efforts in this field are directed to fashion, the world of "kansei" itself. And what I must emphasize here is that the researches for fibers as the raw material of clothes have also been directed to satisfy the requirements related to fashion.

I believe that the similar situation is coming also in the field of plastics and its applications. The customers have been requiring the plastics manufacturers to give support to make "high-touch" products. In most cases the plastics manufacturers probably recommend some grades or some processing technologies which they suppose to be the closest to the requirements, picked up from their line-up developed for the other purposes.

But I think it will be useful for us to look over some examples related to "kansei", in order to consider what kind of materials or technologies could be closer to that, even if they were not developed for that.

3. EXAMPLES OF RECENT DEVELOPMENTS IN ASAHI CHEMICAL

(1) Thermoplastic Elastomer, "TUFTEC_TM"

The interior parts of automobiles require for tender feeling by touch. Among the existing materials vulcanized rubber has a relatively close characteristics to this requirement. But rubber has some defects in color and smell, both the important factors related to "kansei". Plasticized PVC is often applied to this use, but its feeling of resilience is not enough. TPU and TPO are used sometimes, but much softer feeling is wanted.

Asahi has a series of styrene and olefin block copolymers, named "TUFTEC$_{TM}$", which belong to a so-called thermoplastic elastomer, where polystyrene block is the hard segment and polyolefin block composes the soft segment. It is prepared by hydrogenation of styrene and butadiene block copolymer catalyzed by metallocene.

Polystyrene block   Elastomer block

Fig.3 STRUCTURE MODEL OF "TUFTEC$_{TM}$"

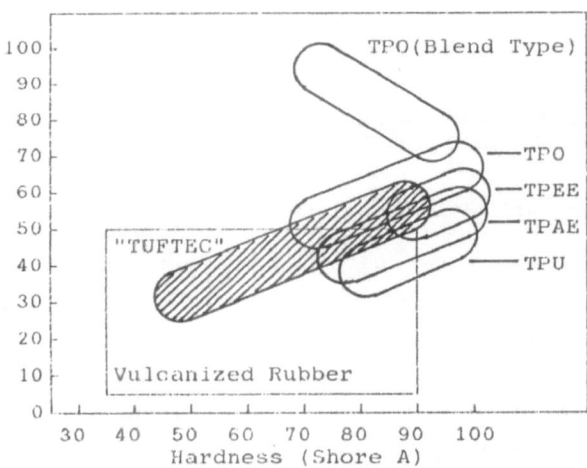

Fig.4 COMPARISON IN RUBBERY PROPERTIES OF ELASTOMERS

Various thermoplastic materials are located on the map in Fig.4, showing the relation between hardness and compression set. The lefter and lower site of this map is thought to be favorable for the interior parts with their feeling by touch, and vulcanized rubber covers this area.

The balance of TUFTEC is the closest to that of rubber among the other thermoplastic elastomers.

Besides, it can be colored clearly, does not have disagreeable smell and has a good weatherability. Since it adheres well to styrenic and olefinic polymers, a double-layered articles having rigid skelton covered with TUFTEC can be molded easily. Fig.5 and 6 are the examples.

TUFTEC has still a defect that its molded surface is a little sticky, but I expect it to become one of the "high-touch" materials for car interior parts, when we can improve the defect.

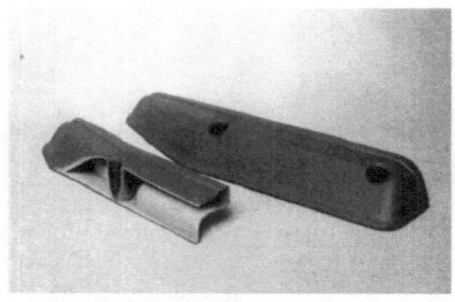 

Fig.5 BLOW-MOLDED DOUBLE-WALLED   Fig.6 DOUBLE-INJECTION-MOLDED
ARM REST (POLYETHYLENE/TUFTEC)   TURNING SWITCHES (ABS/TUFTEC)

(2) Highly-Lubricated Polyacetal Resin, "TENAC$_{TM}$LA501"

Asahi has developed a new ultra-lubricated polyacetal grade, named "TENAC$_{TM}$LA501".

Polyalkyleneglycol-monoalkylate is used as a chain transfer agent during polymerization of formaldehyde, and it is combined with polyoxymethylene chain as a small block (sample A of Table 2).
TENAC LA501 (sample B) is a compound of this block copolymer and a free lubricant which is compatible with said block.

For the reference, ordinary acetal homopolymer (sample C) and its compound containing the same free lubricant as used in sample B (sample D) was prepared.

A thrust-type rotative friction tester was used for evaluating wear rates of the four samples. In the first series of the tests each

Table 2 WEAR RATES IN POM/POM FRICTION

	vs.A	vs.B	vs.C	vs.D
Sample A	MELT	--	0.5	--
Sample B	--	0.5	1.0	---
Sample C	4.5	<0.1	MELT	4.0
Sample D	--	----	0.5	4.5

Unit: mg/km

Test Conditions:
Friction Speed:15cm/sec
Load: 2kg/sq.cm
Temperature: 23°C
Humidity: 50%R.H.

A: Block Copolymer Containing Lubricative Unit
B: A + Free Lubricant
C: Ordinary Homopolymer
D: C + Free Lubricant

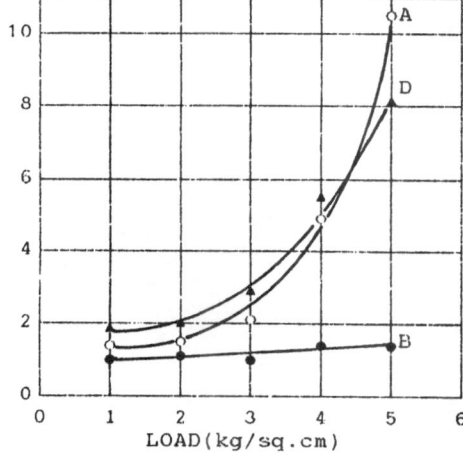

Test Conditions:

Friction Speed:60cm/sec

Temperature: 23°C

Humidity: 50%R.H.

Fig.7 WEAR OF VAROUS POM BY STEEL

material was matched to the sample C. The wear rates of each samples
as results are shown in the table horizontally, while those of the
sample C itself are arranged vertically.

In the second series each sample was matched to the same one, of
which results can be seen as diagonal data.
It shows that the sample B has a very small wear rate of itself and
it does not damage much the opposite material.
Fig. 7 shows their wear rates against steel. The sample C's data is
not drawn here because of scaling out.
When the load is higher, the sample B's wear rate is as low as one
tenth of the others'.

120

The mechanism of its superior friction properties is supposed that the lubricative blocks in the polymer chain might assist the bleeding-out of the free lubricant and maintain it on the surface, as illustrated in Fig.8.

Fig.9 to 11 show the examples of its application. They are slide switches of a pocket-sized cassette player, a tape-loading mechanism of VTR and a laser disk rack, respectively.

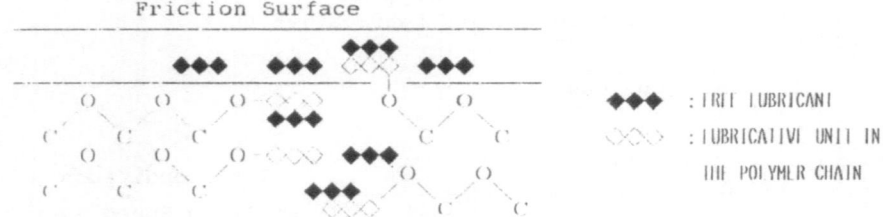

Fig.8 ASSUMED ILLUSTRATION OF THE SURFACE OF "TENAC LA501"

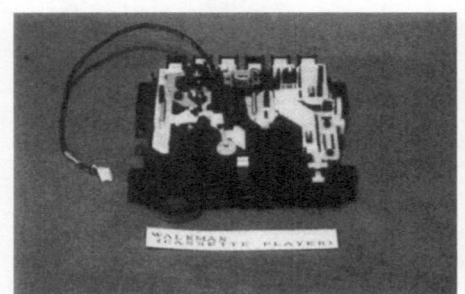

Fig.9 SLIDE SWITCHES OF CASSETTE PLAYER MADE OF "TENAC LA501"

Fig.10 TAPE-LOADING MECHANISM OF VTR MADE OF "TENAC LA501"

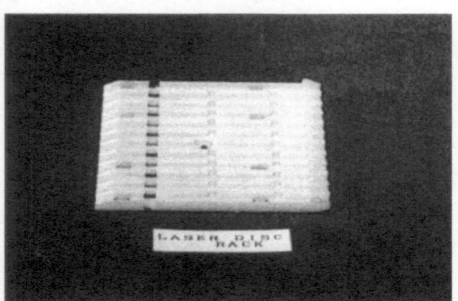

Fig.11 LASER DISK RACK MADE OF "TENAC LA501"

The white parts in every figure are TENAC LA501, and they work smoothly for much longer time and with less friction noise than those made of conventional grades of polyacetal.

Though these applications seem to be quite functional, I believe that this plastics almost satisfies the condition to be the "real" material, because its characteristics is functional, favorable and also intrinsic --- quite hard, smooth and slippery.

(3) Bright Surface Injection Molding Technology

Referring the case of fibers, "high-touch" is achieved not only by the development of polymer itself, but also by that of fabrication technology. Exotic-profile fibers and crimped fibers are the examples.

I introduce two molding technologies of plastics here and in the following section.

It is difficult to get an extremely high-gloss surface of the molded articles even in the injection molding, especially in the case of glass fiber-filled plastic compounds, because solidification of the surface preceeds the enough contact of the resin to the mold surface, prevented by the filler.

Very high mold temperature could avoid this trouble, but it would cause extraordinary retardation of molding cycle time, because it would take so long a time for the hot mold to be cooled down for ejection.

In the BSM or Bright Surface Molding Process, only the surface layer of the mold is heated up at the start of every molding cycle, therefore it can easily be cooled down by heat conduction.

High-frequency induction is applied for realising this. Before the injection step the mold is opened slightly and induction coil is inserted. High frequency induces current in the adjacent electric conductive material, namely, the mold, and generates heat. Conveniently, the induction takes place only at the surface.

Fig.12 shows the apparatus of BSM process. Fig.13 shows the comparison of two molded articles both made of glass-filled ABS resin, the right one is molded by BSM process while the left one is conventionally injection-molded.

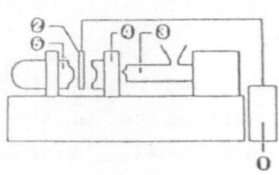

(1) HIGH FREQUENCY OSCILLATOR
(2) INDUCTOR
(3) INJECTION CYLINDER
(4)(5) MOLD

Fig.12 BSM APPARATUS

Fig.13 BSM-MOLDED SAMPLE COMPARED
WITH CONVENTIONAL MOLDING

Fig.14 BSM-MOLDED PLATE WITH
WELL-PRINTED GRAIN

Fig.15 CONVENTIONALLY MOLDED
PLATE WITH DULL GRAIN

Another samples are shown in Figs.14 and 15, both using a grain-surfaced mold. The BSM-molded plate (Fig.14) has the expected grain, but the other is still transparent because the grain is not printed well. You will also find that the seam-line is eliminated in the BSM-molded piece.

(4) Gas Injection Molding

In the conventional injection molding process, surplus resin melt is fed into the mold cavity in order to compensate the volume reduction of filled resin by cooling down. This cannot avoid the surface defects such as sink marks, especially when the molded article has ribs or thicker-walled parts.

In the gas injection molding process, compressed nitrogen gas is blown into the molding instead of the secondary injection. The gas breaks the gate and runs into the inner side of the molding, especially at thicker-walled sites, where the temperature is higher and the melt viscosity is lower, and a hollow-shaped article is formed as a result.

Fig.16 shows a typical apparatus for gas injection. An ordinary injection molding machine is equipped with a specially designed valve nozzle and a gas unit which generates and accumulates pressurized gas. The gas is introduced to the mold and recovered, synchronized with the molding cycles under the control of a sequencer.

Figs.17 and 18 are the samples showing its hollow cross-section.

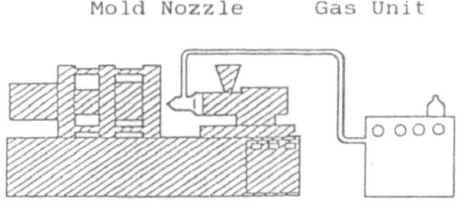

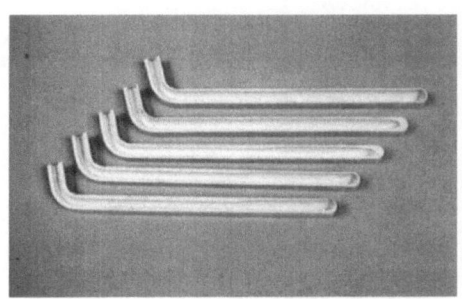

Fig.17 CURVED PIPES MOLDED BY
GAS INJECTION

Fig.16 GAS INJECTION MOLDING SYSTEM

Fig.18 CASING MOLDED BY GAS
INJECTION WITH HOLLOW RIBS

Fig.19 COMPARISON IN WARPAGE
(LEFT ONE IS GAS-INJECTED)

Sink marks are eliminated because the inner pressure of the molding is held until solidification is completed. Warpage, which is often seen in the ordinary injection molding, becomes much smaller in this process, because no overpacking of resin is needed. Fig.19 shows the comparison in warpage.

With above four examples I have tried to find a clue to the "high-touch" materials and technologies. As I mentioned repeatedly, the start of developing "high-touch" plastics will be in the satisfication of functions. And, it will be more important to be conscious of consumers' "kansei" and to keep efforts to make our materials more favorable.

To make plastics a "real" material, that is my perpetual desire.

References:

1 Nagamachi M (1989) Jidosha-Gijutsu 43(1):94
2 Shinozaki S (1989) Nikkei Mechanical Jun.26:51
3 ibid. Jul.17:82
4 ibid. Aug.7:67

Solution Associations in Hydrophobically-Modified Polyethylene Oxides

J. C. Salamone, A. M. Thompson, C. H. Su, Y. Ke,
and A. C Watterson

Polymer Science Program
Department of Chemistry
University of Lowell
Lowell, Massachusetts 01854

ABSTRACT

The solution associations of some hydrophobically modified polyethylene oxides have been investigated. It was found that as the hydrophobe level increased in the polymers, the apparent molecular weight and intrinsic viscosity increased, indicative of increased intermolecular association. Additionally, the solution associations appear to be augmented in salt solution.

INTRODUCTION

The modification of water soluble polymers for tailored solution behavior has been of interest to many research groups seeking different end uses. Enhanced oil recovery, fracturing fluids, flocculation, associative thickeners, personal care products and many other applications exist for polymers of unique solution behavior. Although the ultimate applications may differ, a common goal of researchers in the water soluble polymer field is a solution viscosity enhanced through polymer-polymer intermolecular associations, often with the stipulation that associations must remain intact in the presence of mono- and divalent salts. To this end our research group has extensively studied the properties of several classes of water soluble polymers. While the majority of our studies have involved ampholytic systems: ampholytic ionomers [1], polyampholytes [2,3], polysulfobetaines [4,5] and liquid crystalline ionomers [6], we recently have been investigating the solution properties of some nonionic, associative water soluble polymers based on polyethylene oxide.

Hydrophobic associations in aqueous solution have been recognized in polymeric systems for some time. Their ability to enhance viscosity has led to research directed toward modifying existing water soluble polymers with alkyl substituents capable of associating in solution. These have included studies by Strauss et al. on polysoaps from polyvinylpyridine [7], Schulz [8] and McCormick

[9] and their respective coworkers on hydrophobically modified polyacrylamides and this laboratory concerning hydrophobically modified polyvinylimidazolium salts [10] and more recently, hydrophobically modified polyampholytes and ampholytic ionomers [11-14]. Hydrophobically modified polyethylene oxides have received relatively little attention. Landoll has reported the incorporation of alkyl epoxides into nonionic cellulose ethers [15] and the polymerization of ethylene oxide with *n*-alkyl epoxides [16].

In this account we detail the initiation of research into hydrophobically modified polyethylene oxides. Ethylene oxide has been copolymerized with 1,2-epoxyalkanes to produce water soluble polymers with various amounts of hydrophobic side chains. Through control of hydrophobe level in the copolymers, highly water soluble copolymers can be obtained. The presence of solution associations are investigated through the use of light scattering and capillary viscosity. As solution associations are formed or destroyed, apparent molecular weight (light scattering) and hydrodynamic volume (intrinsic viscosity) are affected. Also, the effects of added salt on these nonionic copolymers is examined.

EXPERIMENTAL

Catalyst Preparation

The preparation of the polymerization catalyst was based on the method of Landoll [16]. To a 100 ml three necked round bottom the following was successively charged: 24 ml of a 25% solution of Et_3Al in toluene (Aldrich), 20 ml tetrahydrofuran, 0.07 ml water, 2.3 ml 2,4-pentanedione and 4.7 ml tetrahydrofural alcohol. Each addition was made dropwise after dissolution of the previous charge. The mixture was kept at 0°C for 1 hour followed by four hours at 80°C, after which the catalyst solution was ready for use.

Polymerization of Ethylene Oxide with 1,2-Epoxyalkanes

Although the feed ratios of 1,2-epoxyalkane to ethylene oxide were varied, the total monomer concentration in solution was maintained at approximately 10 (w/w)%. One mole% catalyst, relative to the total monomer, was used in all polymerizations.

A desired amount of the alkylene epoxide was charged to a 500 ml round bottom flask. The flask was placed on a vacuum line and degassed. The appropriate amount of anhydrous toluene (approx. 90 (w/w)%) was distilled directly into the reaction flask followed by the distillation of ethylene oxide, from lithium aluminum hydride, into the flask. Lastly, the appropriate amount of catalyst solution was charged and the temperature of the solution was brought to 40-45°C for approximately 12 hours. The result was a highly viscous polymer solution. The crude polymer was collected by precipitation in hexane followed by vacuum drying. Purification was carried out by redissolving the polymer in a 5% HCl solution and dialyzing (MWCO = 12,000 - 14,000) against distilled water for one week. The final product was isolated by lyophilization.

Polymer Characterization

The quantity of hydrophobe incorporated in the copolymers was determined with ^{1}H NMR using the peak height ratios between the appropriate protons. A Bruker WP-270 SY Spectrometer was employed at 270 MHz. The molecular weights of the copolymers in various solvents were determined using a Wyatt Technologies DAWN Model F Laser Light Scattering Spectrophotometer. All solutions were made dust free by ultrafiltration through 0.22 micron filters. The refractive index increments were determined using a Baush and Lomb Differential Refractometer. When determinations were made in salt solution, all solutions were dialyzed against solvent to ensure Donnan membrane equilibrium. Data was evaluated by the Zimm technique with the aid of an interfaced IBM microcomputer and software developed in this laboratory. Solution viscosities were determined using a Cannon-Ubbelohde single bulb dilution viscometer for intrinsic viscosity measurements and a Brookfield Model HATDCP cone and plate digital viscometer. All viscosity measurements were made at $25.0 \pm 0.1°C$.

RESULTS AND DISCUSSION

As seen from Table 1, the amount of 1,2-epoxy alkane monomer incorporated into the copolymers is substantially less than the original feed ratio. This is considered to stem from either a lower reactivity of the sterically hindered monomer or a high degree of association of the hydrophobic monomer in the polymerization mixture preventing rapid propagation.

The association of the hydrophobically modified polyethylene oxides in aqueous solution is apparent from the light scattering and intrinsic viscosity data listed in Tables 1-3. As the hydrophobe incorporation level increases within a copolymer, the apparent molecular weight increases dramatically (see 1,2-epoxy-octadecane, x=15, copolymer, Table 1). The radius of gyration increases accordingly. This effect has been observed in several other associating polymer systems studied in this laboratory. As the hydrophobe level increases, the solvent-segment interactions decrease as evidenced by the decreasing second virial coefficient (A_2, Table 1). The polymers thus tend to associate in inter-

128

Table 1. Hydrophobic Monomer Incorporation and Light Scattering Results for Several Hydrophobically Modified Polyethylene Oxides in Aqueous Solution.

x^*	Feed Ratio (mole%)	Resultant Ratio (mole%)	$M_w \times 10^{-6}$ (g/mole)	R_g (nm)	$A_2 \times 10^6$ (ml·mole/g)
15	0.4	0.16	3.5	123	14.4
	1.1	0.20	7.5	140	6.7
7	1.5	0.08	3.2	66	32.5
3	1.5	0.15	3.6	116	31.6

*Refers to the number of side chain methylene carbons:

$$-(CH_2CH_2O)---(CH_2CHO)-$$
$$(CH_2)_x$$
$$CH_3$$

and intramolecular micellar-like complexes. When the amount of intermolecular associations is significantly greater than the intramolecular associations, significant solution viscosity enhancements occur. The solution viscosity has also been seen to be solvent polarity dependent. Solutions of poly(ethylene oxide-co-1,2-epoxyoctadecane) have shown five fold viscosity increases as the solvent polarity increases from toluene to chloroform. Additionally, time dependent behavior is observed for solutions of the copolymer in chloroform. This solution behavior of the copolymers is under detailed investigation.

Table 2. Light Scattering Results For a PEO Copolymer Containing 0.16 mole% 1,2-Epoxyoctadecane (X=15) in Aqueous Salt Solutions.

[NaCl] (mole/l)	$M_w \times 10^{-6}$ (g/mole)	R_g (nm)	$A_2 \times 10^6$ (ml·mole/g)
0	3.5	123	14.4
0.5	6.5	108	2.3
1.0	5.6	105	-5.1
2.0	5.2	104	-9.8

When electrolytes are added to aqueous solution, the copolymers tend to associate to a larger degree. The effect of added electrolyte on the molecular dimensions of the copolymers can be seen in Tables 2 and 3. While the apparent molecular weights increase with increasing salt concentration, the hydrodynamic volume of the complexes decreases, evidenced by both a decreasing A_2 and intrinsic viscosity. The higher salt concentrations yield negative second virial coefficients indicative of a very compact structure.

In the determination of the intrinsic viscosity of the copolymers, the familiar Huggins Equation:

$$n_{sp}/C = [n] + k'[n]^2C \qquad (1)$$

was used which was derived for polymers in dilute solution, completely isolated from one another by a solvent atmosphere [17]. The correlation for the copolymers is not linear, as reflected by the correlation coefficients in Table 3. However, a modified Huggins Equation (equation 2) developed in this laboratory for application to highly associating polymeric systems yielded very linear results when relative viscosity is plotted against concentration (Table 3).

$$n_r = 1 + [n]_m C \qquad (2)$$

Table 3. Intrinsic Viscosities for a PEO Copolymer Containing 0.16 mole% 1,2-Epoxyoctadecane (X=15) in Aqueous Salt Solution Determined by Both the Huggins and Modified Huggins Equation.

	Huggins Equation		Modified Huggins Equation		
[NaCl] (mole/l)	[n] (dl/g)	r	$[n]_m$ (dl/g)	Intercept	r
0.0	2.1	0.84	1.93	1.01	0.99
0.5	1.5	0.98	1.59	0.99	0.99
1.0	1.4	0.96	1.56	0.99	0.99

Note: $[n]_m$ was determined using the Modified Huggins Equation

The linearity of the results using the modified Huggins Equation suggests the polymers are highly associated in salt solution. The degree of association appears to increase with increasing salt concentration.

Investigation into the rheological implications of these findings is currently underway and will be reported in the future.

130

ACKNOWLEDGEMENT

Helpful discussions with L. M. Landoll are gratefully acknowledged.

REFERENCES

1. J. C. Salamone, I. Ahmed, E. L. Rodriguez, L. Quach and A. C. Watterson, *J. Macromol. Sci.-Chem.*, **A25**(5-7) 811(1988)

2. J. C. Salamone, I. Ahmed, M. K. Raheja, P. Elayaperumal, A. C. Watterson and A. P. Olson, "Water Soluble Polymers for Petroleum Recovery," G. A. Stahl and D. N. Schulz, eds., Plenum Publishing, 1988, pp. 181-94

3. J. C. Salamone, C. C. Tsai, A. P. Olson and A.C. Watterson, "Ions in Polymers," A. Eisenberg, ed., Adv. Chem. Ser., 187, Chap. 22 (1980)

4. J. C. Salamone, W. Volksen, A. P. Olson and S. C. Israel, *Polymer*, **19**, 1157 (1978)

5. J. C. Salamone, W. C. Rice and A. C. Watterson, *Am. Chem Soc. Div. Polym. Chem., Polym. Prepr.*, **29**(1), 279 (1988)

6. J. C. Salamone, C. K. Li, S. B. Clough, S. L. Bennett and A. C. Watterson, *Am. Chem. Soc. Div. Polym. Chem., Polym. Prepr.*, **29**(1), 273 (1988)

7. U. P. Strauss and E. G. Jackson, *J. Phys. Chem.*, **6**, 649 (1951)

8. D. N. Schulz, J. J. Kaladas, J. J. Maurer, J. Block, S. J. Pace and W. W. Schulz, *Polymer*, **28**, 2110 (1987)

9. C. L. McCormick, T. Nakamasa and C. B. Johnson, *Polymer*, **29**, 731 (1988)

10. J. C. Salamone, S. C. Israel, P. Taylor and B. Snider, *J. Polym. Sci., Polym. Symp.* **45**, 65 (1974)

11. J. C. Salamone, A. M. Thompson, M. K. Raheja, C. H. Su and A. C. Watterson,*Am. Chem. Soc. Div. Polym. Chem., Polym. Prepr.*, **29**(1), 281 (1988).

12. J. C. Salamone, A. M. Thompson, W. C. Rice, K. T. Lai, M. W. Boden, Y. M. Luo, M. K. Raheja and A. C. Watterson, IUPAC MACRO '88, 32nd Int. Symp. on Macromol., 138 (1988)

13. J. C. Salamone, A. M. Thompson, C. H. Su and A. C. Watterson, *Am. Chem. Soc. Div. Polym. Chem., Polym. Prepr.*, **30**(1), 326 (1989)

14. J. C. Salamone, A. M. Thompson, C. H. Su and A. C. WAtterson, *Am. Chem. Soc. Div. Polym. Chem., Polym. Prepr.*, **30**(1), 333 (1989)

15. L. M. Landoll, *J. Polym. Sci., Polym. Chem. Ed.*, **20**, 443 (1982)

16. L. M. Landoll, U.S. Patent 4,304,902, Dec. 8, 1981

17. M. L. Huggins, G. Natta, V. Desreux and H. Mark, *J. Polym. Sci.*, **56**, 153 (1962)

Preparation of Random, Block and Graft Copolymers of High Stereoregularity and their Characterization

Koichi HATADA, Tatsuki KITAYAMA, and Koichi UTE

Department of Chemistry, Faculty of Engineering Science, Osaka University,
Toyonaka, Osaka 560, Japan

ABSTRACT: Highly isotactic and syndiotactic polymerizations of methacrylates with
t-C$_4$H$_9$MgBr and t-C$_4$H$_9$Li-R$_3$Al, respectively, in toluene at low temperature were utilized for the preparation of highly stereoregular random and block copolymers.   The
detail structure of these copolymers were analyzed by ^{1}H and ^{13}C NMR spectroscopy.
Molecular weight dependence of the copolymer composition was studied by an on-line
GPC/NMR. Stereoregular PMMA macromonomers having styrene group were prepared from
the living PMMA anions.   Radical polymerization of these macromonomers and copolymerization with styrene clearly indicated that the isotactic macromonomer had higher
reactivity than the syndiotactic one.   Termination rate constant of propagating
radical of the macromonomers was directly determined by ESR spectroscopy, and was
found to be much smaller than that for styrene polymerization.   Some of the properties of the stereoregular copolymers and the poly(PMMA macromonomer)s were also
described.

INTRODUCTION

Block, random and graft copolymerizations have provided us great opportunities not
only for modification of the properties of the respective homopolymers but also for
creation of the new properties. We recently found two living systems for the highly
stereoregular polymers of methyl methacrylate(MMA); t-C$_4$H$_9$MgBr prepared in diethyl
ether gives highly isotactic(it-) poly(methyl methacrylate) (PMMA) with narrow molecular weight distribution (MWD)[1] and t-C$_4$H$_9$Li-R$_3$Al gives highly syndiotactic (st-)
PMMA with narrow MWD[2]. These two types of stereoregular polymers have the same
chemical structure; t-C$_4$H$_9$ group at the left-(initiating) end and methine proton at
the right-(terminating) end, and can be most suitably utilized for testing the tacticity dependence of the polymer properties [3].   These living polymerizations
permitted us to prepare various kinds of stereoregular block and random copolymers
of various kinds of methacrylates.   Stereoregular PMMA macromonomers could also be
prepared from the living PMMA anions, and were used for the synthesis of graft and
comblike polymers with stereoregular branches [4].   The control of stereoregularity
of block, random and graft copolymers will provide us a new way of controlling the

B. C. Anderson · Y. Imanishi (Eds.)
Progress in Pacific Polymer Science
© Springer-Verlag Berlin Heidelberg 1991

properties of the copolymers. The detailed structural analysis of the copolymers by means of NMR spectroscopy including a newly developed on-line GPC/NMR[5-8] is also described.

STEREOREGULAR BLOCK AND RANDOM COPOLYMERS OF MMA AND ALKYL METHACRYLATE

Highly isotactic copolymers prepared with t-C_4H_9MgBr in toluene: Anionic polymerization of methacrylate gives a wide variety of stereoregular polymers [9,10]. However, the polymerization reaction often involves side reactions and has been difficult to control [10]. In the polymerization of MMA with Grignard reagent such as n-C_4H_9MgCl, two types of reactions occur between the initiator and the monomer at the initial stage of polymerization; the attack of the initiator on the C=C double bond of MMA forms the initiating species, and the attack on the C=O double bond leads to the formation of butyl isopropenyl ketone and CH_3OMgCl [11].

$$C_4H_9MgCl + CH_2=\overset{\overset{\displaystyle CH_3}{|}}{\underset{\underset{\displaystyle OCH_3}{|}}{\underset{\displaystyle C=O}{C}}} \longrightarrow C_4H_9-CH_2-\overset{\overset{\displaystyle CH_3}{|}}{\underset{\underset{\displaystyle OCH_3}{|}}{\underset{\displaystyle C=O}{C^- MgCl^+}}}$$

$$\overset{\overset{\displaystyle CH_3}{|}}{\underset{\underset{\displaystyle OCH_3}{|}}{\underset{\displaystyle C_4H_9-C-O^-MgCl^+}{CH_2=C}}} \longrightarrow \overset{\overset{\displaystyle CH_3}{|}}{\underset{\underset{\displaystyle C_4H_9}{|}}{\underset{\displaystyle C=O}{CH_2=C}}} + CH_3O^-MgCl^+$$

The ketone is more reactive than MMA and reacts with some of the propagating MMA anions to form the less reactive anions. The alkoxide coordinates with the propagating species to affect their stereospecificity and reactivity. Therefore, the carbonyl attack makes the polymerization reaction complex and uncontrollable.

Another factor which affects the stereospecificity of the polymerization is Schlenk equilibrium of Grignard reagent; $2\ RMgX \rightleftharpoons R_2Mg + MgBr_2$.
Generally, RMgX gives an it-PMMA and R_2Mg a st-PMMA [12,13], and thus the equilibrium mixture of RMgX and R_2Mg may result in the coexistence of active species with different stereoregularities.

Polymerization of MMA in toluene at −78°C by t-C_4H_9MgBr prepared in diethyl ether proceeds in a living manner to give a highly it-PMMA with narrow MWD [1]. The initiator solution contains an excess amount of $MgBr_2$ which forms through Wurtz type coupling reaction between t-C_4H_9MgBr and t-C_4H_9Br, and the Schlenk equilibrium in this Grignard reagent shifts toward RMgBr side, leading to isotactic-specific polymerization. Bulky t-C_4H_9 group prevents the initiator from being involved in the carbonyl addition reaction. Under these conditions, a single type of active species forms without side reaction, and a highly isotactic living PMMA is formed.

Living and highly it-PMMA anion obtained with t-C_4H_9MgBr could be utilized for the preparation of highly block it-copolymers[14,15]. Figure 1 shows the GPC curve of the block copolymer of MMA and ethyl methacrylate(EMA) with the degree of polymerization DP of each block of 59 and that of PMMA formed at the same $[MMA]_0/[t$-$C_4H_9MgBr]_0$

Table 1.   Preparation of *it*-PMMA-block-poly(EMA) with *t*-C$_4$H$_9$MgBr in toluene at -60°C[a]

MMA (mol)	EMA (mol)	[M]$_0$/[I]$_0$ (mol/mol)	Yield (%)	Mn Obsd[b]	Mn Calcd[c]	Mw[d]/Mn	MMA/EMA[e] in polymer	Tacticity(%)[f] mm	mr	rr
0.70	0.70	100	97	12500	10800	1.29	59/59	97	2	1
0.50	0.90	280	88	26900	26900	2.11	97/151	95	3	2
0.30	0.91	300	99	27200	30900	1.76	64/182	95	3	2
1.18	0.25	285	100	29700	29200	1.42	35/50/200[g]	97	2	1
0.80	0.40	90	100	8200	9400	1.17	25/28/25[g]	95	3	2

[a] [MMA+EMA]$_0$/toluene = 2.0 (mol/L).   [b] Determined by VPO.
[c] Estimated from the copolymer composition and the yield.
[d] Determined by GPC.   [e] Number of monomeric units per chain.
[f] Determined from ^{1}H NMR spectra by peak elimination method
[g] Triblock copolymer; PMMA-block-poly(EMA)-block-PMMA.

ratio as the block copolymerization. The chromatogram of the block copolymer indicated the narrow MWD of the copolymer and did not show any peak in the range of the elution volume where the control PMMA showed its peak, indicating that all the living PMMA anions add EMA to form the block copolymer. Polymerization of EMA with the living *it*-PMMA anion prepared at -60°C was examined under several conditions. The results are shown in Table 1. All the block copolymers were highly isotactic and the Mn agreed well with the calculated values, although the MWD's became broad as the poly(EMA) block length became long. Triblock *it*-copolymers with fairly narrow MWD and high isotacticity were also obtained as shown in Table 1.

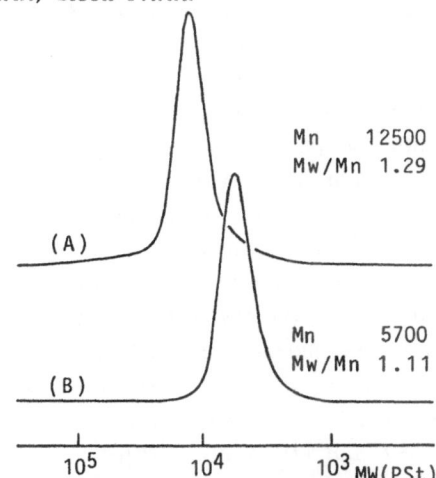

Figure 1 GPC curves of PMMA-*block*-poly(EMA) (A) and PMMA (B) prepared with *t*-C$_4$H$_9$MgBr in toluene at -60°C.
(A) MMA 700 mmol, EMA 700 mmol, *t*-C$_4$H$_9$MgBr 14 mmol, toluene 700mL.
(B) MMA 10 mmol, *t*-C$_4$H$_9$MgBr 0.2 mmol, toluene 5 mL.

The homopolymerization of EMA by *t*-C$_4$H$_9$MgBr in toluene at -60 and -78°C also yielded highly *it*-polymer but the polymer showed mostly bimodal MWD. Numbers of polymer molecules in both higher and lower molecular weight fractions were found to be almost constant during the polymerization, and Mn's for both the fractions increased linearly with conversion. The results suggest that the species giving these fractions were both living and highly isotactic-specific [15]. Since the Mn of the higher molecular weight fraction was about ten times as large as that of the lower molecular weight fraction, one can assume that the rate constant for propagation at the active species giving the former fraction is about ten times as large as that at the species giving the latter fraction [15]. The difference in structure of these active species is not clear at present, and the extensive study is under way to understand the structure not only for controlling the MWD but also for producing highly active and highly isotactic-specific species.

The polymerization of MMA by the poly(EMA) anions formed with t-C$_4$H$_9$MgBr was carried out in toluene at -60°C to obtain further information on the living nature of the poly(EMA) anion. Figure 2 shows GPC curve of the block it-copolymer obtained and that of poly(EMA) formed at the same [EMA]$_0$/[t-C$_4$H$_9$MgBr]$_0$ ratio. Both higher and lower molecular weight peaks in the chromatogram of the control poly(EMA) shifted to higher molecular weight side upon the block copolymerization. The number of polymer molecule in each fraction was found to be almost constant. The results indicate the living character of both the higher and lower molecular weight poly(EMA) anions at -60°C [15].

Molecular weight dependence of the copolymer composition in the block copolymer was analyzed by using an on-line GPC/NMR instrument in which a 500 MHz NMR spectrometer is linked to GPC chromatograph as a detector [5-8]. ^{1}H NMR data were collected over the entire chromatographic peak with time resolution of 24 s. Figure 3 shows the plots of the signal intensities for OCH$_3$ protons in MMA units (broken line) and OCH$_2$ protons in EMA units (dotted line) against elution time, which correspond to the chromatograms representing the MWD's of the PMMA and poly(EMA) blocks, respectively. The chromatographic peak intensities were normalized to represent the mole ratio of MMA and EMA units. The solid line was obtained by adding these two chromatograms and corresponds to the chromatogram for the whole polymer. The result of the analysis indicates that both higher and lower molecular weight fractions contain comparable amounts of MMA and EMA units. Since the amount of active species giving the higher molecular weight fraction is much smaller than that of the active species giving the lower molecular weight fraction, it is concluded that the former species add much more MMA molecules than the latter; the difference between the reactivities of the higher and lower molecular weight poly(EMA) anions is retained even when the anions turn to MMA anions.

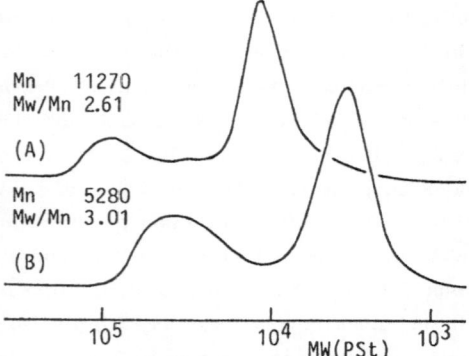

Figure 2 GPC curves of poly(EMA)-$block$-PMMA(A) and poly(EMA)(B) prepared with t-C$_4$H$_9$MgBr in toluene at -60°C.
(A) MMA 50 mmol, EMA 50 mmol, t-C$_4$H$_9$MgBr 1.0 mol.
(B) EMA 10 mmol, t-C$_4$H$_9$MgBr 0.2 mmol.

In Figure 2: (A) Mn 11270, Mw/Mn 2.61; (B) Mn 5280, Mw/Mn 3.01.

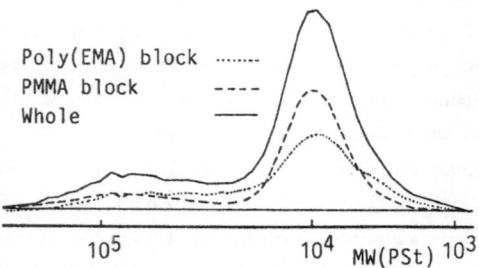

Poly(EMA) block ········
PMMA block   - - - -
Whole   ————

Figure 3 GPC/NMR analysis of poly(EMA)-$block$-PMMA prepared with t-C$_4$H$_9$MgBr in toluene at -60°C.

Conventional copolymerization of MMA and EMA with t-C$_4$H$_9$MgBr in toluene at -60°C also gave highly it-copolymer with bimodal MWD [15]. The copolymer was found to have random comonomer sequence distribution as described in the next section. The amount of higher molecular weight fraction increased as EMA content in the initial monomer mixture increased. The result suggests that the active species giving higher molecular weight copolymer is formed mainly in the initiation reaction of EMA with t-C$_4$H$_9$MgBr.

Structural analysis of highly isotactic copolymers by NMR spectroscopy:

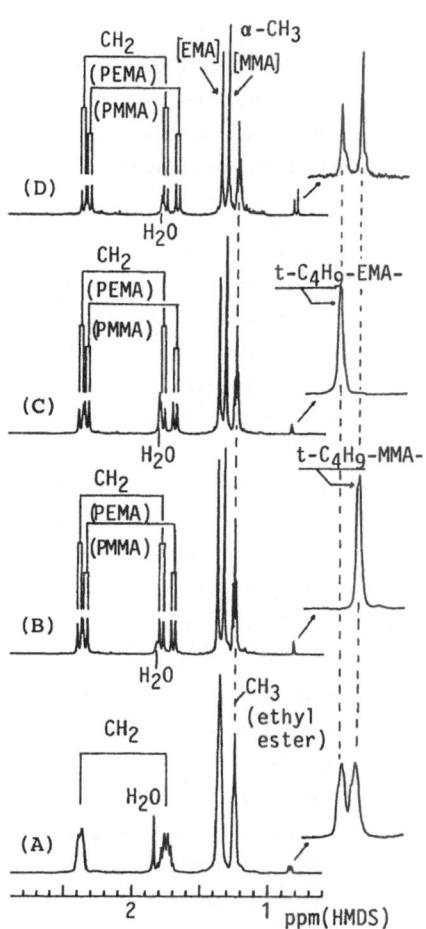

^{1}H NMR spectra of poly(MMA-ran-EMA), poly-(EMA)-$block$-PMMA, PMMA-$block$-poly(EMA) and a mixture of it-PMMA and it-poly(EMA) prepared with t-C$_4$H$_9$MgBr were measured in nitrobenzene-d$_5$ at 110°C (Figure 4)[15]. The spectra of the block copolymers are very similar to that of the mixture of it-PMMA and it-poly(EMA) except for the signals due to the initiator fragments incorporated at the left-end of the polymer chain (0.8-0.9ppm). Methylene protons of the block copolymers showed two sets of quartets due to the highly it-PMMA and it-poly(EMA) blocks. Methylene protons of the random copolymer showed more complicated splittings due to the presence of different monomer sequences. Similar differences in the spectra between the block and random copolymers are also observed for α-methyl proton signals. Thus these ^{1}H NMR signals are useful to distinguish the block copolymers from the random copolymer.

Both the block copolymers show singlet signal at 0.823 or 0.852 ppm, which is assigned to the initiator fragments attached to the left-ends of PMMA and poly(EMA) blocks, respectively. The signals thus provide the information whether the block copolymer was prepared from PMMA anion or poly(EMA) anion. Since the mixture of it-PMMA and it-poly(EMA) both prepared with t-C$_4$H$_9$MgBr shows signals of two

Figure 4 500MHz ^{1}H NMR spectra of poly(MMA-ran-EMA) (A) PMMA-$block$-poly(EMA) (B), poly(EMA)-$block$-PMMA (C) and a mixture of PMMA and poly(EMA) (D) prepared with t-C$_4$H$_9$MgBr in toluene at -60°C. The spectra were measured in nitrobenzene-d$_5$ at 110°C.

types of t-C$_4$H$_9$- group(Figure 4D), the block copolymers can be distinguished from the mixture by ^{1}H NMR spectroscopy.

The random copolymer of MMA and EMA shows two singlet signals at 0.826 and 0.845 ppm assignable to the t-C$_4$H$_9$- groups attached to MMA and EMA units, respectively (Figure 4A). The intensity ratio of these two signals (1:1) was close to monomer feed ratio, indicating that the reactivities of MMA and EMA toward t-C$_4$H$_9$MgBr in the initiation process are almost the same. Although the mixture of it-PMMA and it-poly(EMA) also shows two t-C$_4$H$_9$ signals, the signals due to the monomeric units of the random copolymer are apparently different from those of the mixture. Therefore, by the inspection of the signals due to initiator fragment as well as monomeric units, all of these four polymer samples can be distinguished from their ^{1}H NMR spectra.

^{13}C NMR chemical shifts of carbonyl carbons in the copolymer of MMA and EMA are sensitive to triad comonomer sequence as well as pentad tacticity. Figure 5 shows carbonyl carbon spectra of the it-PMMA-$block$-poly(EMA), it-poly(MMA-ran-EMA), and radically prepared poly(MMA-co-EMA), all of which have 1:1 composition [15]. High stereoregularity of the block and random copolymers made the spectra simpler as compared with that of the radically prepared poly(MMA-co-EMA). The signals of MMA- and EMA-centered sequences in $mmmm$ configurational pentad showed splittings due to the triad monomer sequences. The peak assignments were made by comparing the spectra of the it-copolymers with different compositions and are shown in the Figure. The relative peak intensities of MMM:(MME + EMM) :EME and EEE:(EEM + MME):MEM (M;MMA unit, E; EMA unit) were 1:2:1, indicating that the monomer sequence distributions of both MMA- and EMA-centered triads are almost random.

The block copolymer shows two strong signals at 176.51 and 176.37 ppm due to MMA and EMA sequences in $mmmm$

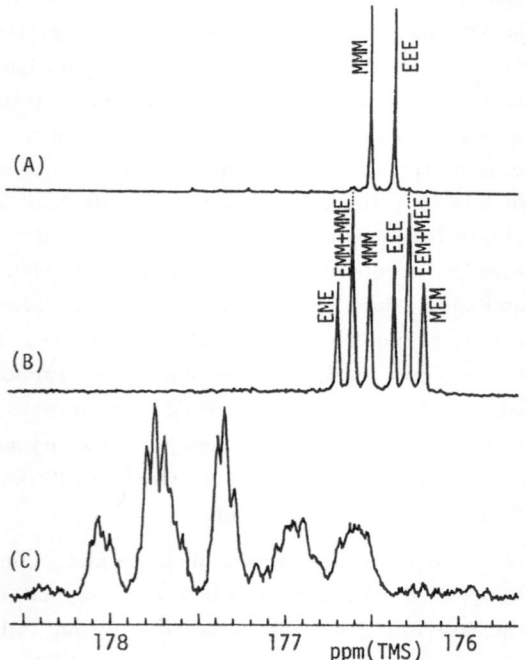

Figure 5 125MHz ^{13}C NMR spectra of PMMA-$block$-poly(EMA) (A), poly(MMA-ran-EMA) (B) prepared with t-C$_4$H$_9$MgBr in toluene at -60°C, and poly(MMA-co-EMA) prepared with AIBN in toluene at 60°C (C), measured in CDCl$_3$ at 55°C. (M and E denote MMA and EMA units, respectively.)

configuration, respectively. Weak signals of equal intensity observed at 176.60 and 176.28 ppm were assigned to MME and MEE triads in *mmmm* configuration, which should exist at the switching point of PMMA block and poly(EMA) block. Thus the block copolymer could be distinguished from the mixture of PMMA and poly(EMA) by ^{13}C NMR spectroscopy. Some other small peaks due to chain end units were also observed in the spectrum of the block copolymer and the details will be published elsewhere.

Highly syndiotactic copolymer prepared with *t*-C$_4$H$_9$Li-R$_3$Al: Preparation of highly *st*-PMMA is also an important subject because PMMA with high glass transition temperature (Tg) is needed for industrial materials such as optical disks and fibers and the Tg increases with the syndiotacticity. We recently found that combinations of *t*-C$_4$H$_9$Li and several trialkylaluminums (R$_3$Al) gave highly syndiotactic living PMMA with narrow MWD [2]. The polymerization of MMA in toluene with *t*-C$_4$H$_9$Li alone gave an isotactic-rich PMMA with broad MWD. Addition of R$_3$Al (R=C$_2$H$_5$, *n*-C$_4$H$_9$, *n*-C$_8$H$_{17}$) increased the syndiotacticity of the polymer formed, and at the ratios of Al/Li≥3, highly *st*-PMMAs with narrow MWD were obtained. The Mn's were close to the values calculated from the amounts of *t*-C$_4$H$_9$Li used and the monomer consumed. The polymerization is initiated with *t*-C$_4$H$_9$ anion and proceeds in a living manner to give *st*-PMMA, the chemical structure of which is identical with that of the *it*-PMMA prepared with *t*-C$_4$H$_9$MgBr;

it-PMMA   *st*-PMMA

Highly syndiotactic and living polymers of various alkyl methacrylates, such as ethyl, isopropyl, butyl and isobutyl methacrylates, could also be obtained by the polymerization in toluene with *t*-C$_4$H$_9$Li-(*n*-C$_4$H$_9$)$_3$Al at -78°C [2]. Thus the highly syndiotactic random and block copolymers of various methacrylates can be prepared easily in toluene with *t*-C$_4$H$_9$Li-(*n*-C$_4$H$_9$)$_3$Al at -78°C. Polymerization of EMA with

Table 2. Preparation of *st*-PMMA-block-poly(EMA) with t-C$_4$H$_9$Li/(*n*-C$_4$H$_9$)$_3$Al in toluene at -78°C[a]

MMA (mmol)	EMA (mmol)	[M]$_0$/[I]$_0$ (mol/mol)	Yield (%)	Mn Obsd[b]	Mn Calcd[c]	Mw/Mn	MMA/EMA[d] in polymer	mm	mr	rr
6.7	13.8	93	99	12300	10020	1.07	37/75	0	11	89
10.2	10.2	93	100	11000	10100	1.12	51/51	0	9	91
13.7	6.8	93	100	12000	9950	1.22	77/38	0	9	91
10.0[f]	10.0	100	83	8900	8900	1.25	42/41	0	9	91

a [MMA+EMA]$_0$/toluene=1.0mol/l, Al/Li=3mol/mol.   b Determined by VPO.
c Estimated from the copolymer composition and the yield.
d Number of monomeric units per chain.   e Determined from ^1H NMR signals of α-methyl protons by the peak elimination method.
f Random copolymerization with t-C$_4$H$_9$Li-(C$_2$H$_5$)$_3$Al(1/3).

the living *st*-PMMA gave *st*-PMMA-*block*-*st*-poly(EMA) with narrow MWD as shown in Table 2. The Mn's of the block copolymers agreed with the calculated values. Copolymerization of MMA and EMA afforded *st*-copolymers with narrow MWD. Comonomer sequence distribution of the copolymer was found from ^{13}C NMR analysis to be random [2], similarly to that of the *it*-copolymers of MMA and EMA formed with t-C_4H_9MgBr.

Several *st*-copolymers of MMA and alkyl methacrylates with narrow MWD were also prepared with t-C_4H_9Li-(n-C_4H_9)$_3$Al. Glass transition temperatures for the *st*-copolymers of MMA and butyl methacrylate (n-BuMA) are shown in Figure 6. Recently, block and random *st*-copolymers of MMA and benzyl methacrylate as well as *st*-poly(benzyl methacrylate) prepared with t-C_4H_9Li-R_3Al were found to form stereocomplex with *it*-PMMA in solid and in solution [16].

Glass transition temperature of stereoregular copolymer: Glass transition temperature (Tg) of copolymer generally varies with its composition. It is also known that Tg of polymethacrylate strongly depends on its tacticity; *it*-PMMA has lower Tg than *st*-PMMA. Therefore, the control of stereoregularity of the copolymer may provide another handle for the control of Tg of the copolymer. Figure 6 shows Tg's of stereoregular block and random copolymers of MMA and EMA determined by DSC* [15]. The Tg's of the random *it*-copolymers lay between 49 and 8°C (Mn = 1.13~3.55x10^4) and decreased with an increase in the content of EMA units in the copolymer. The block *it*-copolymers (Mn=0.82~2.97x10^4) showed single Tg's which also decreased as EMA content increased and were slightly lower than those of the random copolymers with similar compositions. A 1:1 mixture of *it*-PMMA (Mn=16400, Tg=49°C) and *it*-poly(EMA) (Mn=13000, Tg=8°C) showed Tg's of 44°C and 4°C. The single Tg of the block copolymer may indicate that the copolymer does not show microphase separation,

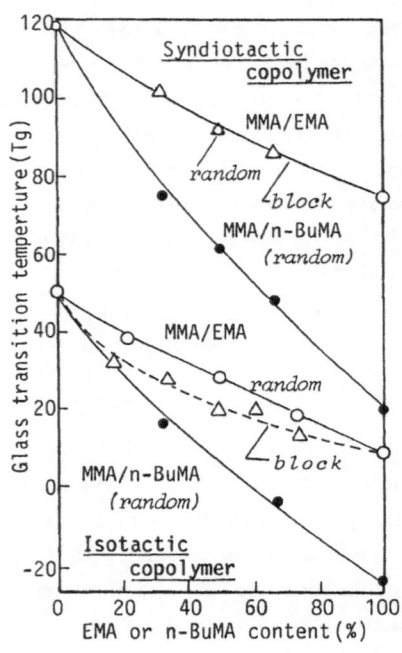

Figure 6   Glass transition temperature (Tg) of stereoregular copolymers of MMA and EMA or *n*-BuMA of various compositions

---

* Tg value was determined according to ASTM D3418-82. When Tg's of the block and random copolymers of MMA and EMA were determined by the procedure reported by Richardson and Savill (*Polymer* (1975) **16**:753), the difference between Tg's of the block and random copolymers became smaller (Suzuki H, private communication).

probably because of too short chain length of either block to form microphase. The Tg's of the *st*-copolymers lay between 120 and 75°C and decreased with increasing EMA content. Therefore, Tg's of the copolymer could be controlled in the temperature range of 120 to 8°C by changing the tacticity and copolymer composition. The *it*-copolymers whose Tg's are around room temperature showed shape-memory ability; the copolymer sample stretched 3 times as much as the original sample in length at 60°C and cooled to room temperature, shrank immediately to the original length when warmed again at 60°C.

The Tg's of poly(MMA-*ran-n*-BuMA) can be altered in the range of 120 to -22°C, by changing the composition and tacticity as shown in Figure 6. Tg values of *st*-poly-(MMA-*ran*-lauryl methacrylate)s were controllable in the range of 120 to -60°C.

PREPARATION AND POLYMERIZATION OF STEREOREGULAR PMMA MACROMONOMERS

Preparation of stereoregular PMMA macromonomer: Macromonomer method has been regarded as one of the promising ways for the preparation of graft copolymer with well-defined structure. The graft copolymer with stereoregular branches can be prepared from stereoregular macromonomer. However, there have been only a few examples of the preparation of stereoregular macromonomer [17,18]. Living nature of the *it*- and *st*-polymerization of MMA with *t*-C$_4$H$_9$MgBr and *t*-C$_4$H$_9$Li-R$_3$Al, respectively, permitted us to prepare stereoregular PMMA macromonomers. *it*- and *st*-PMMA macromonomers of high functionality were prepared from the corresponding living PMMA anions and *p*-bromomethylstyrene in the presence of 1,8-diazabicyclo[5.4.0]undec-7-ene (or hexamethylphosphoramide) and N,N,N',N'-tetramethylethylenediamine, respectively[4]. Both the *it*- and *st*-PMMA macromonomers have the same chemical structure from the left-end to right-end;

$$CH_3-\underset{\underset{CH_3}{|}}{\overset{\overset{CH_3}{|}}{C}}\left(CH_2-\underset{\underset{\underset{OCH_3}{|}}{\overset{|}{C=O}}}{\overset{\overset{CH_3}{|}}{C}}\right)_n CH_2-\langle\bigcirc\rangle-CH=CH_2 .$$

The number of styrene group in a macromonomer molecule could be accurately determined by measuring the intensities of the signals due to *t*-C$_4$H$_9$ group at the left-end and the vinyl group of the styrene group at the right-end, since the *it*- and *st*-PMMA anions have one *t*-C$_4$H$_9$- group per chain.

Tacticity dependence of reactivity of PMMA macromonomer: Polymerization of *it*-(Mn= 2900) or *st*-(Mn=2600) PMMA macromonomer was carried out with AIBN in toluene-d$_8$ at 60°C. The reaction was monitored by ^{1}H NMR spectroscopy [4], and the conversion of the macromonomer was determined from the relative intensity of vinyl methylene proton signals to t-C$_4$H$_9$- signal. The plots of -ln([M]/[M]$_0$) versus polymerization time (Figure 7) clearly indicate that the reactivity of the *it*-PMMA macromonomer is slightly but meaningfully higher than that of the *st*-PMMA macromonomer. A similar results were also obtained in the polymerization at 80°C.

The *it*- and *st*-macromonomers (M_1) were copolymerized with styrene (M_2) in toluene at 60°C using AIBN as an initiator.   Figure 8 shows copolymer composition curves for the copolymerizations [4].   The copolymer compositions could be accurately determined over a wide range of the composition from the intensity measurements of the ^{1}H NMR signals due to the t-C_4H_9- group in the macromonomer unit and the phenyl groups. Monomer reactivity ratios were determined as follows; r_1=0.79, r_2=1.29 for *it*-macromonomer, r_1=0.40, r_2=1.62 for *st*-macromonomer.   The r_2 values indicate that the reactivity of *it*-macromonomer toward styrene radical is higher than that of the *st*-one.   Judging from r_1 and r_2 values, the reactivities of the styrene group of the macromonomers are only a little lower than that of styrene itself, indicating that the introduction of PMMA chain at the *p*-position of styrene does not strongly alter the reactivity of the vinyl function in the propagation process.

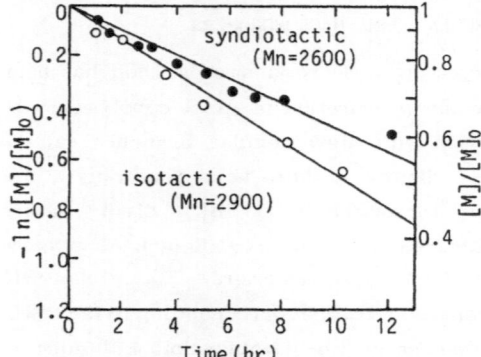 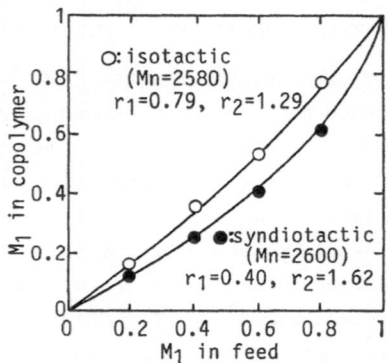

Figure 7   Time-conversion plots for the radical polymerization of *it*- and *st*-PMMA macromonomers with AIBN in toluene-d_8 at 60°C.

Figure 8   Copolymer composition curves for radical copolymerization of PMMA macromonomer(M_1) and styrene with AIBN in toluene at 60°C.

Copolymerization of the *it*- and *st*-macromonomers in toluene at 50-80°C, in chloroform at 60°C and in dimethylsulfoxide at 60°C also confirmed higher reactivity of the *it*-macromonomer[19].   Higher reactivity of *it*-PMMA macromonomer may be due to the higher segmental mobility of *it*-PMMA chain than that of *st*-PMMA chain, which was evidenced by the measurement of ^{13}C NMR spin-lattice relaxation times in solution[20].

ESR studies on the propagating radicals of macromonomer: ESR spectroscopy is the most useful tool for the study on the nature of propagating radicals.   Under the usual polymerization conditions, the stationary concentration of propagating radicals is too low to observe the ESR signals with a conventional ESR instrument. However, there is a possibility of the ESR observation of the radicals when the rate of termination reaction is lowered.

When a benzene solution of *st*-macromonomer(Mn=5380) was irradiated by UV light from a 1 kW high pressure mercury lamp at -20°C in the presence of AIBN, the ESR signal of propagating radical could be observed [21]. The signal decay was followed in the temperature range of at 30-70°C. Typical examples of the decay curves are shown in Figure 9. From the second order plot of the signal intensity and the initial radical concentration, k_t was estimated to be 54 L/mol·s at 30°C. A st-macromonomer of Mn=2720 had the k_t value of 94 L/mol·s at 30°C. These k_t values are 10^5 times smaller than the value for styrene polymerization, indicating that the termination reac-

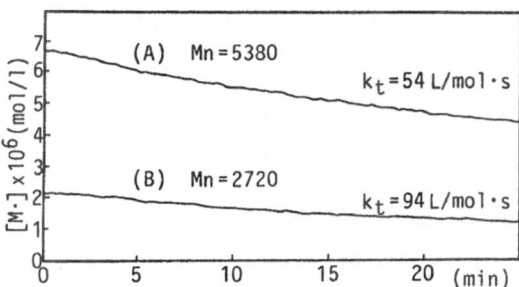

Figure 9   Decay curves of ESR signals observed in the polymerization of st-PMMA macromonomers in benzene at 30°C. The decays were followed at 30°C after the samples were irradiated by UV light at -20°C for one min.

tion is extremely restricted in the polymerization of macromonomer.   The smaller k_t value for the higher molecular weight macromonomer is explained by the higher segmental density around the propagating radicals than that of the lower molecular one.

The k_t values for the it-macromonomer could also be determined at 30°C and was found to be about 100 times as large as that for the st-macromonomer [22].   Since the rate of polymerization (R_p) for it-macromonomer was a little larger than that for the st-macromonomer and R_p is proportional to $k_p/k_t^{0.5}$, k_p for the it-macromonomer should be at least about 10 times as large as that for the st-macromonomer. Higher segmental mobility of it-PMMA chain than that of st-chain [20] should give rise to both larger k_p and k_t values for the it-PMMA macromonomer.   Although the increases of k_p and k_t affect R_p in opposite ways, the larger R_p for it-macromonomer suggests that the positive contribution of increase of k_p to R_p overcomes the negative contribution of increase of k_t.

Properties of polymers of stereoregular PMMA macromonomers: Table 3 shows the [η]'s and Tg's of the it- and st-PMMA macromonomers, the poly(PMMA macromonomer)s derived therefrom, and it- and st-PMMAs with molecular weights similar to the corresponding

Table 3   Characteristics of polymers of stereoregular PMMA macromonomers
in comparison with the starting macromonomer and linear PMMA

Polymer	Tacticity(%)			Mn		$\frac{Mw^a}{Mn}$	$[\eta]^b$ (dL/g)	k'^c	T_g (°C)
	mm	mr	rr	VPO	GPC				
Isotactic									
Macromonomer	95	4	1	2520	2350	1.11	0.005		21
Poly(macromonomer)	95d	4d	1d	75000	43700	1.45	0.090	1.70	44
Linear PMMA	97	3	0	56500	61600	1.24	0.393	0.43	50
Syndiotactic									
Macromonomer	1	11	88	2600	2120	1.10	0.005		94
Poly(macromonomer)	1d	11d	88d	61100	31100	1.26	0.074	1.48	114
Linear PMMA	0	8	92	53900	59600	1.19	0.260	0.59	124

a Determined by GPC.   b Measured in toluene at 30.0°C.
c Huggin's coefficient.   d Tacticities for the PMMA chains.

polymacromonomers[4]. The Tg's of the polymacromonomers lay between those of the starting PMMA macromonomers and the linear PMMAs. The [η]'s of the polymacromonomers were lower than those of the corresponding linear PMMAs with a similar molecular weight, and the Huggins' coefficients were larger than those of the linear PMMAs. Elution volumes of the polymacromonomers in GPC chromatogram were larger than those of the linear PMMAs, indicating that the polymacromonomers have smaller hydrodynamic volume in solution than linear PMMAs. From GPC-LALLS experiments, the *a* values in Mark-Houwink-Sakurada equation for the polymers of *it-* and *st-* macromonomers were found to be less than 0.5, suggesting that the molecules take compact spherical form in solution[23].

ACKNOWLEDGMENT

A part of this work was supported by Grant-in-Aids for Scientific Research from the Ministry of Education, Science, and Culture of Japan (No. 6143022 and No. 01550717).

REFERENCES

1. Hatada K, Ute K, Tanaka K, Okamoto Y, Kitayama T (1986) *Polym J* 18:1037
2. Kitayama T, Shinozaki T, Sakamoto T, Yamamoto M, Hatada K (1989) *Makromol Chem Supplement* 15:167
3. Kitayama T, Horii H, Hatada K, Kashiwagi T (1989) *Polym Bull* 21:433
4. Hatada K, Kitayama T, Ute K, Masuda E, Shinozaki T, Yamamoto M (1989) *Polym Bull* 21:165
5. Hatada K, Ute K, Okamoto Y, Imanari M, Fujii N (1988) *Polym Bull* 20:317
6. Hatada K, Ute K, Kitayama T, Yamamoto M, Nishimura T, Kashiyama M (1989) *Polym Bull* 21:489
7. Hatada K, Ute K, Kashiyama M, Imanari M (1990) *Polym J* in press
8. Ute K, Kashiyama M, Oka K, Hatada K, Vogl O (1990) *Makromol Chem Rapid Commun* 11:xxx
9. Yuki H, Hatada K (1979) *Adv Polym Sci* 31:1
10. Hatada K, Kitayama T, Ute K (1988) *Prog Polym Sci* 13:189
11. Ute K, Kitayama T, Hatada K (1986) *Polym J* 18:249
12. Matsuzaki K, Tanaka H, Kanai T (1981) *Makromol Chem* 182:2905
13. Allen PEM, Bateup BO (1978) *Eur Polym J* 14:1001
14. Hatada K, Kitayama T, Ute K, Masuda E, Shinozaki T, Yamamoto M (1988) *Polym Prep Am Chem Soc Div Polym Chem* 292:54
15. Kitayama T, Ute K, Yamamoto M, Fujimoto N, Hatada K (1989) *Polym J* submitted
16. Kitayama T, Fujimoto N, Terawaki Y, Hatada K (1989) *Polym Bull* submitted
17. Hatada K, Nakanishi H, Ute K, Kitayama T (1986) *Polym J* 18:581
18. Hatada K, Ute K, Shinozaki T, Kitayama T (1988) *Polym Bull* 19:231
19. Hatada K, Kitayama T, Masuda E (1989) *Polym Prep Jpn* 38:348
20. Hatada K, Kitayama T, Okamoto Y, Ohta K, Umemura Y, Yuki H (1978) *Makromol Chem* 179:485
21. Hatada K, Kitayama T, Masuda E, Kamachi M, *Makromol Chem Rapid Commun* in press
22. Hatada K, Kitayama T, Masuda E, Kamachi M, unpublished result
23. Masuda E, Kitayama T, Hatada K, unpublished result

Condensation Block Copolymer Synthesis via Decarboxylation

E. P. Woo*, M. J. Mullins, S. E. Bales and S. P. Crain

Central Research, The Dow Chemical Company,
Midland, MI 48674

INTRODUCTION

Block copolymers comprising carbonate and arylene ether blocks have been prepared by the catalytic decarboxylation of random or alternating copolycarbonates. This novel route to block copolymers is catalyzed by a number of catalysts and requires only that one of the bisphenols in the starting copolycarbonates be activated by electron-withdrawing substituents, e.g. 4,4'-dihydroxydiphenyl sulfone (bisphenol S), or 4,4'-dihydroxybenzophenone (bisphenol K).

McGrath and coworkers (1, 2) had investigated extensively the three syntheses of polycarbonate-polysulfone block copolymers. The first is where the polycarbonate segment is formed in the presence of the hydroxyl-terminated polysulfone oligomer. The second technique is to preform both oligomers with hydroxy groups and then extend to high molecular weight by phosgenation in a common solvent, e.g., dichloromethane. The most rigorous synthesis of these block copolymers employs the pyridine catalyzed reaction of chloroformate-terminated polysulfone with hydroxy-terminated polycarbonate in dichloromethane. Since these are all solution polymerzations, copolymers in which one of the blocks is semi-crystalline would be inaccessable.

Our synthetic procedure uses high molecular weight copolycarbonates which are easily prepared by phosgenation in solution instead of low molecular weight oligomers or monomers. This novel procedure, although may be viewed formally as carbon dioxide extrusion, in fact comprises a series of reactions: the cleavage of a carbonate group by the catalyst, the decarboxylation of the resulting carbonate anion to the corresponding phenolate, and the recombination of the reactive ends. Inspite of this multiplicity of reactions where in each occurrance the polymer chain is broken, the resulting polymers suffer no overall molecular weight degradation. More surprisingly is the observation that blocks are formed as a consequence of the reaction sequence from either random or alternating copolycarbonates. To the best of our knowledge these unique features are unprecedented

Although nucleophilic aromatic displacement chemistry is well known (3) and has been the basis of a number of commercial engineering thermoplastics, the reactivity of activated aromatic carbonates toward nucleophiles has not been applied to polymers. Witt and coworkers (4) discovered that diphenyl carbonates with at least one electron-withdrawing substituent at either

B. C. Anderson · Y. Imanishi (Eds.)
Progress in Pacific Polymer Science
© Springer-Verlag Berlin Heidelberg 1991

144

the ortho or para positions could be decarboxylated to the corresponding diphenyl ethers when heated at ≥180°C with a catalytic amount of potassium carbonate or acetate. The yields were dependent on the position and the nature of the substituents and ranged from 17 to >90%. A later work (5) showed that alkali metal fluorides and cyanides were also catalytically active and that potassium salts were ≥20 times more active than the corresponding sodium ones. However, the yields of diphenyl ethers were still considerably less than quantitative. The 80% yield of 4-phenylsulfonylphenyl ether had perhaps discouraged the exploration of the utility of this chemistry in polymers.

RESULTS AND DISCUSSION

The random copolycarbonates were prepared by the standard phosgenation procedure in dichloromethane with pyridine as the acid acceptor. Because of the slight difference in reactivity, these polymers may have a very small degree of blockiness which should not affect the results of this work. The alternating copolycarbonates were prepared in a similar fashion except that bisphenol A was first converted to the bis-chloroformate before the addition of bisphenol S or bisphenol K. All polymers were capped with tert-butylphenol. This work was conducted on 1:1 copolycarbonates of bisphenol A with either bisphenol S or bisphenol K. The chemistry of this novel procedure is illustrated in Scheme 1.

Scheme 1

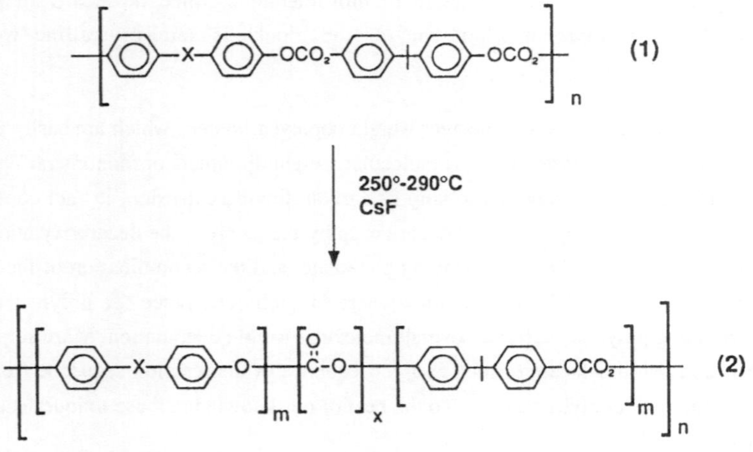

(1)

250°-290°C
CsF

(2)

a, X = SO$_2$ b, X = CO

Since the starting copolycarbonates were all of high molecular weight, diphenyl sulfone was used as a diluent to faciliate mixing. A variety of alkali salts may be used but cesium fluoride (≤1% by weight) is preferred because of its solubility in the reaction medium. There was an induction period of 10 minutes of less and a color change from pale yellow to red before carbon dioxide evolution was observed. The rate of carbon dixoide evoluation was, within the experimental reproducibility, dependent on the concentration of the catalyst and the reaction temperature. Higher catalyst concentrations and higher temperatures led to higher reaction rates. There appeared, however, an upper temperature limit. Reactions conducted at 320°C gave gel-containing polymers. The carbon dioxide extrusion typically proceeded at a fairly linear rate and became very slow when the total amount evolved approached 50% of theory. For copolycarbonate of bisphenol A and bisphenol S, **1a**, this occurred at about 30-190 minutes at 250°-290°C. Reactions were terminated either at or before this point. The polymers were treated with acetic anhydride to capped any phenolic groups and freed of diphenyl sulfone and cesium fluoride by washing with methanol and water. When care was taken to exclude air and water, tough, nearly colorless films could be made by compression molding. For those cases where the polymers were soluble in *syn*-tetrachloroethane, their inherent viscosities were actually higher than those of the starting copolycarbonates.

The decarboxylation of random **1a** was studied under different reaction conditions and several experiments were repeated using alternating **1a**. Since the results from random and alternating **1a** were identical within the experimental uncertainties, no further reference will be made to this aspect. The results are given in Table 1. In all cases except the last (entry 9) the polymer films were transparent and exhibited a single T_g at about 160°-180°C. Surprisingly, the polymers from the last run gave translucent films and exhibited glass transition temperatures at 175°C and 195°C. Since these results were indicative of block structures, which were entirely unexpected from the non-block nature of **1a**, further analysis was conducted.

Table 1. Results of the Decarboxylation of Copolycarbonate **1a**

Entry	CO_2 Loss (% Theory)	M_n of PES[a] (g/mole)	Bis S/ Bis A	CsF (wt %)	DPS[b] (wt %)	Temp (°C)	Time (min)	Inh. Vis.[c] (dL/g)
1	13.2	656	27.0	0.09	0	290	30	0.56
2	27.0	769	61.9	0.29	17.0	290	40	0.88
3	31.3	1236	28.8	0.36	70.0	290	17	0.46
4	45.5	1561	51.9	0.34	69.0	290	26	0.58
5	46.4	1653	73.2	0.30	55.0	290	30	0.66
6	40.3	1289	74.2	0.40	55.0	250	85	0.51
7	42.3	1531	44.3	0.33	69.0	250	113	0.62
8	43.0	1711	63.9	1.00	69.0	250	37	0.51
9	48.4	2639	8.4	0.92	67.0	250	92	0.69

[a] Poly(ether sulfone) block determined by NMR.
[b] Diphenyl sulfone as the inert diluent.
[c] Determined in *sym*-tetrachloroethane.

It was reasoned that PES blocks should survive mild hydrolyis while carbonate groups should be hydrolyzed. Accordingly, polymer **2a** was hydrolyzed at room temperature with methanolic KOH; the product mixture was acidified, and extracted with boiling water to remove the bisphenols. The residual arylene ether oligomers were acetylated and analyzed by proton NMR from which the number average molecular weight and the ratio of bisphenol S to bisphenol A were obtained by integration. The block copolymer nature of **2a** was thus unequivocally established. Three interesting trends can be identified from the results summarized in Table 1. First, there is a definite relationship between the extent of decarboxylation and the length of the PES block: the greater is the extent of decarboxylation, the longer is the PES block. Second, the PES blocks contain both bisphenols, the ratio of bisphenol S to bisphenol A----which may be viewed as block purity----increasing with increasing block length. Thus extending the extrusion of carbon dioxide increases the block length but decreases the block purity. Third, increasing the catalyst concentration usually gives faster reaction rate. Finally, it should be pointed out that although the rate of decarboxylation was fairly linear until about 20-30% of the theory, these rates were far from reproducible from experiment to experiment. We believe that this is primarily due to the difficulty in mixing for the reaction mixtures were still extremely viscous even with added inert diluent.

If this unique reorganization of molecular structures could be applied to the copolycarbonates of biphenol A and bisphenol K, then it would be a simple way to prepare the hitherto unknown polycarbonate poly(ether ketone) (PEK) block copolymers. As might be expected from the difference in the 'activiting power' of ketone and sulfone, the decarboxylation of **1b** (1:1 bisphenol A:bisphenol K) was considerably slower, approaching the 50% point after about 4 hours at 250°-280°C. Upon cooling at the end of the reaction, crystallization of the polymer was occasionally observed. The resulting polymers, **2b**, were not soluble in most organic solvents and the hydrolyzed segment of greater than four repeating units were also similarly insoluble so that analysis of block length was not feasible. Strong acid solvents, such as sulfuric acid and methanesulfonic acid, commonly used for semi-crystalline poly(aromatic ketones) rapidly degraded the copolymers. However, it was found that pentafluorophenol was a suitable solvent for viscosity measurements. The results are shown in Table 2.

A film of **2b** made by standard compression molding was typically opaque, indicative of some degree of crystallinity. A film molded at 370°C and quenched immediately in ice-water was amorphous and showed a T_g at 147°C, a crystallization exotherm at 232°C (16 J/g), and a broad melting endotherm at 300°-340°C (14 J/g). Annealing the amorphous film at 250°C for 15 min brought about crystallization. Assuming that the heat of crystallization of the PEK block of **2b** is similar to the 130 J/g reported for poly(ether ether ketone) (6), then the crystallinity is about 10% overall or about 20% for the PEK blocks which comprises a maximum of 50% of **2b**. This result was confirmed by wide-angle X-ray study.

Table 2. Decarboxylation of Copolycarbonate **1b**

Entry	CO$_2$ Loss (% Theory)	CsF (wt %)	Temp. (°C)	Time (min)	Inh. Vis.[a] (dL/g)
1	40.4	0.70	250	262	---
2	39.4	0.52	250	272	---
3	53.5	0.58	250-280	326	---
4	41.3	0.37	250	137	0.40
5	37.5	0.40	250	268	0.71
6	46.0	0.09	250	214	0.48
7	---	0.15	250	174	0.57
8	---	0.02	290	172	0.54

[a] Determined in pentafluorophenol at 25°C, 0.5 g/dL.

While the more detailed morphological picture of **2b** would have to await further study, the benefit engendered by the crystalline nature is clearly evident in the resistance of the polymer to aqueous base and organic solvents. A film (about 0.4 mm in thickness) of **2b** immersed in 10N NaOH for 200 hours showed no change in weight while a film of bisphenol A polycarbonate of the same thickness had lost about 50% of its original weight. To evaluate solvent resistance, loops of films of about 0.25 mm thickness prepared from **1b** and **2b** were immersed in a number of organic solvents. The loop diameter was chosen to achieve an 1% strain (% strain = 100 x film thickness/loop diameter). Although not a very rigorous test, the time to rupture, nevertheless, is reflective of the solvent resistance of the film. The data summerized in Table 3 show that films of **2b** is vastly superior in solvent resistance than films of **1b.**

Table 3. Rupture Time of Polymer Films

Solvent	Copolymer **1b**	Copolymer **2b**
Dichloromethane	< 1 sec	2 min
Toluene/hexane	3 sec	> 3 min
Toluene	3 sec	> 3 min
Tetrahydrofuran	6 sec	> 3 min
Acetone	24 sec	> 3 min
DMF	> 3 min	> 3 min
Carbon tetrachloride	> 3 min	> 3 min

It is evident from the formation of block polymers that the mechanism of the reaction must be more complex than we had originally envisaged. Since the earlier work on the decarboxylation of substituted diphenyl carbonates shed no light on the subject, we decided to investigate the mechanism by studying the decarboxylation of phenyl 4-benzoylphenyl carbonate 3, which should give 4-phenoxybenzophenone, 4. Surprisingly, the major products observed, in about equimolar amounts, were 4,4'-dibenzoyldipheny ether, 5, and diphenyl carbonate, 6, while the expected 4 was present only in trace amounts (see Scheme 2). Compound 3 may be viewed as the model for the copolymer 1b in that only one of the two phenylenes of the carbonate group is activated by a ketone. If 4 were the predominant product in the model decarboxylation experiment, the polymer resulting from alternating 1b would be the poly(ether ether ketone) of bisphenol A and benzophenone. The fact that 5 and 6 were the major products are entirely consistent with the formation of 2b. Additional information concerning the reaction mechanism was derived from monitoring the decarboxylation of 3, at 250°C with CsF as the catalyst and diphenyl sulfone as the solvent, by liquid chromatography. The composition of the reaction mixture as a function of time is graphically represented in Figure 1.

Scheme 2

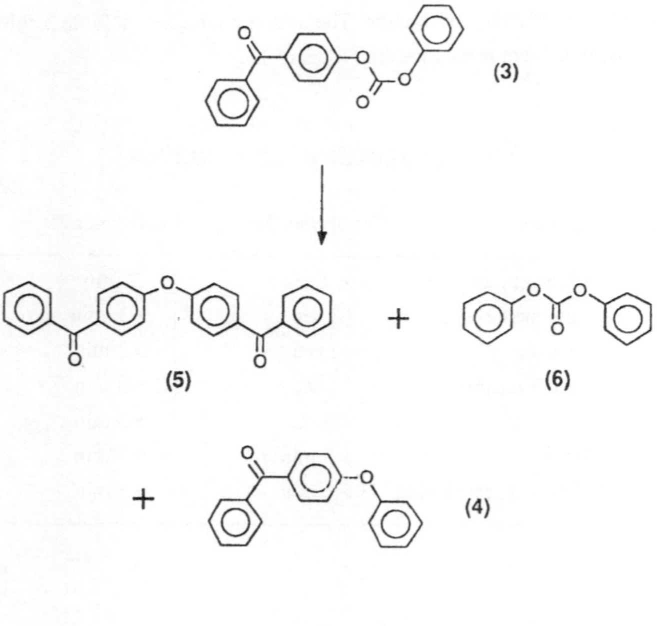

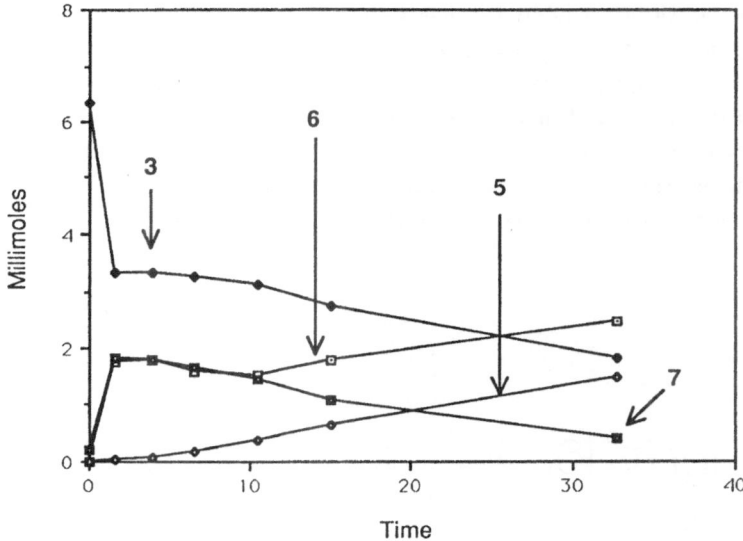

Figure 1. Decarboxylation of phenyl 4-benzoylphenyl carbonate.

As can be seen, an immediate transesterification occurred which was essentially complete before the first aliquot could be removed. This fast carbonate exchange transforms the starting unsymmetrical **3** into a statistical (1:2:1) mixture of bis(4-benzoylphenyl) carbonate (**7**), **3**, and **6** before decarboxylation can occur to any detectable extent. Carbonate **7** being doubly activated is expected to be more reactive towards nucleophiles than carbonate **3** and should react at a significantly faster rate. The depletion of **7** upsets the ester exchange equilibrium and more **3** would be converted to **7** and the inert **6**. Thus, one would expect the rate of disappearance of **3** to parallel that of **7** while the concentration of **6** would build up at the same rate. Indeed, this scenario is exactly what is shown in Figure 1.

The overall reaction mechanism may be represented by the sequence shown in Scheme 3. The main feature is that carbonate exchange is much faster than the nucleophilic attack of any nucleophile at the carbon para to the activating benzoyl group. Thus the attack of a nucleophile (B⁻) at the carbonyl carbon of **3** generates phenolates **8** and **9**; in other words, there is an equilibrium between **8** and **9** through the carbonate exchange process. Phenolate **8** is expected to be predominant since it is the conjugate base of a stronger acid. The total concentration of phenolates should be the same as that of the catalyst barring any significant concentration of contaminants. Nucleophilic attack by **8** on carbonate **3** produces the observed disubstituted ether **5** and phenyl carbonate anion (**10**). Decarboxylation of **10** gives **9** which reenters the fast carbonate exchange process to regenerate **8** and more diphenyl carbonate. Applying this mechanism to polymer **1b**, it may be envisaged that the initial reaction is an equilibration of the polymers in a manner akin to the model compound study. After the first cleavage of a carbonate group, the active nucleophile is the one derived from bisphenol K which then attacks a carbonate junction that is flanked on at least one side by a bisphenol K moiety.

Scheme 3

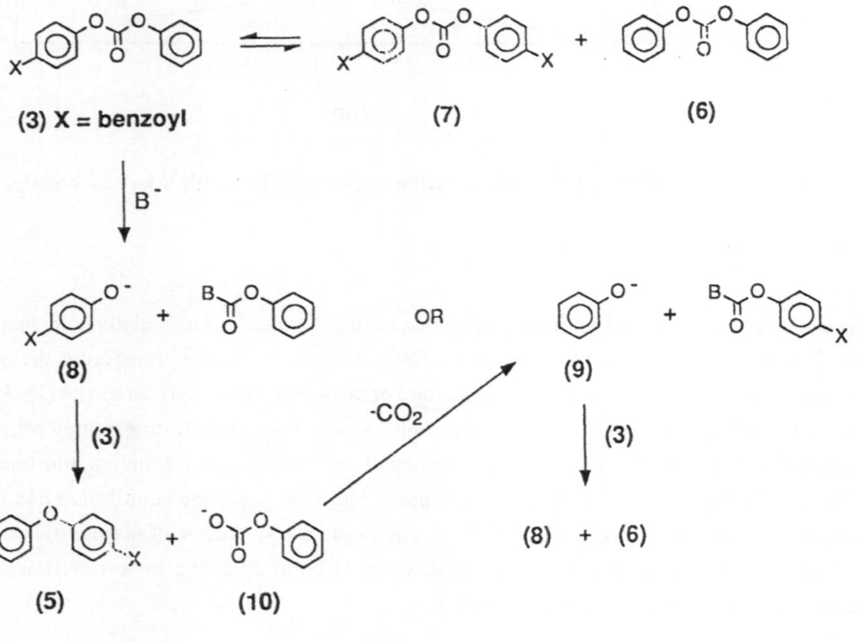

(3) X = benzoyl   **(7)**   **(6)**

(8)   **(9)**

(5)   **(10)**

-CO$_2$

(8) + (6)

CONCLUSION

This work has demonstrated that copolycarbonates with one bisphenol that is activated by electron-withdrawing groups can be decarboxylated into copolymers containing carbonate and arylene ether blocks. Although the conversion involves multiple sequences of cleavage of the polymer backbone, extrusion of carbon dioxide and nucleophilic aromatic displacement, the resulting polymers suffer no apparent molecular weight degradation. Particularly noteworthy is the conversion of a bisphenol A-bisphenol K copolycarbonate to a semi-crystalline block copolymer of polycarbonate and poly(ether ketone).

REFERENCES

1   McGrath JE, Ward TC, Shchori E, Wnuk AJ (1977) Polym Eng Sc 17: 647

2   McGrath JE, Matzner M, Robeson LM, Barclay R (1977) J Polym Sci Polym Symp 60: 29

3   Bunnett JF, Zahler RE (1951) Chem Rev 9:273

4   Witt H, Holtschmidt H, Muller E (1970) Angew Chem Int Ed Engl 9:67

5   Jost P, Forestiere A, Sillion B (1982) Tetrahedron Letts 23:4311

6   Blundell DJ, Osborn BN (1983) Polymer 24:953

Polymeric Solid Electrolyte and Ion-Conduction

E. Tsuchida

Waseda University
Tokyo 169

Introduction

The study of polymeric solid electrolyte was started in my laboratory ten years ago. Recently, it collects keen interests from not only pure scientific research fields but also industrial application fields. Since it was firstly recognized by Dr. Wright in Univ. of Sheffield, England in 1975 [1], it has been studied vigorously and has also been applied in several fields. It is very easy to think about the benefits when an electrolyte solution is replaced with completely dry polymer films.

Polymeric solid electrolytes mean that the polymeric systems which enable efficient ion migration even in solid state comparable to solution state. Compared with inorganic ion conductive materials, the polymer systems have great advantages for mechanical properties as well as wide variety of chemical design. The selective ion conduction can be performed by design of functional polymer matrix. There are more advantages of polymeric solid electrolyte such as, flexibility, transparency, excellent processibility, light weight and complete dry system. If we succeeded to prepare polymeric solid electrolytes satisfying the above mentioned requirements, lots of devices could be changed completely.

In general, polymeric solid electrolytes can be classified into four groups.

(1) porous polymer + solvent + salt
(2) porous polymer + oligomer + salt
(3) polar polymer + salt
(4) polar polymer with charged sites

The first case is the most simple system. Organic solvent and inorganic salt are mixed and incorporated into the polymer matrix which has micro porous structure. For the second case, oligomers are used instead of organic solvent for the case 1. This system

B. C. Anderson · Y. Imanishi (Eds.)
Progress in Pacific Polymer Science
© Springer-Verlag Berlin Heidelberg 1991

contains no volatile solvent, and is expected to have superior stability than systems classified into the case 1. These polymeric solid electrolytes are developed from solution system. Ionic conduction is realized through solution column incorporated in polymer matrix. The third one composes polymer and inorganic salts without solvent or oligomer. This therefore can be called as the exact polymeric solid electrolyte. Ions migrate in polymer matrix and the molecular design should be very important to enable such a conduction. The last one is rather advanced one. As the charged sites are fixed on the polymer structure, the mobile carriers are only counter ions of the fixed charges. This system is expected to be a system which is extremely important for devices driven under direct current.

(1) Porous Polymer + Solvent + Salt System

This is a very simple model, and is composed of organic salt solution and polymer matrix. The solution is infiltrated into porous matrix with high dielectric constant which is required to show high conductivity. This is therefore classified into apparently a solid state but should be classified into a solution system microscopically. Cations and anions are surrounded by a number of solvents. Ion conduction obeys the mechanism for the ionic conduction in solution, so called percolation model.

The conductivity of polymeric solid electrolyte of case 1 is shown in Table 1 [2-4]. In the second column, the conductivity of about 10^{-6} S/cm can be confirmed for PVdF-LiClO$_4$ system containing 20 mol% of propylene carbonate. This system shows the conductivity of 10^{-7} S/cm without propylene carbonate.

Table 1 Lithium ionic conductivity of hybrid films

Base polymer (mol%)	Dielectric constant ε (25°C)	LiClO$_4$ (mol%)	additive (mol%)	Conductivity $10^7 \cdot \sigma_i$ (S/cm)
PVdF 70.0	9.2	30.0	0	2.30
PVdF 58.6	9.2	25.1	PC*)20.5	10.0
PAN 58.1	8.0	39.0	EC**)2.9	0.14
PAN 56.8	8.0	24.4	EC 18.8	2.80
PVAc 64.0	3.3	36.0	0	0.08
PVAc 52.1	3.3	29.2	PC 18.7	1.10
PS 70.0	2.5	30.0	0	0.01
PS 62.9	2.5	27.0	PC 10.1	0.25
PMMA 68.0	3.3	22.0	PC 10.0	0.60
PVC 60.0	2.5	40.0	0	0.001

*) propylene carbonate   **) ethylene carbonate

The effect of a polar solvent such as propylene carbonate, is understandable to promote salt dissociation and to construct ionic conduction column in PVdF matrix.

(2) Porous Polymer + Oligomer + Salt System

This is performed in principle by substituting organic solvents with oligomers. One example is Nafion film incorporating oligomer(PEO_{400E})/$LiClO_4$ solution [5]. Cylindrical ionic conduction column is expected to be revealed in the Nafion matrix. The conductivity of 10^{-6} S/cm is obtained (Table 2). For the same example, Flemion is used as polymer matrix. Flemion has ionic group at higher concentration than Nafion. We therefore succeeded to incorporate PEO_{400E}/$LiClO_4$ solution at higher content than Nafion. The lithium ion conductivity of more than 10^{-5} S/cm was obtained (Table 3).

$$-(CF_2CF_2)_{0.88}-(CFCF_2)_{\overline{0.12}}$$
$$\overset{|}{O}$$
$$CFCF_3$$
$$CF_2$$
Nafion 117 $OCF_2CF_2SO_3Li$

$$-(CF_2CF_2)_{\overline{0.74}}-(CFCF_2)_{\overline{0.26}}$$
$$\overset{|}{O}$$
$$(CF_2)_3$$
Flemion $COOLi$

Table 2
Lithium ionic conductivity of hybrid films a)

Composition (wt%)			Conductivity	Pore size
Nafion	$LiClO_4$	PEO_{400E}	$10^7 \times \sigma_i$ (S/cm)	(Å)
93.8	0.4	5.8	11.1	
98.1	0.3	1.6	6.7	~10
97.7	0.2	2.1	6.1	
99.2	0.1	0.7	1.7	

a) $LiClO_4$·PEO_{400E} solutions immersed Nafion film.

Table 3
Ionic conductivity of hybrid films

Composition (wt%)			Conductivity
Flemion	$LiClO_4$	PEO_{400E}	$10^6 \times \sigma_i$ (S/cm)
50.0	6.0	44.0	0.14
50.0	9.0	41.0	0.31
50.0	12.0	38.0	0.25
50.0	15.0	35.0	3.98
40.0	10.0	50.0	6.14
40.0	15.0	45.0	13.7
40.0	17.0	43.0	3.18
40.0	20.0	40.0	4.10
40.0	25.0	35.0	.1.20
40.0	30.0	30.0	1.37

Poly(methacrylic acid) interacts with polyoxyethylene(POE) to form polymer assembly so-called polymer complex [6]. In this complex system, POE constructs ionic conduction column and PMAA supports this conduction column by the three dimensional polymer complex frame. This system is expected to provide an unique polymer matrix.

Fig. 1 shows the effect of component concentration on the conductivity of PMAA-PEO/$LiClO_4$ system[6]. The conductivity increases with

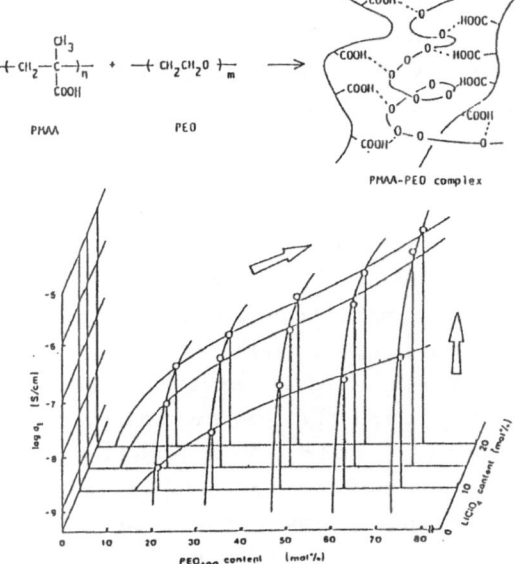

Fig. 1
Effect of PEO_{400} and $LiClO_4$ content on ionic conductivity for PMAA-PEO_{400}/$LiClO_4$ system at 25 °C.

increasing LiClO$_4$ and PEO oligomer content. The increase in LiClO$_4$ and PEO oligomer content realizes the increase in the carrier number and a number of conduction column, respectively.

The interpenetrating polymer network, IPN, is also known to be an unique material to provide ion conduction matrix.

(3) Polar Polymer + Salt System

Completely dry polymeric solid electrolyte can be presented by polyoxyethylene (POE). POE can interact with ions by the similar manner to the water molecules as seen in Fig. 2. In an aqueous solution, inorganic salts dissociate by solvation of water molecules and generated cation can migrate with solvated state. In case of POE, salt can also dissociate by pseudo solvation of ether oxygen. Cation can be transported in turn by segmental motion of POE and exchange of cation between ether oxygens. Therefore, high segmental motion of polyoxyethylene and high exchange rate are key points to design high ionic conductor.

To understand the importance of lower glass transition temperature, the ion conduction mechanism in the inorganic salt crystals is compared with that in polymer systems (Fig. 3). In the inorganic salt crystals, potential field generated inevitably by the crystal field. Ion migration is explained by the ion hopping mechanism on the energy level low enough to jump the carrier potential. On the other hand, it seems not easy to migrate the ions in the solid matrix provided by

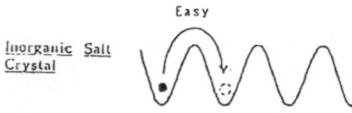

in Aq   vs.   in POE

Fig. 2
Cation-dipole interaction in aqueous solution and in polyoxyethylene.

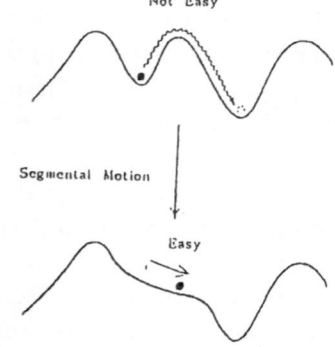

Fig. 3
Continuous potential field for ion conduction in solid state.

polymers. However, it becomes easy when the sufficient segmental motion is given in the matrix. Ions can be transferred along with the segmental motion of matrix polymer, just like surfing. Higher segmental motion can be available at room temperature when it has lower glass transition temperature. The linear POE, however, isn't the best material for the design of ion conductor. The linear and high molecular weight POE has high crystallinity and poorer mechanical property than any other common plastics.

The segmental motion of the polymer is important to design polymeric solid electrolytes with high ionic conductivity. Lower T_g of the polymer is required to increase the segmental motion. Therefore, polymer matrix such as shown in Fig. 4 was designed to satisfy these requirements. In this polymer system, main chain possesses good processibility and side chain constructs the ionic conduction column with low T_g. Salt dissociation is promoted in high dielectric environment and/or by pseudo solvation. Consequently, formed ions migrate in this conduction column in the solid. Further, degree of crystallinity decreases considerably by grafting

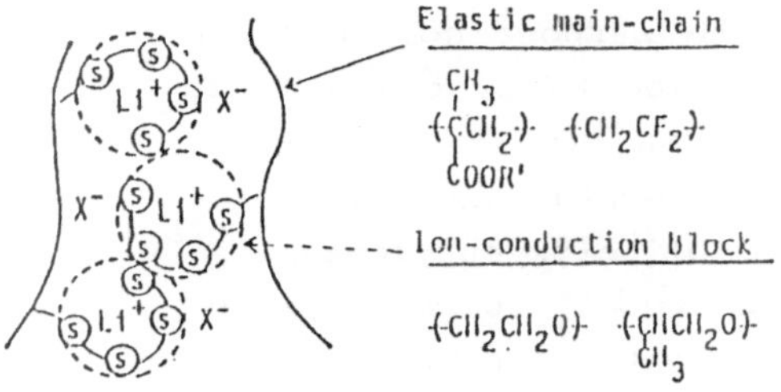

Fig. 4
Design of polymeric solid electrolyte with processibility and high conductivity.

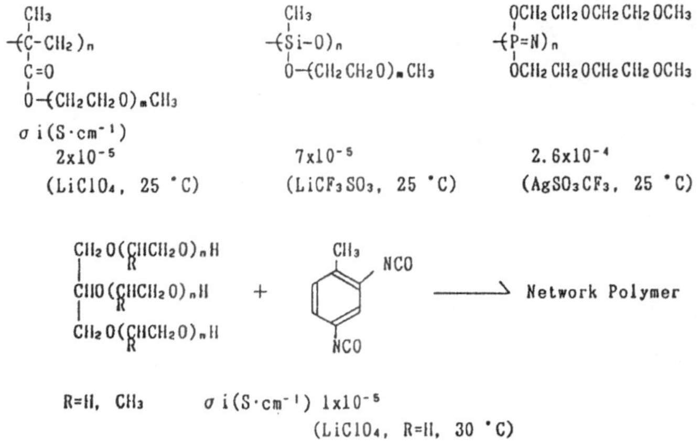

Fig. 5
Designed polymer matrix

oligo(oxyethylene) chains. Figure 5 shows typical samples of other works designed to satisfy the requirements, such as low T_g and good processibility. All systems containing inorganic salt such as $LiClO_4$ showed the conductivity of 10^{-5} S/cm. The main component for these is POE because of its excellent characteristics. Our system is shown in the top of the left hand side compound, called "poly [oligo(oxyethylene) methacrylate]" derivatives [7].

The chemistry of this system is briefly mentioned here. ω-Methoxy polyethylene oxide was esterified with methacrylic acid chloride and product was isolated by alumina column. Structure of the purpose compound was confirmed by proton NMR, as shown in Fig. 6, infra-red, and elemental analysis.

Oligo(oxyethylene) methacrylate compounds with different number of

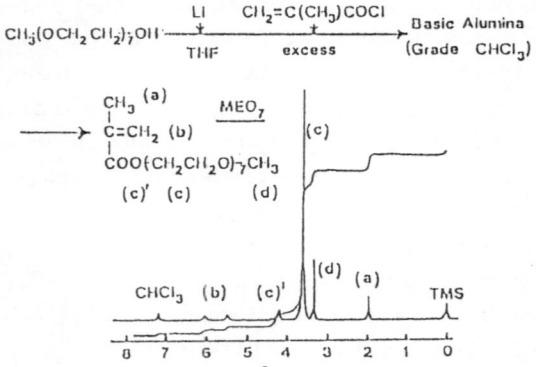

Fig. 6
Synthetic scheme and ^{1}H-NMR spectrum of MEO$_7$

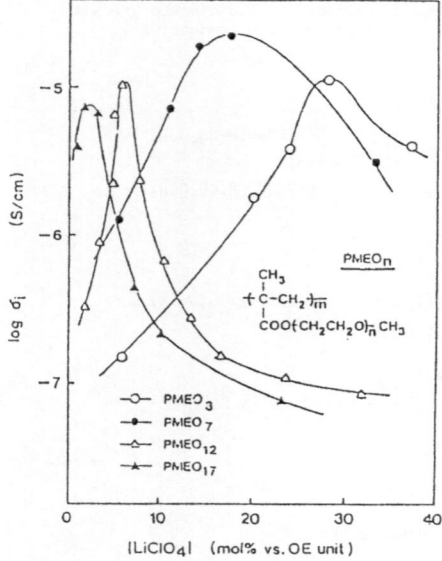

Fig. 7
Salt content dependence of the ac(1 V) ionic conductivity for P(MEO$_n$)/LiClO$_4$ hybrid films.

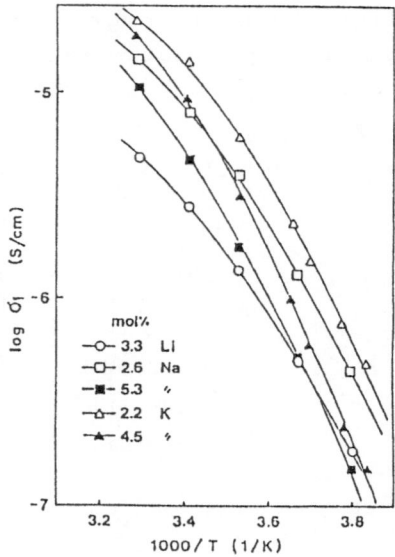

Fig. 8
Temperature dependence of the ionic conductivity for P(MEO$_7$)/MSCN hybrid films.

oxyethylene unit were polymerized and mixed with LiClO$_4$. The ac ionic conductivity was measured by complex impedance technique. Salt content dependence of the ionic conductivity was evaluated for these hybrid films(Fig. 7). The maximum conductivity is observed in every system. This is attributed to the compensation of two factors, namely the carrier number increases with increasing the salt content, leading to an increase in ionic conductivity. While a glass transition temperature at the same time is raised by salt addition. This lowers the ionic conductivity. Therefore, the maximum conductivity can be observed. The peak is shifted to lower salt content side by increasing the side chain length. This suggests that the interaction between cations and polymer segments is affected deeply by the side chain length.

The conductivity of P(MEO$_7$)/MSCN hybrid film was measured at different temperatures ranging from 0 to 80 $^{\circ}$C, and the Arrhenius plot for each system is shown in Fig. 8 [8]. The all of these could be drawn as curved line rather than linear. In other words, the ionic conduction mechanism in each system is considered to obey the Williams-Landel-Ferry(WLF) behavior, in which ionic movement is influenced by the segmental motion of the polymers.

(4) Polar Polymer with Charged Sites

This system can carry only one ion species in the matrix such as only lithium ion. The single ionic conduction is impossible in solution system. In inorganic system, ion conduction is always single but ion species cannot be changed easily in comparison with those of polymeric solid electrolyte system. The most important characteristics for the single ionic conduction is excellent stability in current or conductivity under DC conductive field.

Copolymer system: A simple example for the single ion conductor can be proposed by the copolymerization of oligo(oxyethylene) methacrylate and methacrylic acid salt [9-12]. The present copolymer is a typical single ion conductor without any added salt. Generally, the conductivity for single ion conductor is very low, in the range of 10^{-8} - 10^{-9} S/cm. This can be improved by this copolymerization.

P(MEO$_n$-MAM)

Figure 9 shows the effect of copolymer composition on the single ion conductivity for this copolymer [11]. The conductivity more than 10^{-7} S/cm is seen at 25°C depending on the cation species. Composition effects considerably on the conductivity. The same explanation can be applied for this. Namely, an increase in salt monomer content increases number of mobile cations but decreases mobility because of crosslinking effect of cations to the polymer matrix. This effect can be seen directly as the changes of T_g with DSC.

T_g increases considerably with an increase in methacrylate salt content in copolymer [11]. The cation species dependence can also be confirmed in Fig. 10. Sodium and potassium are rather effective to increase the T_g than lithium.

The time dependence of dc conductivity for the P(MEO$_7$-MALi) copolymer shows almost no change as shown in Fig. 11 [12]. Namely, a constant conductivity is provided easily. As reference, that for the P(MEO$_7$)/LiClO$_4$ hybrid system was also plotted in this figure. This hybrid system shows higher initial conductivity but it decreases considerably with

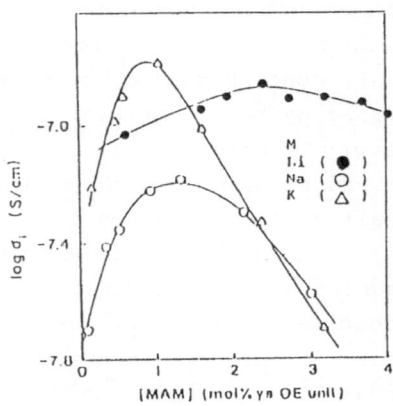

Fig. 9
Comonomeric electrolyte content dependence of ionic conductivity for P(MEO$_7$-MAM) films.

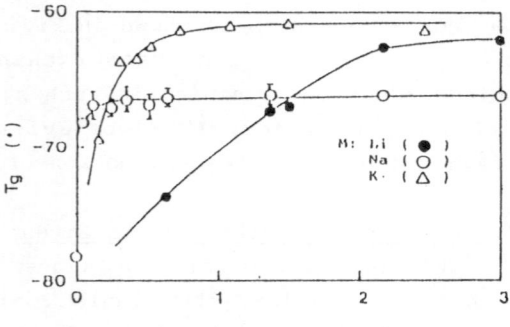

Fig. 10
Relation between comonomeric electrolyte content and T_g of the P(MEO$_7$-MAM) films.

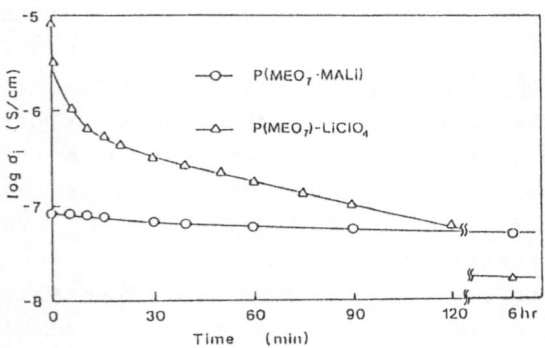

Fig. 11
Time dependence of dc(3 V) ionic conductivity of polymeric solid electrolytes.

dc polarization. Finally, it reaches lower value than that for the single ionic system. This excellent stability should be required for the ionic devices driven under dc such as battery.

Homopolymer: The most sophisticated system is a homopolymer single ion conductor [13,14]. This homopolymer contains three different roles such as flexible structure holder, ion conductive pathway, and carrier ion source in one repeating unit(Fig. 12). This (ω-carboxy) oligo (oxyethylene) methacrylate homopolymer is an unique material for not only application but also basic physico-chemical researches.

In polymeric solid electrolyte systems, there are three different factors concerning ion conduction (Fig. 13). A is the segmental motion, B is the dissociation energy, and C is the Ion-Dipole Interaction. It has been generally quite difficult to illuminate only one factor because of considerable difficulty to set all of other factors the same. However, this system provides a series of ion conductors having different

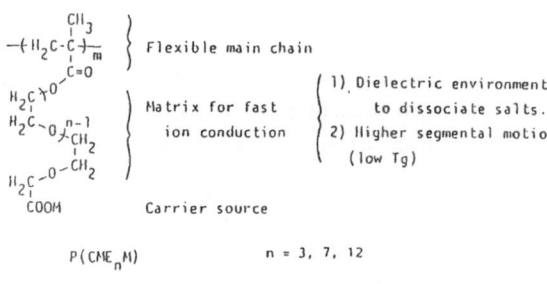

Fig. 12
Structure of P(CME$_n$M) and expected roles of segments.

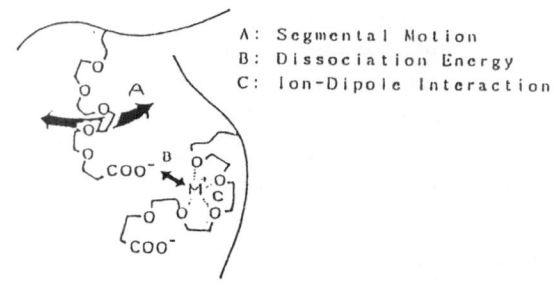

A: Segmental Motion
B: Dissociation Energy
C: Ion-Dipole Interaction

Fig. 13
Factors of ion conduction for P(CME$_n$M) system.

Table 4 Conductivity and WLF parameters for P(CME$_n$M).

Polymer	Salt	Conductivity (S/cm, 30 °C)	Tg (°C)	C_1 (17.74)	C_2 (51.6)
P(CME$_3$M)	Li	< 10^{-11}	32	12.0	75.1
	Na	3.2×10^{-11}	27	10.2	44.5
	K	2.0×10^{-10}	33	9.34	62.9
	Rb	4.5×10^{-10}	32	9.72	65.2
	Cs	7.9×10^{-10}	32	9.92	50.4
P(CME$_7$M)	Li	4.3×10^{-10}	-23	12.3	51.2
	Na	2.2×10^{-8}	-24	12.6	72.4
	K	1.1×10^{-7}	-24	11.4	46.3
	Rb	2.7×10^{-7}	-23	12.2	45.6
	Cs	4.8×10^{-7}	-23	10.5	45.6
P(CME$_{12}$M)	Li	8.9×10^{-9}	-59	----	----
	Na	6.5×10^{-8}	-58	----	----
	K	4.5×10^{-7}	-57	----	----
	Rb	5.9×10^{-7}	-55	----	----
	Cs	3.3×10^{-6}	-54	----	----

cation species without changing any other structural characteristics such as molecular weight, polyether content, charged site density and so on. Further, the conductivity measurement under constant $T-T_g$ provides information about the factors B and C.

Table 4 summarizes all of the conductivity data. T_g and WLF parameters for a series of $P(CME_nM)$ with different cation species and different oligo(oxyethylene) chain length. The T_g decreases with increasing oligo(oxyethylene) chain length. The higher conductivity is seen for systems having larger cations and/or longer oligo(oxyethylene) side chain. It should be noted here that the system in the lower most, $P(CME_{12}Cs)$, showed the highest ionic

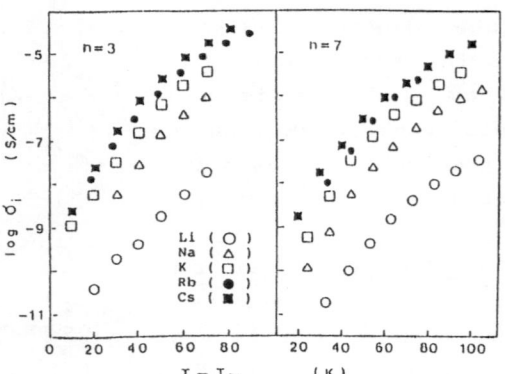

Fig. 14
Relation between $T-T_g$ and ionic conductivity of $P(CME_nM)$.

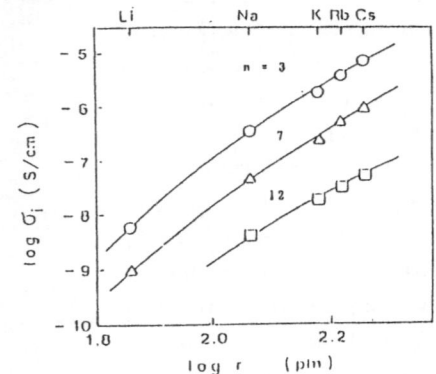

Fig. 15
Relation between ionic radius and conductivity of $P(CME_nM)$ at $T-T_g = 60\ °C$.

conductivity at room temperature for the polyelectrolyte homopolymer system. Over 3×10^{-6} S/cm was found at 25 °C.

The conductivity is plotted against $T-T_g$ as seen in Fig. 14. It surely shows that larger cations can provide higher conductivity. In Figure 15, the ion radius is plotted with conductivity for a series of monovalent cation. There is an easy explanation for this tendency. Namely, larger cation salt has smaller dissociation energy, and therefore provides much more ions in number. It seems reasonable, but it can not explain such wide range of difference. Further, sodium ion was analyzed by ESCA to be dissociated about 15-20% in this solid system. The most possible explanation for this is the different force of Ion-Dipole interaction. Larger ions have smaller surface charge density and therefore receive weaker interaction with ether oxygens. Accordingly larger cations can move faster in the matrix through the fast exchange between sites.

Until quite recently, most scientists believe that the Stokes-Einstein equation can be applied even in solid polymer system. That means, larger cations receive larger resistance because of larger size. However, it was clarified that the fast ion migration could be achieved by the use of not strong but moderate Ion-Dipole interaction. In near future, polymers with more excellent ion conductive characters will be designed. The control of interaction between carrier ions and polymer matrix is the key point for this purpose.

References

1 Wright PV (1975) Br Polym J 7:319

2 Watanabe M, Kanba M, Tsunemi K, Mizoguchi K, Tsuchida E, Shinohara I (1981) Macromol Chem, Rapid Commun 2:741

3 Ohno H, Matsuda H, Mizoguchi K, Tsuchida E (1982) Polym Bull 7: 271

4 Tsuchida E, Ohno H, Tsunemi K (1983) Electrochim Acta 28:833

5 Shigehara K, Kobayashi N, Tsuchida E (1984) Solid State Ionics 14:85

6 Tsuchida E, Ohno H, Tsunemi K, Kobayashi N (1983) Solid State Ionics 11:227

7 Kobayashi N, Uchiyama M, Shigehara K, Tsuchida E (1985) J Phys Chem 89:987

8 Kobayashi N, Ohno H, Tsuchida E, Hirohashi R (1987) Koubunshi Ronbunshu 44:317

9 Kobayashi N, Uchiyama M, Tsuchida E (1985) Solid State Ionics 17:307

10 Kobayashi N, Hamada T, Tsuchida E (1986) Polymer Journal 18:661

11 Tsuchida E, Kobayashi N, Ohno H (1988) Macromolecules 21:96

12 Ohno H, Tsuchida E (1989) J Macromol Sci -Chem A26:551

13 Tsuchida E, Ohno H, Kobayashi N, Ishizaka H (1989) Macromolecules 22:1771

14 Ohno H, Kobayashi N, Takeoka S, Tsuchida E (1990) Solid State Ionics, in press

Cure of High Performance Epoxy Resin Systems

C.E.M. Morris, P.J. Pearce, B.C. Ennis and V.T. Truong
DSTO Materials Research Laboratory
Ascot Vale, Vic., 3032 Australia

Introduction

High performance, epoxy-based systems are widely and increasingly used in structural aerospace and other applications as both adhesives and the matrix of fibre-reinforced composites. Many studies have been conducted on epoxy systems directed towards elucidating structure-property relationships. This has resulted in a greatly increased level of understanding of these complex systems but has also indicated the complexity of the interrelations between composition, thermal history and properties.

This paper focusses on the role of cure conditions on properties and performance using examples from the extensive studies at this Laboratory. While temperature is the major influence in cure conditions, other factors, such as the degree of confinement of the reactants, can have significant effects.

Temperature Effects

During structural adhesive bonding, circumstances may arise where the use of a temperature different from that recommended by the adhesive manufacturer is either deliberately sought or inadvertently applied. An understanding of the relationship between chemical composition, cure temperature, extent of reaction and glass transition temperature (Tg) is especially important if a product suitable for the application is to be obtained.

A study has been conducted on the effect of cure temperature for three nominally 120°C curing, structural film adhesives. All three were toughened systems based on resins of the diglycidyl ether of bisphenol A (DGEBA) but containing different curing agents and toughness modifiers. The curing agents were (1) dicyandiamide/the adduct of toluene diisocyanate and dimethylamine (TDI-DMA), (2) dicyandiamide/Monuron and (3) the TDI-DMA adduct alone. Curing temperatures ranged from 76°C to 170°C. Thermal analysis showed clearly that the first two systems require the recommended cure temperature and at lower temperatures are substantially undercured, the reactivity being effectively terminated by vitrification before cure is complete. However, for the third adhesive the Tg increases with cure temperature to a maximum at about 100°C before decreasing again at higher temperatures (Fig 1). Further, measurements on Al-Al joints showed that the maximum lap-shear strength was obtained when the adhesive was cured at 100°C. Water uptake by the adhesive was also a minimum for material cured at 100°C. Thus in this case optimum properties are obtained from a cure temperature of 100°C rather than the recommended 120°C.

The effects of vitrification before complete cure can be exemplified with the curing agent diaminodiphenyl sulfone (DDS), commonly used in high temperature systems. Figure 2 shows the dynamic mechanical thermal analysis (DMTA) of two tetraglycidyl methylene dianiline (TGMDA) systems, one cured with DDS and the other with piperidine. In the latter instance the system was almost fully cured and hence Tg was well defined and virtually unaffected by further heating. In contrast the TGMDA/DDS reaction has no well defined end point and further heating results in additional reaction and higher Tg, as shown by the upturn in the modulus curve and the width of the tan delta peak (Fig. 2), and increased brittleness.

Cure temperature conditions can also have a substantial effect on the morphology of rubber-toughened epoxies and thus on their fracture behaviour. DGEBA systems containing carboxy-terminated butadiene acrylonitrile reactive rubber can show major increases in fracture toughness, compared with unmodified analogues, through the formation of rubber particles in the cured epoxy matrix. Different cure temperature conditions can induce a variety of second phase morphologies from the same epoxy formulation. Table 1 shows fracture toughness (K_IC) data for some rubber modified, fully cured DGEBA/piperidine systems. Two values of K_IC indicate stick-slip behaviour. The higher temperature cure has led to higher values of K_IC and a wider range of

B. C. Anderson · Y. Imanishi (Eds.)
Progress in Pacific Polymer Science

compositions where fracture occurs in a ductile manner while other mechanical properties, such as Young's modulus and Tg, remain almost unchanged.

Open vs Closed Containers

Epoxy systems used in structural applications, whether as adhesives or the matrix of fibre-reinforced composites, are normally cured under some pressure and can be regarded as in closed containers. Studies on the kinetics and mechanisms of cure chemistry are often conducted without pressure in containers essentially open to the atmosphere. Extensive thermal analysis examinations at this Laboratory on a range of epoxy formulations have shown that for such fundamental quantities as the heat of reaction substantially different values can be obtained by using open or hermetic pans (Table 2). These differences are apparent with both the TDI-DMA adduct and dicyandiamide as curing agent but not with DDS.

Conclusions

Despite extensive studies understanding of the interactions between chemical composition, thermal history and properties is still inadequate for predictive purposes and individual studies on specific systems are still required.

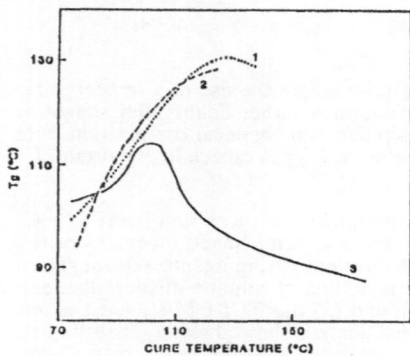

Fig 1: Tg vs cure temperature for three 120°C curing adhesives.

TABLE 1

EFFECT OF RUBBER CONTENT AND CURE CYCLE ON K_{IC} OF DGEBA/PIPERIDINE

Rubber Content (phr)	K_{IC} (MNm$^{-3/2}$)	
	Low Temp Cure (16h at 120°C)	High Temp Cure (2.5h at 150°C + 16h at 120°C)
0	1.2-0.8	1.2-0.8
5	2.3-0.8	3.0
10	2.5	3.5
15	2.5	4.0

Loading rate 1 mm/min; Test Temp. 25°C

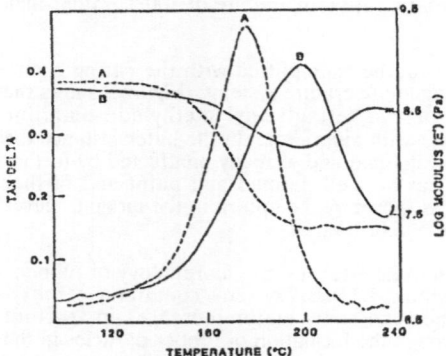

Fig 2: DMTA curves.
A: TGMDA/piperidine
B: TGMDA/DDS

TABLE 2

EXAMPLES OF THE EFFECT ON DSC DERIVED DATA OF OPEN VS CLOSED PANS

System	Heat of Reaction J/g	
	Open	Closed
DGEBA/TDI-DMA 6 phr		
5°C/min	300	590
isothermal 150°C	325	550
DGEBA/dicy 5.5 phr		
20°C/min	219	366

Functional Polymers Derived from 2-Oxazolines

Takeo Saegusa and Yoshiki Chujo

Department of Synthetic Chemistry,
Faculty of Engineering, Kyoto University,
Yoshida, Sakyo-ku, Kyoto 606, JAPAN

Ring-opening polymerization of a family of 2-oxazoline monomers 1 produces poly(N-acylethylenimine)s 2 [1-5] which are regarded as a series of polymer homologues of N,N-dimethyl amides of
various carboxylic acids. Thus, the polymer homologues of N,N-di-
methylformamide (DMF) and N,N-dimethylacetamide (DMAc) are readily
prepared by the polymerizations of the 2-oxazoline monomers having
R=H and R=CH$_3$, respectively. DMF and DMAc, which are called "apro-
tic polar solvent", are characterized by the strong hydrogen-bonding property. They are very hydrophilic or even hygroscopic. The
characteristic properties of DMF and DMAc are observed also in
their respective polymer
homologues of 1 of R=H
and R=CH$_3$ which are
readily soluble in water
and absorb moisture from
the atmosphere.

The reaction of the polymerization of 2-oxazoline is very
clean, which is not disturbed by chain-transfer and termination.
In the following scheme which exemplifies the 2-oxazoline polymerization, the propagating species having a structure of 2-oxazolin-
ium salt is not fragile, which is conveniently utilized from the
syntheses of block copolymers and end-reactive polymers.

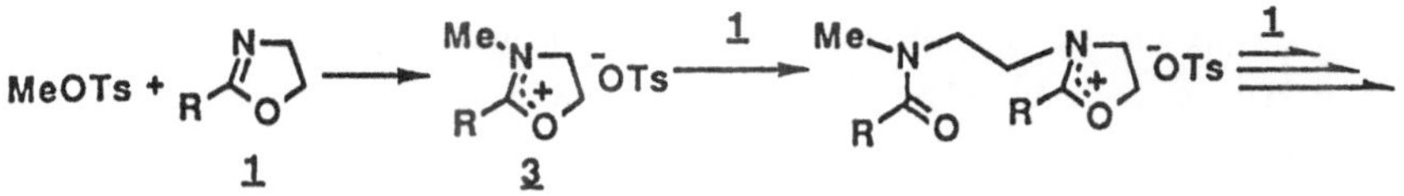

On the basis of the high hydrophilicity of the polymers of
R=CH$_3$ as well as the cleaness of the polymerization reaction

B. C. Anderson · Y. Imanishi (Eds.)
Progress in Pacific Polymer Science
© Springer-Verlag Berlin Heidelberg 1991

(living polymerization mechanism), three novel functional polymers
have been explored.

1. Non-ionic surface active agents
2. Non-ionic hydrogels
3. An organic/inorganic hybrid polymer

NON-IONIC SURFACE ACTIVE AGENTS

A polymeric segment of 2-methyl-2-oxazoline (MeOZO) having high hydrophilicity has been combined with various hydrophobic groups to produce a new group of surface active agents which are characterized by the non-ionic nature.

Scheme 1

The new group of surfactants is exemplified by the AB and ABA types block copolymers between MeOZO and one of 2-oxazoline derivatives having medium to higher alkyl substituent at the 2-position.

The AB and ABA type block copolymers involving MeOZO and another derivative of 2-oxazoline are readily prepared by the so-called "One-pot Multistage Block Copolymerization" [6].

(m, n=5-15, R= alkyl>C$_4$, phenyl)

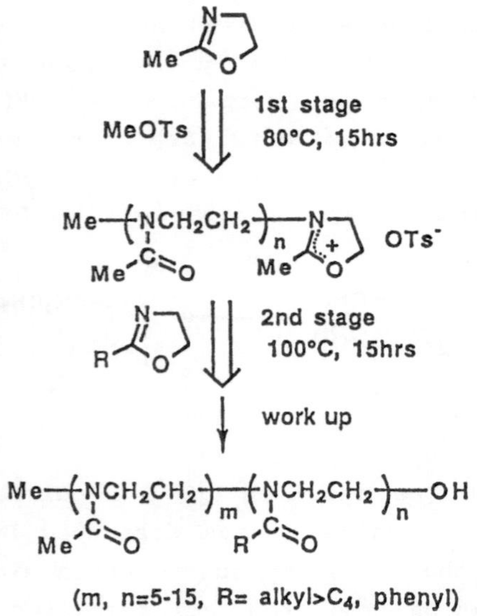

Scheme 1 shows the "One-pot Two-stage" process to produce AB type block copolymer, in which the first monomer is MeOZO and the second one is a 2-oxazoline derivative having a medium to higher alkyl or phenyl group.

Figure 1 shows GPC curves of the products of the one-pot two-stage block copolymerization between MeOZO (the first monomer) and 2-phenyl-2-oxazoline (PhOZO) (the second monomer). Curve A represents the first stage product of the homopolymer of MeOZO. Curves B, C and D indicate the second stage products with a varying

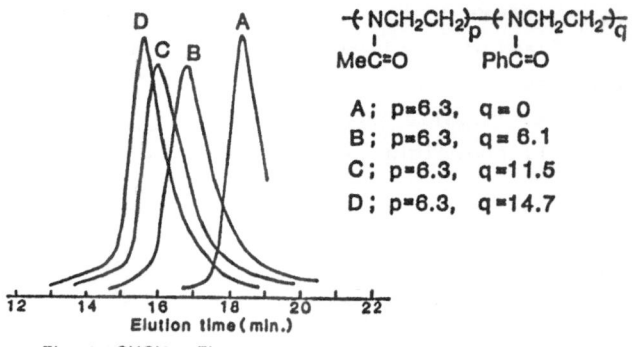

A; p=6.3, q=0
B; p=6.3, q=6.1
C; p=6.3, q=11.5
D; p=6.3, q=14.7

Eluent : CHCl3, Flow rate 1.0 ml/min,
Column : Shodex A803.

Figure 1. GPC Curves of One-pot
Two-stage Polymerization.

amount of the second monomer of 2-phenyl-2-oxazoline. The results of GPC analysis are taken to confirm that the second stage products consist exclusively of block copolymer. Neither the homopolymer of the first (MeOZO) nor that of the second monomer (PhOZO) is observed. All the second stage products in Figure 1 are completely water-soluble. The homopolymer of PhOZO, if it were present in the product, should have been detected as the water-insoluble fraction. Table 1 shows some data of surface tension values of aqueous solutions of block copolymers at a concentration of 1.0% at 29°C.

Table 1. Surface Tension Values of
Block Copolymers.

$$Me + \left(NCH_2CH_2 \atop MeC{=}O \right)_m \left(NCH_2CH_2 \atop RC{=}O \right)_n OH$$

m	R	n	δ (dyne/cm)
11.3	—	—	62.8
5.4	C_2H_5	5.3	52.9
5.4	$n{-}C_3H_7$	5.8	48.4
5.5	$n{-}C_4H_9$	5.3	30.9
14.0	$n{-}C_4H_9$	7.8	28.9
5.5	$n{-}C_8H_{17}$	4.9	32.1
6.3	$n{-}C_{12}H_{25}$	4.2	37.1

1.0% aq. soln. at 29 °C (H_2O 71.3)

It is of deep interest to note that block copolymers with the second monomers having an alkyl group larger than C$_4$ (butyl) group, show excellent surface tension values as low as thirties. The values of critical micelle concentration are quite low, i.e., at a level of 1/10 - 1/100 weight percents (Figure 2).

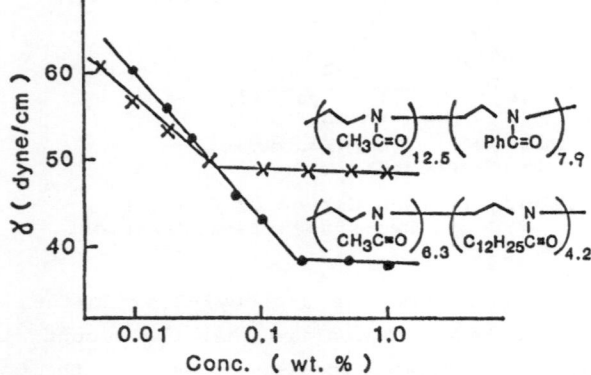

Figure 2. Critical Micelle Concentrations of Block Copolymers.

On the basis of the same principle, ABA type tri-block copolymers have been prepared by the "One-pot Three-stage" process as shown by Scheme 2. The first monomer (R$_1$= a medium to higher alkyl group) is employed also in the third stage. The second monomer of MeOZO produces the hydrophilic segment. The surface tension values of their aqueous solutions at a concentration of 1.0% were in a similar range of the corresponding AB type di-block copolymers, i.e., 30-40 dyne/cm. A series of ABA type tri-block copolymers were prepared also by the "One-pot Two-stage" process starting with a bifunctional oxazoline monomer which was first converted into the corresponding bifunctional oxazolinium salt. The two-directional propagation of the polymerization of the first monomer was followed by that of the second monomer (Scheme 3).

Based on the unit structure of N,N-dimethyl carboxyl amide, the above block copolymers are expected to have useful character-istics such as high compatibility with organic compounds and with synthetic polymers as well as coordination onto transition metal salts and oxides. They may possess excellent performance as a dis-persing agent for organic compounds of antioxidants, stabilizers, UV absorbers and for metal oxides of fillers in synthetic polymers.

Scheme 2

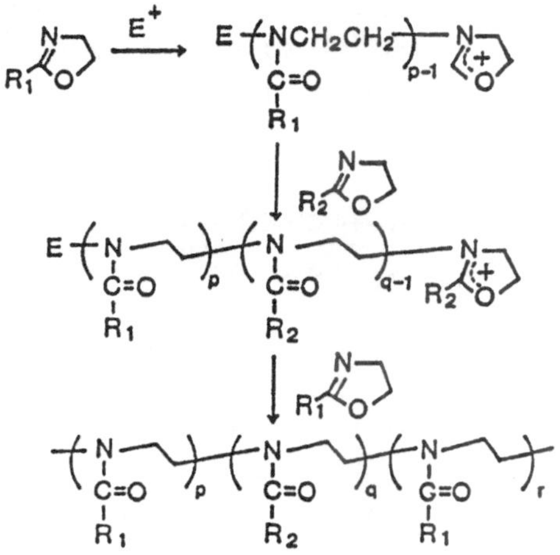

Scheme 3

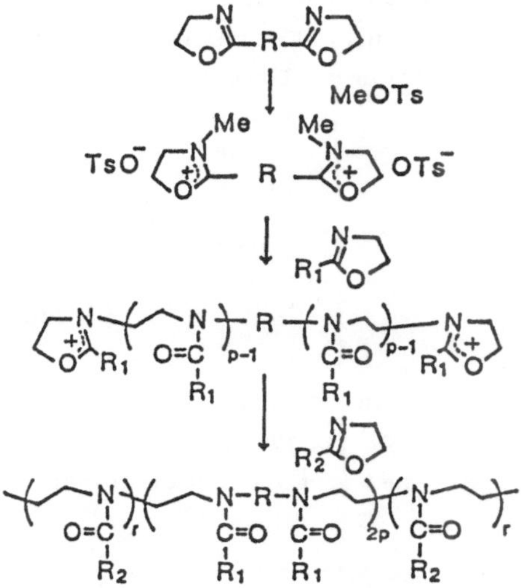

Perfluoroalkyl group is characterized in two ways. First, it is a strong electron-withdrawing group which affects largely the chemical reactivity of the parent compound. Second, it gives a water- and oil-repellency to the compound. A block copolymer was prepared by the "One-pot Two-stage" process using 2-perfluoroethyl-2-oxazoline and MeOZO as shown by Scheme 4 [7]. By GPC analysis, the product of the second stage was water-soluble, which consisted exclusively of a block copolymer. Due to the high hydrophobicity of the segment of 2-perfluoroethyl-2-oxazoline, an aqueous solution of the block copolymer shows an extremely low value of surface tension.

Scheme 4

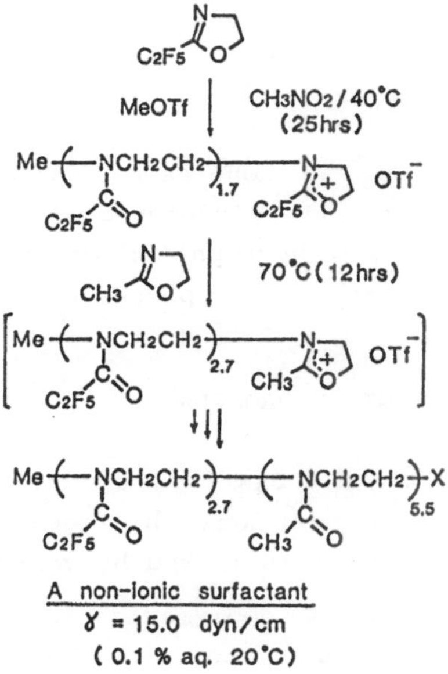
MeOTf
CH3NO2 / 40°C (25hrs)
70°C (12hrs)
A non-ionic surfactant
γ = 15.0 dyn/cm
(0.1 % aq. 20°C)

A newly appeared hydrophilic constituent of polyMeOZO has been combined also with a conventional hydrophobic constituent of a long alkyl group. For example, initiation of the MeOZO polymerization with a higher alkyl tosylate gives rise to a non-ionic surfactant [8].

$$\text{ROTs} + \underset{Me}{\overset{N}{\diagdown}}\!\!\diagup\!\!\overset{O}{\diagup} \;\Rightarrow\; \xrightarrow{\text{NaOH}}\; R\!\!-\!\!\left(NCH_2CH_2\right)_n\!\!-\!\!OH \;\; \underset{Me\,C=O}{}$$

(R: higher alkyl group)

In addition, the propagating species of MeOZO polymerization with the methyl tosylate initiator is reacted with a higher alkyl primary amine to produce another type of non-ionic surfactant [9].

$$Me\!\!-\!\!\left(NCH_2CH_2\right)_{n-1}\!\!\overset{N}{\underset{Me}{\diagup}}\!\!\overset{+}{\diagup}\!\!\overset{O}{\diagup}\; OTs^{\cdot} \;+\; R\text{-}NH_2 \underset{Me\,C=O}{}$$

$$\xrightarrow[\text{resin}]{\text{ion-exchange}}\; Me\!\!-\!\!\left(NCH_2CH_2\right)_n\!\!-\!\!NH\text{-}R \underset{Me\,C=O}{}$$

(R: higher alkyl group)

As the non-ionic hydrophilic constituent, poly(oxyethylene) segment and polyfunctional alcohols have hitherto been employed. Now a new hydrophilic constituent of polyMeOZO segment has been discovered, which possessed additional useful properties.

NON-IONIC HYDROGELS

On the basis of the hydrophilic property of polyMeOZO, a new group of non-ionic hydrogels have been explored, i.e., polyMeOZO has been closs-linked by various ways.

Random copolymerization of MeOZO with a bis-oxazoline monomer produces a hydrogel (Scheme 5) [10, 11]. The swelling degree depends upon the cross-linking density, which is controlled by the feed ratio of the bis-oxazoline monomer. As is shown in Table 2, the copolymer hydrogel shows similar values of swelling degree in pure water and in salt water (5% NaCl). Figure 3 indicates that the swelling degree of the copolymer hydrogel changes very little according to the NaCl concentration.

The second method of the preparation of hydrogel is the partial hydrolysis of poly-MeOZO followed by cross-linking with diisocyanate (Scheme 6) [12]. Some results

Scheme 5

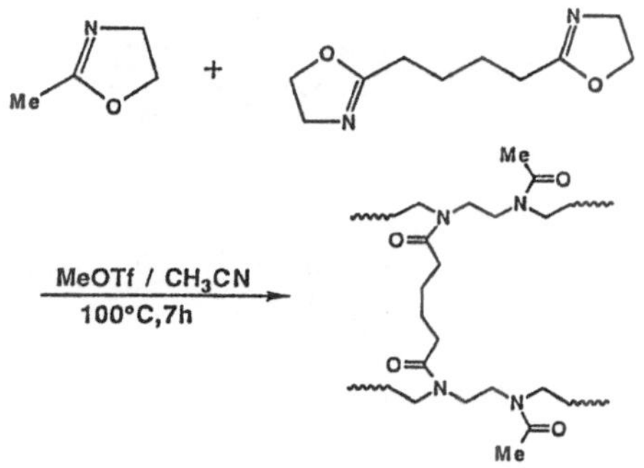

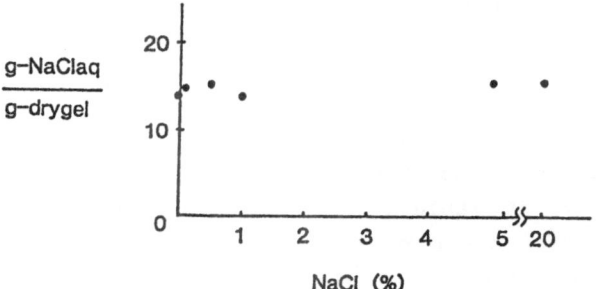

Table 2. Synthesis of PolyMeOZO Hydrogel by Copolymerization.[a]

Run	MeOZO / MeOTf	BisOZO / MeOTf	MeOZO / BisOZO	Yield (%)	H₂O Content[b]	
					in H₂O	5%NaCl aq
1	100	20	5	92	6	6
2	100	10	10	86	15	13
3	100	5	20	68	36	27
4	200	5	40	85	40	31
5	300	5	60	74	45	39
6	90	2.5	36	0[c]	-	-

a) in CH_3CN, 100°C, 7h, in sealed tube
b) g-H_2O / g-dry gel
c) No gelation was observed after 7 days at 100°C

Figure 3. Swelling Degrees of PolyMeOZO Hydrogel in Various Concentrations of Aqueous NaCl.

Scheme 6

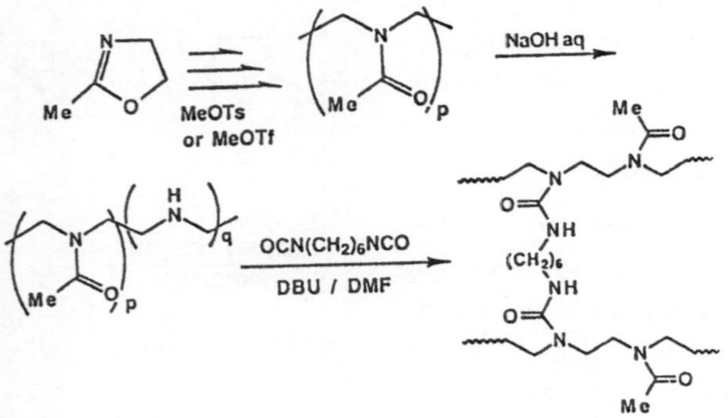

Table 3. Synthesis of PolyMeOZO Hydrogel
by Partial Hydrolysis and
Cross-linking with Diisocyanate.

MW a)	Hydrolysis %	−NCO / −NH−	Mn between cl.pts	H2O Content g H2O/g gel in H2O	5% NaClaq
16,000	3.4	1.53	2,500	14(13)[b]	11
16,000	3.4	0.65	3,850	45(55)[b]	19
16,000	3.4	0.79	3,160	63	29
6,000	7.2	0.53	2,230	48	15
16,000 c)	5.0	0.49	3,470	32	32
Hydrogel of −(CH2CH2O)n−				17	17

a) Mol. wt. of parent poly(MeOZO)

b) The value in parenthesis was obtained after
one cycle of swelling and drying. Swelling
equilibrium is reproducible.

c)

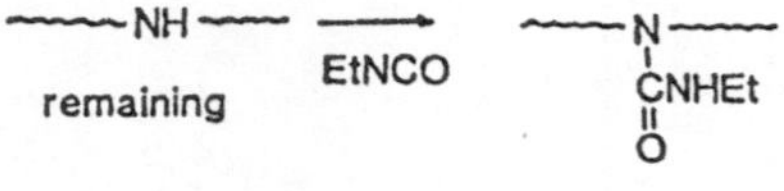

of the preparation of polyMeOZO hydrogel as well as the swelling
degrees are shown in Table 3. The reaction of diisocyanate with
the secondary amino group generated by partial hydrolysis is not
complete. Accordingly, the swelling degree in 5% salt water was
lower than that in pure water. The decrease is ascribed to the
generation of an ammonium group from the remaining secondary amino
group. Conversion of the remaining secondary amino group into the

corresponding urethane group by the reaction with monofunctional isocyanate resulted in similar degrees of swelling in pure water and in salt water.

Non-ionic hydrogel based on polyMeOZO has a useful property of high compatibility with several synthetic polymers. For example, the above copolymer hydrogel is compatible with poly(vinyl chloride), and the blend swells in water in a homogeneous way [13].

On the other hand, partially hydrolyzed polyMeOZO was treated with acid chloride of (7-coumaryloxy)acetic acid in the presence of triethylamine. This coumaryl-polyMeOZO was cast upon a slide glass and irradiated for 3 h (450W high-pressure Hg lamp) to form a gel (Scheme 7), which showed a quite high swelling property in water as a hydrogel. From the results of UV spectroscopy, the gelation was caused by the photo-dimerization of coumarin moieties in polyMeOZO. The swelling degree of the obtained gel can be controlled by the coumarin content or by the irradiation time [14].

Scheme 7

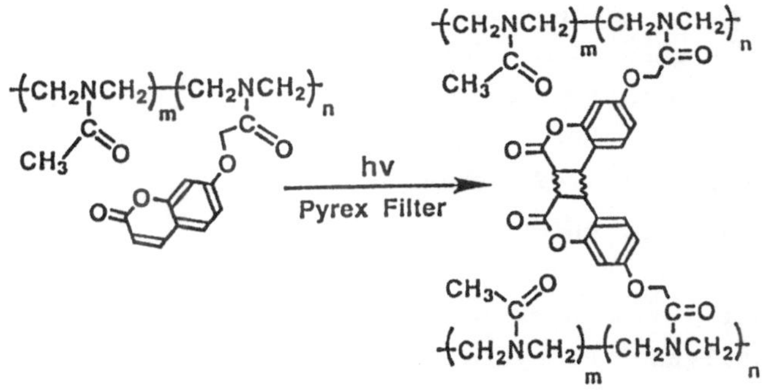

450W High Pressure Hg Lump, Irradiated for 5h in Bulk Film, at r.t.

Scheme 8

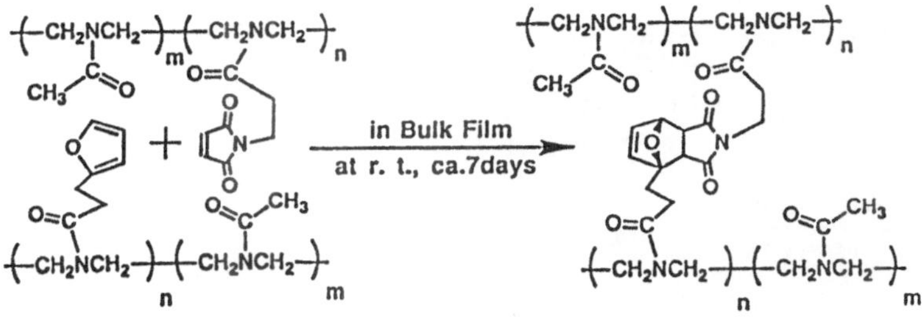

Furan or maleimide group was introduced to polyMeOZO as a diene and a dienophile moiety, respectively. Intermolecular Diels-Alder reaction between these two groups caused the gelation at room temperature (Scheme 8). The obtained gel absorbed water to form a non-ionic hydrogel [15].

The gel having S-S bonds at the cross-linking points (Scheme 9) was prepared. The S-S bonds were cleaved reductively to form -SH groups. This means that the gel was transformed to the soluble polymer after the reductive treatment. This reversible conversion by redox system has a potential to new functional materials [16].

Scheme 9

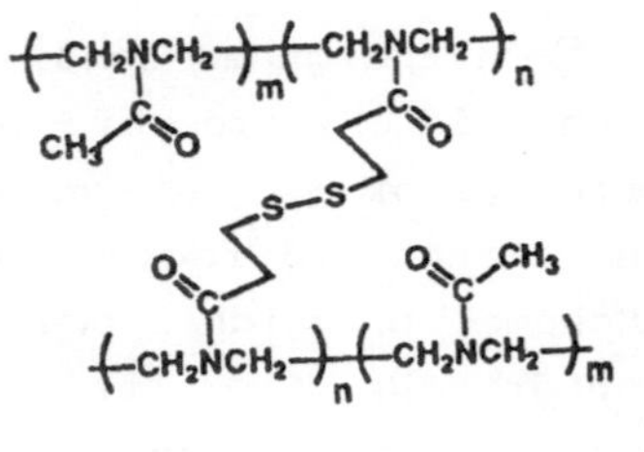

AN ORGANIC/INORGANIC HYBRID POLYMER

Metal oxide in general may be regarded as a three-dimensional network inorganic polymer consisting of metal-oxygen bond. For example, silica gel is formulated by Scheme 10.

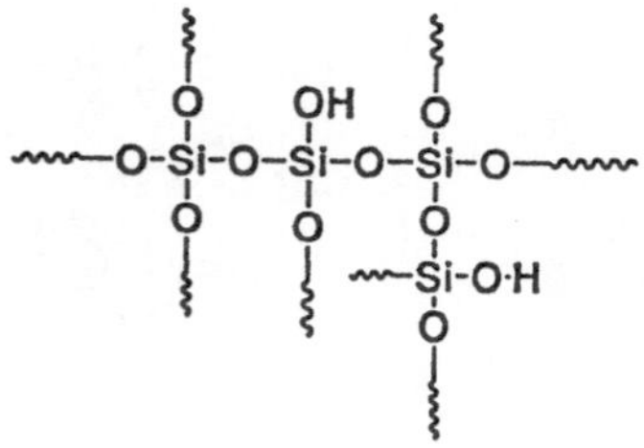

A new material which is expressed by a term of "Organic/Inorganic Hybrid Polymer" has been prepared by the acid-catalyzed co-hydrolysis/co-

Scheme 10

Scheme 11

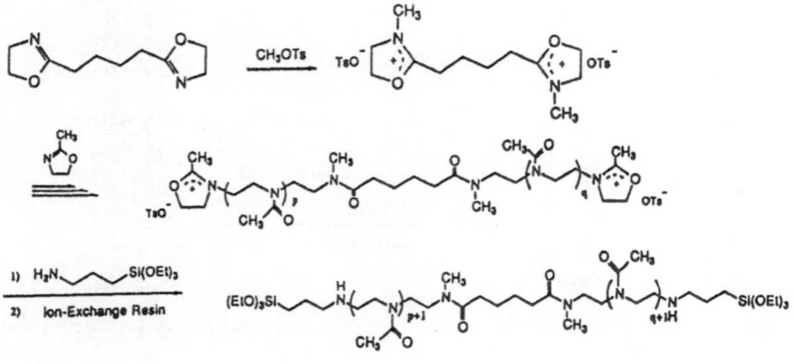

condensation of ethyl orthosilicate with the terminal $-Si(OEt)_3$ group of polyMeOZO.

A typical example of the synthesis of polyMeOZO having a terminal $-Si(OEt)_3$ group is illustrated by Scheme 11 [17].

Co-hydrolysis/co-condensation between ethyl orthosilicate and polyMeOZO having the above terminal group proceeds in a homogeneous solution of EtOH. Evaporation of EtOH, the solvent and the condensation product, gave a colorless and transparent glass. An idea concerning the molecular structure of the glass is formulated in Scheme 12. The composition of the glass depends upon the feed ratio between the two components in the co-hydrolysis/co-condensation. Transparent and homogeneous glassy materials having composi-

Scheme 12

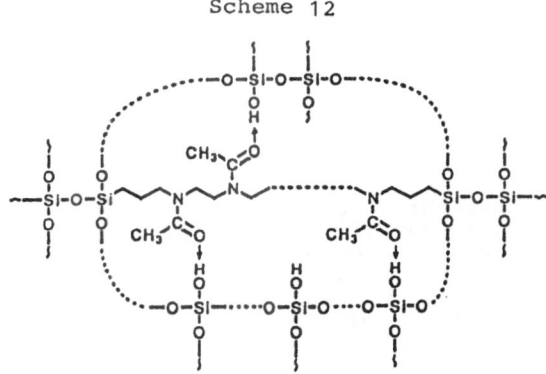

Table 4. Synthesis and Adsorption Property of PolyMeOZO-Silica Hybrid.

Run	POZO (D.P)[a]	$POZO/Si(OEt)_4$	Weight loss (TGA) (%)	POZO(Wt%)[b]	H_2O Content[c]
1	O———[d]	1/2	49.0	50.0	2.98
2	(9.0)	1/10	29.4	15.6	1.87
3	O———	1/2	47.5	47.2	2.26
4	(14.4)	1/10	27.3	14.2	1.63
5	O———O	1/2	50.3	53.1	3.46
6	(16.1)	1/10	28.5	17.9	1.92
7		0	18.2	0	1.53

a) Calculated from feed ratio
b) Calculated from elemental analysis
c) g wet gel / g dried gel
d) O$-Si(OEt)_3$

tions up to 50/50 weight ratio were obtained.   In comparison with the silica gel without polyMeOZO segments, the present composite glass was found to show higher water adsorption property as shown in Table 4 [18].

This material may be defined as a block copolymer consisting of polyMeOZO and poly-SiO$_2$ (three dimensional network). Production of a homogeneous clear glass is ascribed to the chemical bonding between SiO$_2$ and the terminal group of polyMeOZO and the strong hydrogen-bonding between the monomeric units of polyMeOZO and -SiOH group at the interface.

Pyrolysis of the above hybrid at a temperature, e.g. 600°C, below the fusion temperature of silica gel eliminates the component of polyMeOZO to leave a silica gel having micro-pores of a radius of 18 Å.

References

1 Saegusa T, Kobayashi S (1976) Encyclopedia of Polymer Science and Technology, Suppl. Vol. 1:220
2 Kobayashi S, Saegusa T (1984) In: Ivin K, Saegusa T (eds) Ring-Opening Polymerization, Elsevier, London 2:761
3 Kobayashi S, Saegusa T (1985) Makromol. Chem., Suppl. 12:11
4 Saegusa T (1988) Makromol. Chem., Macromol. Symp. 13/14:111
5 Saegusa T, Chujo Y (1989) In: Saegusa T, Higashimura T, Abe A (eds) Frontiers of Macromolecular Science, Blackwell Scientific Publications 119
6 Kobayashi S, Igarashi T, Moriuchi Y, Saegusa T (1986) Macromolecules 19:535
7 Miyamoto M, Aoi K, Saegusa T (1989) Macromolecules 22:3540
8 Kobayashi S, Iijima S, Igarashi T, Saegusa T (1987) Macromolecules 20:1729
9 Kobayashi S, Uyama H, Higuchi N, Saegusa T (1990) Macromolecules 23:54
10 Chujo Y, Sada K, Matsumoto K, Saegusa T (1990) Macromolecules 23: in press
11 Chujo Y, Sada K, Matsumoto K, Saegusa T (1989) Polym. Bull. 21:353
12 Chujo Y, Yoshifuji Y, Sada K, Saegusa T (1989) Macromolecules 22:1074
13 Chujo Y, Yoshifuji Y, Sada K, Saegusa T, in preparation
14 Chujo Y, Sada K, Saegusa T (1990) Macromolecules 23: in press
15 Chujo Y, Sada K, Saegusa T (1990) Macromolecules 23: in press
16 Chujo Y, Sada K, Saegusa T, in preparation
17 Chujo Y, Ihara E, Ihara H, Saegusa T (1989) Macromolecules 22:2040
18 Chujo Y, Ihara E. Kure S, Saegusa T (1990) Macromolecules 23: in press

Conducting Polymer Fibres with Excellent Mechanical Properties and High Electrical Conductivity

A. Andreatta[a], S. Tokito[b], P. Smith[a,c] and A.J. Heeger[a,d]

Institute for Polymers and Organic Solids
University of California at Santa Barbara, CA 93106

Abstract: We present a summary of our recent results on the electrical and mechanical properties of fibers made from poly(2,5-dimethoxy-p-phenylene vinylene), PDMPV and poly(2,5-thienylene vinylene), PTV, using the precursor polymer methodology, and from polyaniline, PANI, using the method of processing as polyblends with poly-(p-phenylene terephthalamide), PPTA, from sulfuric acid. The solubility of both PANI and PPTA in H_2SO_4 presents a unique opportunity for co-dissolving and blending PANI and PPTA to exploit the excellent mechanical properties of PPTA and the electrical conductivity of PANI; we summarize the electrical and mechanical properties of such composite fibers. For PDMPV and PTV fibers, we find a strong correlation between the conductivity and the tensile strength (and/or modulus), and we show from basic theoretical concepts that this relationship is an intrinsic feature of conducting polymers.

I. INTRODUCTION

In conjugated polymers, the π-bonding leads to π-electron delocalization along the polymer chains and to the possibility of relatively high charge carrier mobility, μ, which is extended into three dimensional transport by the interchain electron transfer interactions.[1] The high density of redox sites within the π-electron system (essentially one per monomer) offers the additional advantage of a relatively high density, n, of carriers (charge e) through doping. Thus, high electrical conductivity, $\sigma=ne\mu$, is possible. As a result of the same intrachain π-bonding and relatively strong interchain electron transfer interactions, the mechanical properties (Young's modulus and tensile strength) of conjugated polymers are potentially superior to those of saturated polymers, such as polyolefins. Moreover, because of these two features, it may be possible to achieve exceptional mechanical properties with aligned conjugated polymers at lower chain lengths than required for their saturated counter-

[a]Materials Department
[b]Permanent address: Department of Materials Science and Technology, Kyushu University, Kasuga-shi, Fukuoka 816, JAPAN
[c]Chemical and Nuclear Engineering Department
[d]Physics Department.

B. C. Anderson · Y. Imanishi (Eds.)
Progress in Pacific Polymer Science
© Springer-Verlag Berlin Heidelberg 1991

parts.[2] Thus, conjugated polymers are of special interest because of the potential of a unique combination of electrical and mechanical properties.[3]

Since the electrical and mechanical properties are currently limited by defects and structural disorder, improvement in material quality that will enable the exploration of intrinsic properties has become an important goal of conducting polymer research.[3] It has been long recognized, however, that conjugated polymers tend to be insoluble and infusible. Thus, the question to be answered is whether or not processing methods can be developed that will lead to chain extended and chain aligned materials of sufficient quality. Significant progress has been made; the addition of long alkyl side chains[4] has opened opportunities for processing from solution or from the melt. However, the relatively bulky side chains decrease the π-electron density (and thus the carrier density) and the interchain coupling, thus making it more difficult to achieve the structural coherence needed to obtain high carrier mobility and exceptional mechanical properties.

A promising strategy to approach the intrinsic mechanical and electrical properties is through the use of the versatile precursor route which involves the preparation of a processible precursor polymer and subsequent conversion of the precursor polymer to the conjugated polymer.[5,6] The significant advantage of this route is that the saturated precursor polymers can be processed from solution prior to the thermal conversion to the conjugated final product. The precursor polymers may, therefore, be drawn prior to and during the thermal conversion process so as to yield oriented, homogeneous conjugated polymers.

Poly(p-phenylenevinylene), PPV, and its derivatives can be prepared from a precursor polymer, a polyelectrolyte, which is soluble in water.[6] The dimethoxy-derivative of PPV, poly(2,5-dimethoxy-p-phenylenevinylene), PDMPV, has been prepared via a similar precursor route to PPV and exhibited high conductivities after doping. However, the commonly used aqueous solutions of the PDMPV precursor polymer tend to form gels, and the gradual elimination of the sulfonium group in the solid precursor cannot be avoided even at room temperature; both effects make subsequent processing into highly oriented films and fibers difficult. Recently, the Kyushu University group[7] succeeded in the preparation of dense PDMPV film from a new precursor polymer which is soluble in common organic solvents, easily processible, and stable even at 100ºC. Similarly, poly(2,5-thienylene vinylene) (PTV) is one of the larger class of poly(arylenevinylene) polymers which is attractive as a conjugated polymer and which can be synthesized via the precursor polymer route. As with PDMPV, the Kyushu University group [8,9] and Murase et al[10] have found that PTV can be prepared through a new precursor polymer which is soluble in common organic solvents and chemically stable.

An alternative strategy is to identify stable conjugated polymer systems that can be processed. Of this class, polyaniline (PANI) is certainly a promising example. The use of concentrated acids[11] as solvents for PANI has specific advantages in that both the salt and the base form can be completely dissolved at room temperature, with polymer concentrations ranging from extremely dilute to more than 20% (w/w), in concentrated protonic acids such as H_2SO_4, CH_3SO_3H, and CF_3SO_3H. Perhaps more important is the fact when precipitated from acid solution, PANI comes out in the conducting (protonated) emeraldine salt form.[11] Although the ability to process conducting polyaniline from solution represents genuine progress, fibers and films made from these solutions have mechanical properties which are not adequate for many applications due mainly to the low molecular weight of the polyaniline used.[12] Fibers with significantly enhanced mechanical properties have been obtained by blend processing polyaniline with the rigid chain polymer poly-(p-phenylene terephthalamide), PPTA.[13] It is well known that PPTA is processed from solutions in concentrated H_2SO_4 to yield one of the strongest and stiffest fibers commercially available.[14] The process takes advantage of the fact that PPTA is a rod-like polymer that exhibits a liquid crystalline phase at high solution concentrations which facilitates the formation of highly oriented structures.[15]

In this review, we present a summary of our recent results on the electrical and mechanical properties of fibers made from PDMPV[16] and PTV[17] using the precursor polymer methodology, and from PANI using the method of processing as polyblends with PPTA from sulfuric acid.[13]

II. EXPERIMENTAL METHODS AND TECHNIQUES

A. Preparation of Precursor Polymers for PDMPV and PTV

The preparation of the precursor polymer and conversion to the conjugated polymer are summarized in Figure 1a (PDMPV) and 1b (PTV). Details on the synthesis of the precursor polymer and the conversion to the conjugated polymer are presented elsewhere.[16,17] The DSC thermogram of the PDMPV precursor polymer indicated a glass transition at 110°C, well-separated in temperature from the ≈195°C needed for thermal elimination of the methoxy leaving groups (Scheme 1a). For the PTV precursor polymer, the glass transition is at 50°C, again, well-separated in temperature from the elimination of methoxy leaving groups, which occurred at 188°C.

B. Preparation of the PANI/PPTA Blends[13]

The synthesis of the polyaniline used in this study was reported elsewhere[8,9]. Poly(p-phenylene terephthalamide) with inherent viscosity of 7.43 dl/g was obtained from DuPont as Kevlar® powder.

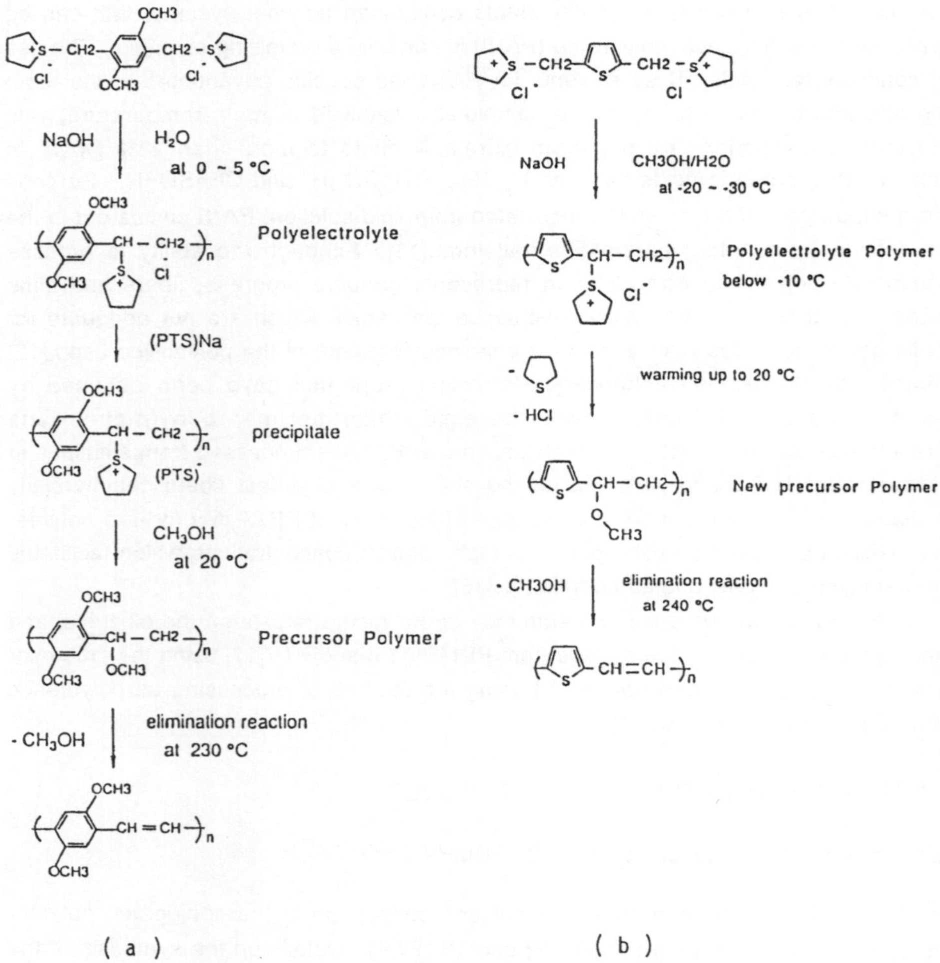

(a)   (b)

Figure 1

a: Synthesis of poly(2,5-dimethoxy-p-phenylenevinylene) from precursor polymer soluble in organic solvents.

b: Synthesis of poly(2,5-thienylene vinylene) from precursor polymer soluble in organic solvents.

A stock solution of 2 wt% PPTA in sulfuric acid was prepared by mixing 3.76 g of PPTA in 100 ml of 98% H_2SO_4 (Fisher). The mixture was mechanically stirred overnight to yield a homogeneous solution. This solution was also used to produce the 100% PPTA fibers. Polyaniline was weighed into a vial and concentrated H_2SO_4 was subsequently added. The mixture was stirred until the polyaniline dissolved and the solution was homogeneous. An amount of 2 wt% PPTA stock solution was added to yield a desired specific PANI/PPTA ratio. The mixture was mechanically stirred for 5 to 12 hours at room temperature, and then allowed to stand for 24 hours before spinning, to let the trapped air escape. In all cases, optically homogeneous solutions were obtained. Since the concentration of PPTA in the solutions was maintained at 1.5 wt% (which is below the onset of the formation of the lyotropic phase; typically at 6-8 wt% PPTA[11]), these polyblend fibers were spun from isotropic solutions.

C. Fiber Spinning, Drawing and Conversion of PDMPV[16] and PTV[17]

Details on the fiber spinning of the purified precursor polymers are given elsewhere.[16,17] For PDMPV and PTV, the precursor polymers were dissolved into chloroform. The solutions were spun using a high precision syringe pump (Sage Instruments, model 355). The viscous solutions were pumped at a speed of 0.013 ml/min through a needle with diameter of 0.5 mm into hexane; the resulting precursor fibers were taken up onto a bobbin at a speed of 30 cm/min. The precursor polymers were dried in a vacuum oven overnight; uniform pale yellow fibers were obtained with diameters of 50 -100 μm. The drawing of the precursor fiber and conversion to fibers of the conjugated polymer were carried out using a temperature controlled, continuous drawing, tube furnace system.[16,17]

The PDMPV and PTV fibers were doped by exposure to the vapor pressure (approximately 1 mm Hg) of iodine at room temperature. Electrical measurements were carried out in-situ during the doping using the four-probe method.

D. PANI/PPTA Fiber Spinning[13]

The polymer blend solutions were wet spun into 1N H_2SO_4 using a high precision syringe pump. Monofilaments were collected onto a take-up spool; applied tension was applied to elongate the fiber during coagulation. The draw down ratio (take up speed/extrusion speed) invariably was as high as possible, for the production of continuous fibers, and it was increased from 7 to 20 with increasing PPTA concentration in the solution. Extrusion speed varied from 0.12 to 0.3 m/min and the windup speed from 1.8 to 4.2 m/min.

The fibers were prevented from drying on the bobbin by continually spraying them with deionized water; this procedure impeded the open structure of the wet spun fibers from collapsing and allowed removal of the residual H_2SO_4. Subsequently, the

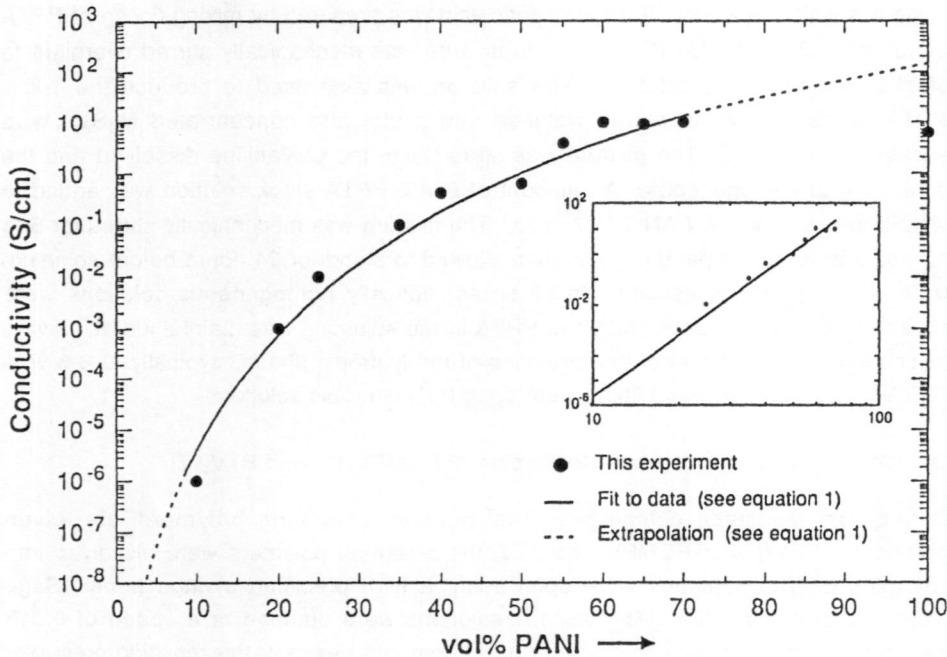

Figure 2: Conductivity versus volume fraction of PANI in the fiber; the inset shows the same data on a log-log plot.

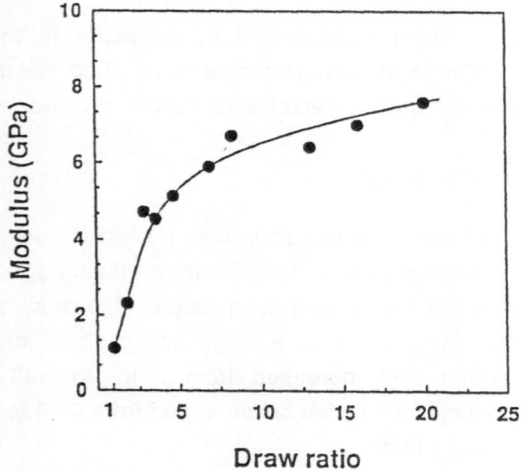

Figure 3: Relationship between draw ratio and modulus of PTV fibers.

fibers were washed with running deionized water for 48 hr. Half of the bobbin was then submerged in 1.5N HCl for 12 hr. This allowed the acid to penetrate the fibers and homogeneously protonate them to the conducting emeraldine salt form (partial reduction of the emeraldine salt occurs during the washing; the HCl treatment restores PANI to the fully protonated form). Finally, the bobbin was placed in an oven and the fibers dried under vacuum at 50°C while maintained at constant length by the bobbin. The pure PPTA fibers were spun with the same method. The pure polyaniline solution was dry-jet wet spun into a 1N H_2SO_4 solution. Washing, HCl treatment, and drying was carried out as described for the blend fibers.

III. RESULTS

A. PANI/PPTA

Wide angle X-ray patterns of the composite fibers consisted of superimposed reflections from the two components, indicating that PANI and PPTA segregated during the coagulation.[13] In Figure 2 (inset), the conductivity vs volume fraction of PANI is shown on a log-log plot; the results indicate that over a relatively wide range of concentrations, the conductivity follows a simple power law,

$$\sigma = \sigma_0 f^\alpha \qquad (1)$$

where f is the weight fraction of PANI. The solid curve in Figure 2 shows the power law fit to the experimental data with $\alpha=8.3$ and $\sigma_0=1.2 \times 10^{-14}$; the conductivity varies with fractional concentration of conducting polymer (from 10-70 % PANI) according to equation 1 over a range of values spanning nearly seven orders of magnitude.

B. PTV[17]

The Young's modulus and the tensile strength of the PTV fibers were measured at room temperature. Figure 3 shows the Young's modulus plotted against draw ratio; Figure 4 displays the tensile strength plotted against draw ratio. The modulus of the undrawn, converted PTV fiber was 1.1 GPa. Values of the Young's modulus of 7 GPa and of the tensile strength of 0.5 GPa were obtained for fibers which had been drawn to 20 times their initial length.

The tensile strength of 0.5 GPa is identical to the value as reported for highly stretched PPV film prepared by a heated roll method[18] and comparable to the values obtained with stretched trans-polyacetylene films.[19] The modulus, however, was lower than the values reported for stretched PPV films[18] and for stretched polyacetylene[19] films.

The conductivities of the iodine-doped PTV fibers were measured at room temperature. Figure 5 shows the conductivity plotted against the draw ratio. The

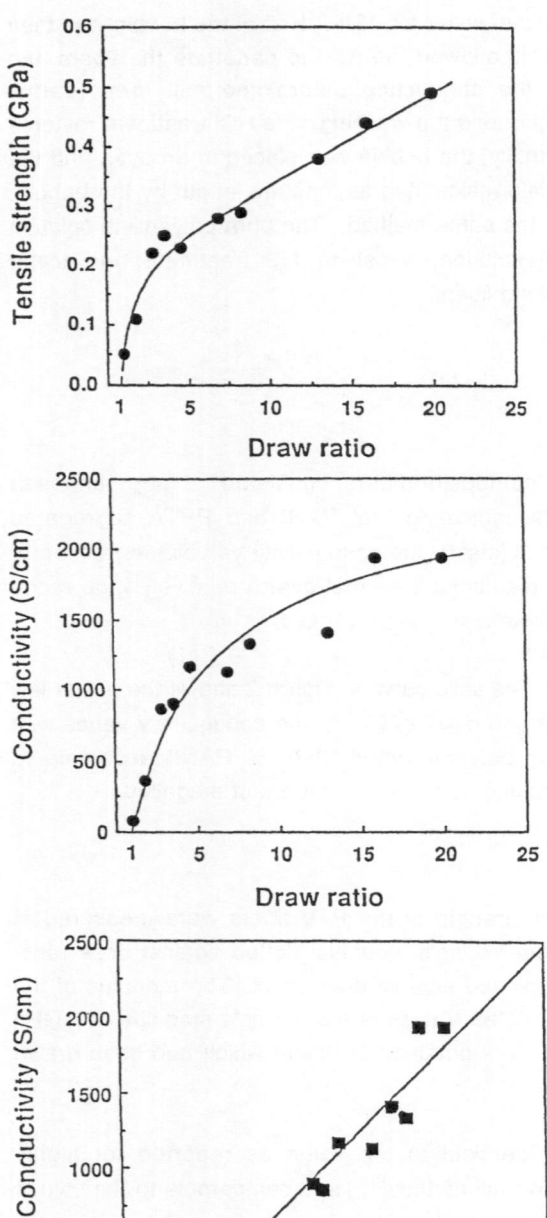

Figure 4:
Relationship between draw ratio and tensile strength of PTV fibers.

Figure 5:
Relationship between draw ratio and conductivity of PTV fibers.

Figure 6:
Correlation between modulus and conductivity of PTV fibers.

undrawn PTV fiber exhibited a conductivity of 80 S/cm. As expected, the conductivity was found to increase with the draw ratio; the maximum conductivity of 2000 S/cm was obtained at the maximum draw ratio. This value was 20 times higher than that of an undrawn fiber and comparable to that of an oriented film reported by Murase et al.[10] The relationship between the conductivity and the draw ratio was similar to that for the modulus and tensile strength (and draw ratio), indicating a strong correlation between conductivity and the mechanical properties.

This correlation between the structural dependence (as induced by the draw ratio) of the mechanical and electrical properties are illustrated more clearly by Figures 6 and 7, where the conductivity of drawn/converted PTV samples is plotted against, respectively, the modulus and the tensile strength. In both cases, an essentially linear relationship is found.

In this initial study of PTV, we have not developed a fully balanced set of processing parameters. This is illustrated by the modest orientation indicated by the X-ray patterns.[17] Nevertheless, the current materials have a Young's modulus of 7 GPa, a tenacity of 0.5 GPa and a conductivity of 2000 S/cm; a combination of properties which are adequate for conductive textile applications. It should be clear that further optimization of the inter-related processing variables will undoubtedly result in materials of superior quality.

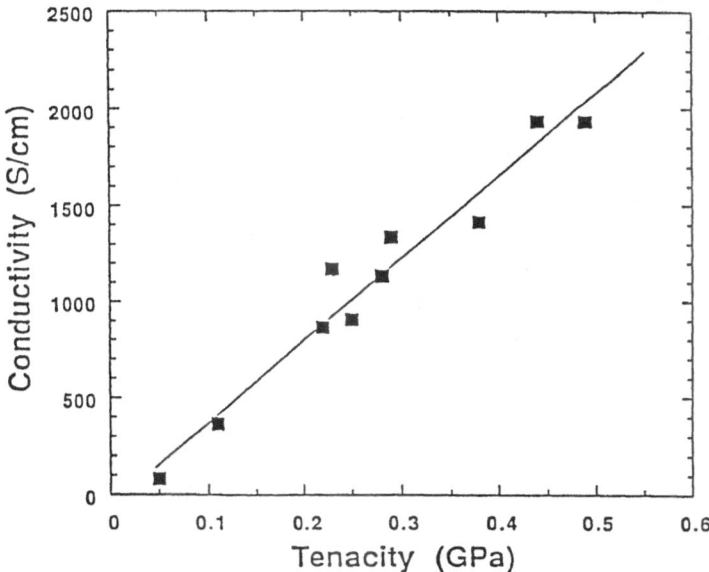

Figure 7: Correlation between tensile strength and conductivity for PTV fibers.

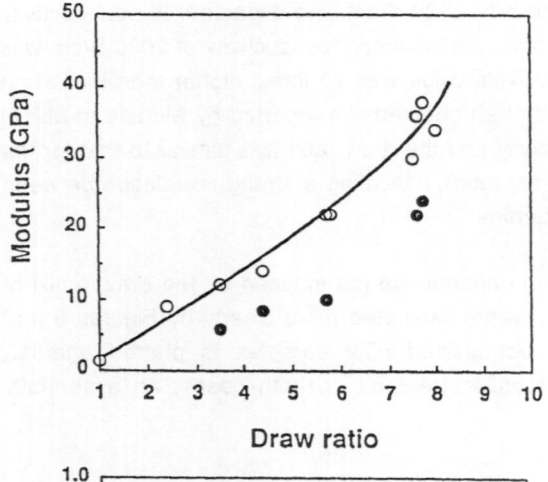

Figure 8:
Young's modulus as a function of the draw ratio for PDMPV fiber (open circles) and for PDMPV fiber doped with iodine (closed circles).

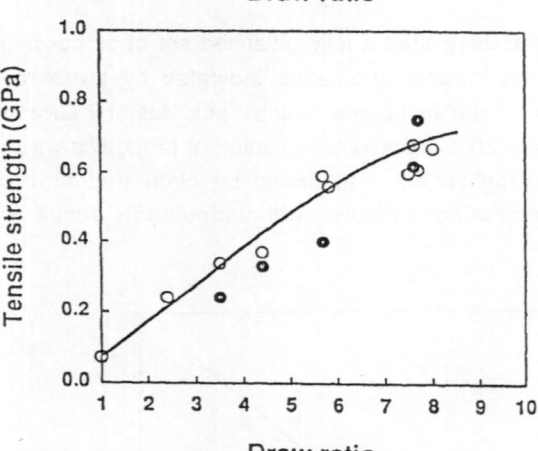

Figure 9:
Tensile strengths as a function of the draw ratio for PDMPV fiber (open circles) and for PDMPV fiber doped with iodine (closed circles).

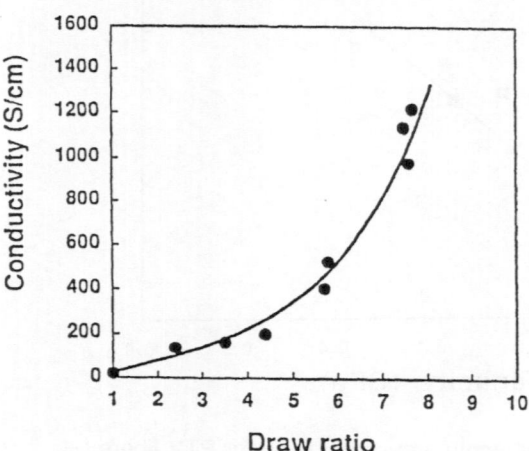

Figure 10:
Electrical conductivity as a function of the draw ratio of PDMPV fiber.

C. PDMPV[16]

The Young's modulus and the tensile strength of PDMPV fibers are shown versus draw ratio in Figures 8 and 9. The undrawn PDMPV fiber exhibited relatively poor mechanical properties: the modulus and tensile strength were 1.3 GPa and 0.07 GPa, respectively. Figures 8 and 9 reveal that fibers which had been drawn to 8 times their initial length had a Young's modulus as high as 35 GPa and a tensile strength of 0.7 GPa. The effects of doping on the mechanical strength are also displayed (filled circles in Figures 8 and 9). The data presented on these graphs indicate that doping caused only a moderate reduction of the modulus and essentially no loss of tensile strength.

Figure 10 shows the conductivity plotted against draw ratio. The undrawn PDMPV fiber exhibited a conductivity of 20 S/cm, with a gradual increase of conductivity seen up to a draw ratio of 5. At draw ratios greater than 5, the conductivity increased dramatically, as reported for stretched films[15] (although different in detail, since in ref. 15 the unstretched films were already about 30% converted). At a draw ratio of 8, the conductivity was 1200 S/cm, 60 times higher than that of the undrawn material.

Figures 11 and 12 show the relationship between Young's modulus and electrical conductivity (Figure 11) and between the tensile strength and electrical conductivity (Figure 12) for drawn PDMPV fibers. In both cases, a strong correlation is observed.

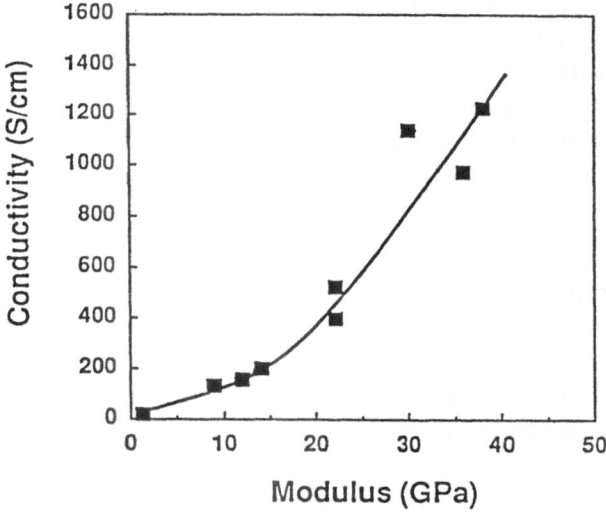

Figure 11: Electrical conductivity as a function of Young's modulus for PDMPV fiber.

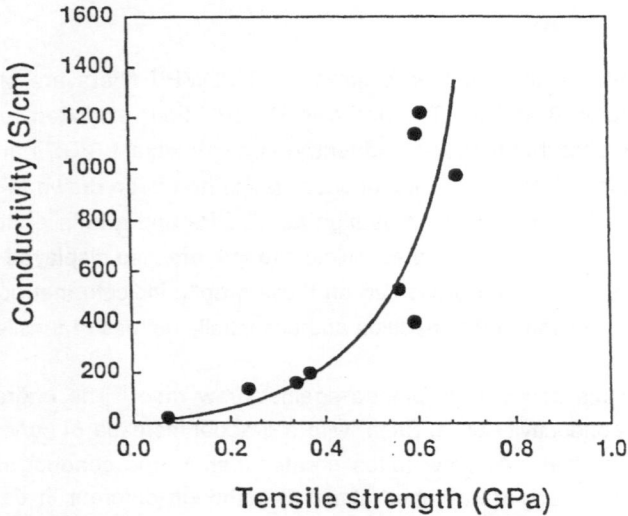

Figure 12: Electrical conductivity as a function of the tensile strength for PDMPV fiber.

IV. DISCUSSION

Although the electrical conductivity of conducting polymers is enabled by intra-chain transport, in order to avoid the localization inherent to one-dimensional systems, one must have the possibility of interchain charge transfer.[1,3] The electrical transport becomes essentially three-dimensional (and thereby truly metallic) so long as there is a high probability that an electron will have diffused to a neighboring chain between defects on a single chain. For well-ordered crystalline material in which the chains have precise phase order, the interchain diffusion is a coherent process. In this case, the condition for extended transport is that [1,3]

$$L/a >> (t_0/t_{3d}) \qquad (2)$$

where L is the coherence length, a is the chain repeat unit length, t_0 is the intra-chain π-electron transfer integral, and t_{3d} is the inter-chain π-electron transfer integral. An analogous argument can be constructed for achieving the intrinsic strength of a polymer. If E_0 is the energy required to break the covalent main-chain bond and E_{3d} is the weaker interchain bonding energy (from Van der Waals forces and hydrogen bonding for saturated polymers), then the requirement is coherence over a length L such that[2,3]

$$L/a >> E_0/E_{3d}. \qquad (3)$$

In this limit the large number (L/a) of weak interchain bonds add coherently such that the polymer fails by breaking the covalent bond. The direct analogy between equations 2 and 3 is clearly evident. In fact, for conjugated polymers, E_0 results from a combination of σ and π bonds (the latter being equal to t_o, see eqn. 2) and E_{3d} is dominated by the interchain transfer integral, t_{3d}. Thus, equations 2 and 3 predict that quite generally the conductivity and the mechanical properties will improve in a correlated manner as the degree of chain alignment is increased, with each approaching intrinsic values when the inequalities of equations 2 and 3 are satisfied. These predictions are in general agreement with the data obtained from PTV and PDMPV as shown in Figures 6,7,11, and 12. In both PTV and PDMPV, a strong correlation is observed, suggesting that major improvements in electrical conductivity can be anticipated as the materials are further improved such that the mechanical properties approach their intrinsic values. However, until a more quantitative understanding of the implied relationships is attained, extrapolation to the intrinsic electrical conductivity is not possible.

V. CONCLUSIONS

We have shown that the modulus and tensile strength derive from a combination of the intra-chain interactions (e.g. strength of chemical bonding, chain conformation, etc.) and inter-chain interaction (e.g. van der Waals forces, interchain transfer interactions, chain conformation, etc.). In conjugated polymers, these same features (band conduction within a polymer chain and efficient electron transfer between polymer chains) determine the carrier mean free path, and thus, the electrical conductivity. Therefore, we conclude that the mechanical and electrical properties of conjugated polymers are intrinsically linked, and we anticipate that in general as the tensile strength (and/or modulus) improve with improved chain orientation, the electrical conductivity will show corresponding improvements until both approach their respective intrinsic theoretical values.

Acknowledgement: The research on PDMPV and on PTV was supported by the office of Naval Research (N00014-83-K-0450). The synthesis of the polyaniline used in this study was funded through a MRG grant from the National Science Foundation (NSF-DMR87-03399). The polyblend fiber spinning and the mechanical and electrical measurements were supported jointly by DARPA-AFOSR and monitored by AFOSR under contract no. F49620-88-C-0138.

REFERENCES:

1. Kivelson S, Heeger AJ (1988) Synth Met 22:371
2. Termonia Y, Smith, P (1988) The Path to High Modulus Polymers with Stiff and Flexible Chains, (eds) Zachariades AE, Porter RS, Marcel Dekker, New York, p. 321
3. Heeger AJ (1989) Faraday Discuss Chem Soc 88:1
4. a. Jen KY, Oboodi R, Elsenbaumer R (1985) Polym Materials: Sci Eng 53:79
   b. Nowak MJ, Rughooputh SDD, Hotta S, Heeger AJ (1987) Macromolecules 20:212
   c. Rughooputh SDD, Hotta S, Heeger AJ, Wudl F (1987) J Polym Sci, Polym Phys Ed 25:1071
   d. Nowak MJ, Spiegel D, Hotta S, Heeger AJ, Pincus P (1989) Macromolecules 22:2917
   e. Sato M, Tanaka S, Kaeriyama K (1986) J Chem Soc Chem Commun 295:873
5. Edwards JH, Feast WJ (1980) Polym Commun, 21:595
6. a. Gagnon DR, Capistran JD, Karasz FE, Lenz RW (1984) Polym Bull 12:93
   b. Murasel I, Ohnishi T, Hirooka M (1984) Polym Commun 25:327
7. Momii T, Tokito S, Tsutsui T, Saito S (1988) Chem Lett 1201
8. Yamada S, Tokito S, Tsutsui T, Saito S, (1987) J Chem Sci, Chem Commun 1448
9. Tokito S, Murata H, Tsutsui T, Saito S (1989) Polymer, to be published
10. Murase I, Ohnishi T, Taniguchi T, Hirooka M (1987) Polym Commun 28:229
11. Andreatta A, Cao Y, Chiang JC, Heeger AJ, Smith P (1988) Synth Met 26:383
12. Cao Y, Andreatta A, Heeger AJ, Smith P (1989) Polymer (in press)
13. Andreatta A, Heeger AJ, Smith P (1989) Polym Commun (in press)
14. Blades H, U. S. Pat. 3,767,756 (1973), 3,869,429 (1975), 3,869,430 (1975)
15. Lewin M, Preston J (eds) (1985) Handbook of Fiber Science and Technology, Marcel Dekker Inc, New York, Vol III, part A, chapter 9
16. Tokito S, Smith P, Heeger AJ Polymer (in press)
17. Tokito S, Smith P, Heeger AJ Synth Met (in press)
18. Machado JM, Karasz FE (1989) Polym Preprints 30:154
19. Akagi K, Suezaki M, Shirakawa H, Kyotani H, Shimomura M, Tanabe Y (1989) Synth Met 28:D1

Nonlinear Optically Active Polymers for Waveguide Application

H. Sasabe, T. Wada, H. Ookawa, M. Hosoda, M. Sekiya,

A. Yamada and A. F. Garito*

Frontier Research Program, RIKEN Institute

2-1 Hirosawa, Wako, Saitama 351-01, JAPAN

Abstract: The macroscopic third order optical susceptibilities $\chi^{(3)}_{1111}(-3\omega;\omega,\omega,\omega)$ of π-conjugated polymeric systems were determined by means of Maker fringe technique at a wavelength of 1907 nm. We demonstrated the propagation of light in these polymeric thin films by mode-line measurements and/or by a direct observation with CCD camera. The possibility of waveguide application was also discussed.

INTRODUCTION

Under the electric field E all the materials exhibit nonlinear polarization as

$$P = \varepsilon_0[\chi^{(1)}:E + \chi^{(2)}:EE + \chi^{(3)}:EEE + \cdots] \qquad (1)$$

where $\chi^{(1)}$ is a linear susceptibility, $\chi^{(i)}$ ($i \geq 2$) is the i-th order nonlinear susceptibility and ε_0 the permittivity of free space. High power laser light (*e.g.*, Q-switched Nd:YAG) induces the nonlinear optical polarization in the material in a similar fashion, which is called the nonlinear optical effect. Nonlinear optics have recently been established its research status as the key issue in the field of optically active applications such as frequency doubling and/or tripling, optical spatial modulation, optical data processing, guided waves, and so forth. For the waveguide application the optical Kerr effect is available; the refractive index n of the material is dependent on the incident light intensity I as

$$n = n_0 + n_2 I \; ; \quad n_2 = \chi^{(3)}_{ijkl}(-\omega;\omega,-\omega,\omega)/(n_0^2 \varepsilon_0 c) \qquad (2)$$

where n_0 is a linear refractive index and c is the light velocity. Therefore the large $\chi^{(3)}$ materials are required for this application. From the viewpoint of materials it is well known that various kinds of ferroelectric crystals and semiconductors show high efficiency of nonlinear optical (NLO) activity but they have many problems in getting large crystals and in processability. On the contrary some organic compounds such as methylnitroaniline (MNA) show larger NLO activity

---

*Permanent Address: Department of Physics, University of Pennsylvania, Philadelphia, PA 19104-6396, U.S.A.

B. C. Anderson · Y. Imanishi (Eds.)
Progress in Pacific Polymer Science
© Springer-Verlag Berlin Heidelberg 1991

than inorganic substances and are processable to some extent, say, in thin films and/or fibers. Moreover the variety of molecular structures enables the suitable design of optical devices.

The intramolecular charge transfer through π-electron conjugation gives large optical nonlinearities in the molecular level, whereas the centro-symmetry of the crystal structure determines the macroscopic second order nonlinearity ($\chi^{(2)}_{ijk}(-\omega_3;\omega_1,\omega_2)$); if the crystal is centrosymmetric, then $\chi^{(2)}$ becomes zero. On the contrary the third order optical nonlinearity $\chi^{(3)}_{ijkl}(-\omega_4;\omega_1,\omega_2,\omega_3)$ does not depend on the crystal symmetry but on the microscopic third order susceptibility $\gamma_{ijkl}(-\omega_4;\omega_1,\omega_2,\omega_3)$ of the constituent molecular unit. From the quantum field theory for low dimensional systems [1] it is suggested that in conjugated linear chain structures such as polyenes and polydiacetylenes the π-electrons are delocalized in their motion only in one dimension along the chain axis. The major contribution to γ_{ijkl} is the dominant chain axis component γ_{xxxx} with all electric fields aligned along the chain x-axis. That is, in the 1-D π-conjugation systems only one tensor component γ_{xxxx} contributes dominantly to the isotropically averaged susceptibility $<\gamma>$ as $<\gamma> = (1/5)\gamma_{xxxx}$. A power law dependence of γ_{ijkl} has been found on the number of carbon atom sites with exponents of 5.4 for the *trans* and 4.7 for the *cis*-polyene conformer, and γ_{xxxx} is more sensitive to the physical length of the chain than to the conformation [1].

When the dimensionality of the π-electron system is expanded from linear to cyclic chains, the theoretical results on cyclic structures such as cyclooctatetr-aene show a decrease of γ_{ijkl} due to an actual reduction in the effective length available for the π-electron to respond to an optical electric field [2]. In the widely spread 2-D π-conjugation systems, on the other hand, other tensor compo-nents also contribute to $<\gamma>$, *i.e.*,

$$<\gamma> = (1/5)[\gamma_{xxxx} + \gamma_{yyyy} + (1/3)(\gamma_{xxyy} + \gamma_{xyxy} + \gamma_{xyyx} + \gamma_{yyxx} + \gamma_{yxyx} + \gamma_{yxxy})].$$

Polydiacetylene obtained by solid state polymerization was firstly reported to have large $\chi^{(3)}_{ijkl}(-\omega_4;\omega_1,\omega_2,\omega_3)$ comparable to that of semiconductors, especially parallel to the conjugated main chain [3]. For the enhancement of processability of materials can be used such techniques as polymerization [4], molecular disper-sion in guest-host systems [5] and so on. In order to elucidate the theoretical predictions concerning the nonlinear optical properties of the two dimensional π-electron systems, we have done systematic studies on various kinds of systems based on the π-conjugation, *i.e.*, conducting polymers such as polythiophene and its derivatives [6], polydiynes [6], molecularly dispersed polymers with macrocyclic π-conjugated system (soluble phthalocyanines)[7] and tetradehydromethanoannulenes [8]. In this paper we will discuss values of $\chi(3)_{1111}(-3\omega;\omega,\omega,\omega)$ (abbreviated as $\chi^{(3)}$ hereafter) of NLO active polymer systems and their possibility to the optical waveguide application.

EXPERIMENTAL

Sample Preparation

Polythiophenes: Polythiophene was polymerized electrochemically in the solution of thiophene monomer/nitrobenzene with tetramethylammonium perchlorate (TMAP) as a supporting electrolyte. The working and counter electrodes were an ITO glass and a Pt plate, respectively. The synthesized polymer film was washed in nitrobenzene, and then perchlorate ions were extracted from the film in the solvent of nitrobenzene/TMAP. After extraction the undoped film was washed in nitrobenzene again and dried up in a vacuum chamber. Soluble polythiophene derivatives [poly(3-alkyloxymethylthiophene)], on the other hand, were prepared chemically, as shown in Figure 1. Details are appeared elsewhere [6]. In this study R=6:poly(3-hexyloxymethylthiophene) and R=12:poly(3-dodecyloxymethylthiophene) (PDTh) were prepared. Thickness-controlled thin films were made by spin coating of polymer solution. From the optical absorption spectra of polythiophene derivatives, it is clearly observed that the absorption maximum shifts towards shorter wavelength by the introduction of alkyloxymethyl pendants (hyposochromic shift). This indicates the shortening of conjugation length due to the rotational motion of thiophene rings caused by longer alkyl chain. The effect of alkyl chain length on hyposochromic shift is almost negligible.

Polydiynes: Two types of polydiynes were prepared: one was an alkane- bridged

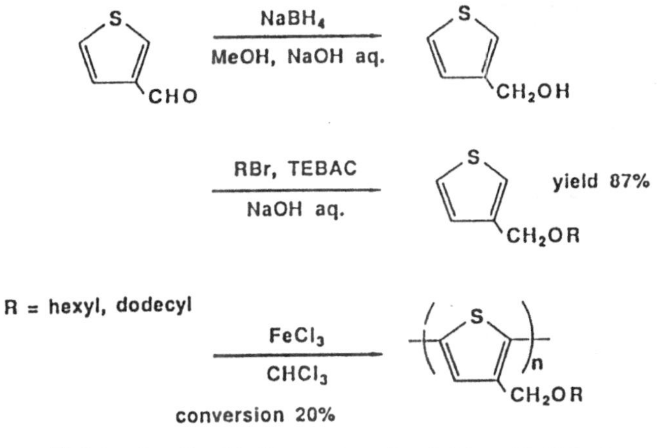

Figure 1. Synthetic route of soluble polythiophenes.

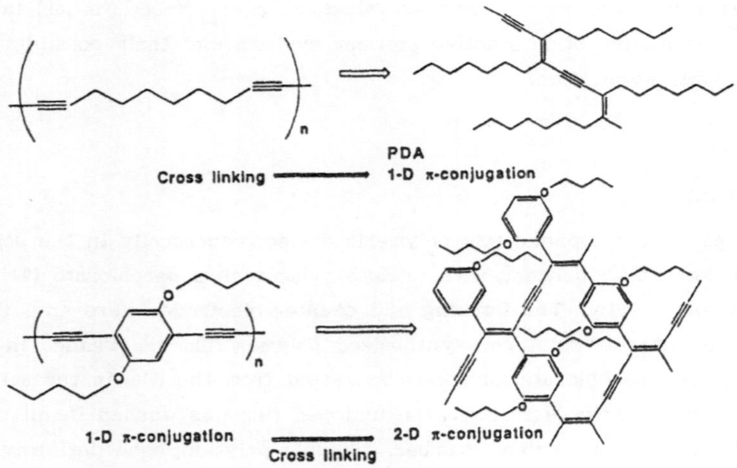

196

Figure 2. Schematic diagram of 1-D and 2-D π-conjugation in polydiynes.

polydiyne and the other was an arene-bridged polydiyne. The former is not a
π-conjugation system, but by the interchain crossliking due to thermal treatment
it becomes a one-dimensional conjugated polymer like PDA as shown in Figure 2.
The latter is, on the other hand, a 1-D π-conjugation system and easily converted into the 2-D system by the interchain crosslinking. Poly(1,9-decadiyne)(PDD)
and poly(1,4-diethynyl-2,5-dibutoxybenzene) (PDEDBB) are examples of alkane- and

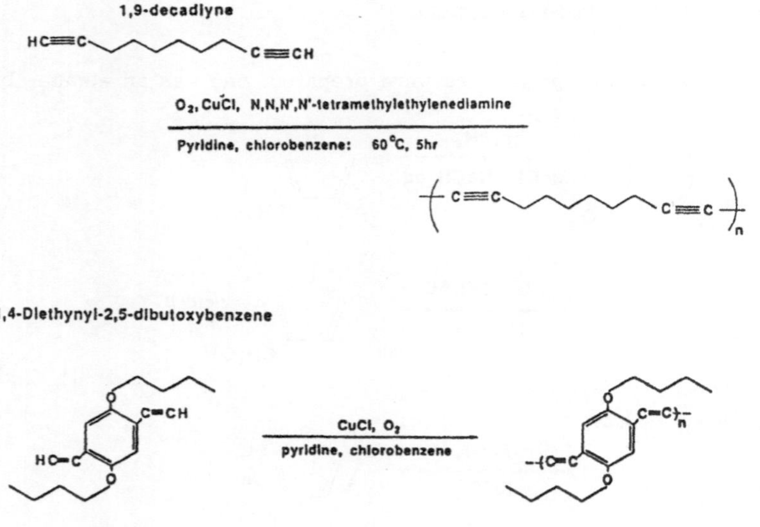

Figure 3. Synthetic route of polydiynes.

arene-bridged polydiynes, respectively, and prepared as shown in Figure 3 [6]. The solution of PDD/dichloroethane was spread over the quartz substrate and then a high quality film could be obtained. Though the as prepared film was opaque, it turned to be colored but transparent by annealing at 150°C for 48 hrs, which indicates the formation of 1-D π-conjugated system. From the IR spectra of this film the formation of C=C bonds can be confirmed. The film was applicable for the third harmonic generation (THG) measurement. Under the UV irradiation, however, the PDD film becomes dark purple in color and less applicable for THG measurement. The film of PDEDBB was also obtained by a spin coating technique. X-ray diffraction pattern indicates that PDEDBB film is completely amorphous, though PDEDBB powders show some indication of crystalline diffraction.

Soluble Phthalocyanines: We have already reported that vanadylphthalocyanine (VOPc) has a large $\chi^{(3)}$ (ca. 10^{-10} esu at 1907 nm) [9]. In order to provide an excellent solubility to polymer matrix, VOPc is modified by substituting peripheral four phenyl rings with *tert*-butyl group (TBVOPc). This compound has been synthesized according to Figure 4. For comparison metal free tetra-*tert*-butylph-thalocyanine (TBH$_2$Pc) has also been synthesized. The materials were thoruoughly purified by column chromatography on silica gel using chloroform as an eluent, followed by precipitation from chloroform to methanol. Optically transparent films containing TBH$_2$Pc or TBVOPc in PMMA (Mn=31,000) were obtained by spin coating of chloroform solutions onto a fused silica substrate. Refractive indices of thin films are measured by ellipsometry and mode-lines methods.

THG Measurement: Maker Fringe Pattern

Transmitted optical third harmonic generation from the thin film deposited on a

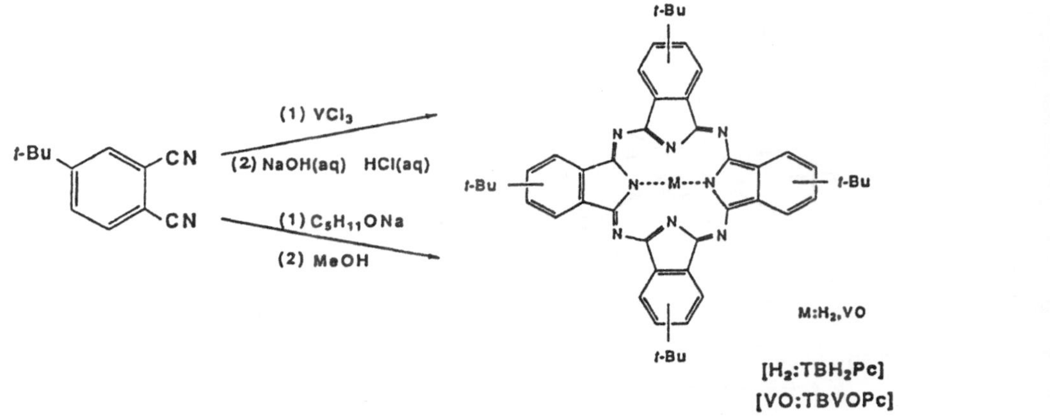

Figure 4. Synthetic route of soluble phthalocyanines.

198

fused silica substrate was observed at a wavelength of 1907 nm generated by the Raman sift of Nd:YAG laser (1064 nm, 10 pulse/sec). The details of measuring system for THG is reported elsewhere [10]. The output intensity of TH (636 nm) from the sample $J_{3\omega}$ is normalized at the same time with that from the reference to eliminate the fluctuation of incident light. $J_{3\omega}$ changes sinusoidally as a function of rotational angle of the plane paralell slab of sample about an axis, and appears as a set of fringes known as Maker fringes. According to Kajzar *et al* [11], $J_{3\omega}$ is given as

$$J_{3\omega} = \frac{256\pi^4 J_\omega^3}{C^2} \left| \frac{A\chi^{(3)}}{\Delta\varepsilon} \right|^2 \sin^2 \frac{\Delta\Phi}{2} \qquad (3)$$

where J and n are respectively the light intensity and refractive index for fundamental (subscript ω) and harmonic (3ω) frequencies, $\Delta\varepsilon = n_\omega^2 - n_{3\omega}^2$, A a factor arising from transmission and boundary conditions, $\Delta\Phi$ a phase mismatch ($=\Delta kl$), l the sample thickness and c the speed of light in the free space. Under the condition of $l \ll l_c$ (l_c: a coherence length), Eq.(3) can be simplified as

$$J_{3\omega} = \frac{2304\pi^6 J_\omega^3}{C^2} \left| \frac{A\chi^{(3)}}{n_\omega + n_{3\omega}} \right|^2 \left(\frac{l}{\lambda_\omega} \right)^2 \qquad (4)$$

where λ_ω is the wavelength of fundamental light.

M-lines Measurement

In order to check the thin film light guide, the prism-film coupling is an essential technique. The coupling between the prism (refractive index of n_p) and the film waveguide (n_f) formed on the substrate (n_s) takes place along the bottom plane of the prism through a thin gap layer (air) of index n_c, as shown in Figure 5. The propagation constant β of the incident light beam along the waveguide film is given by

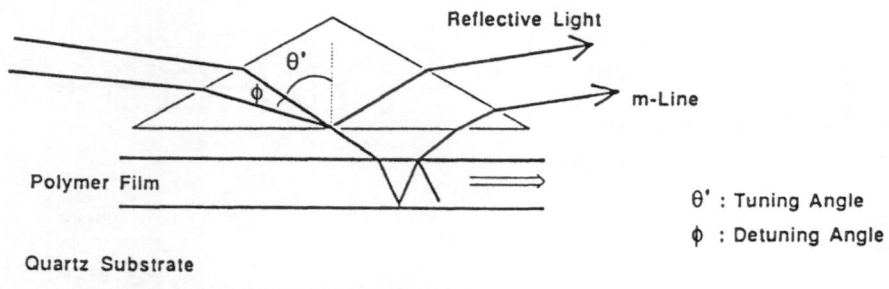

Figure 5. Schematic representation of m-line measurement.

$$\beta = n_p k \sin\theta' \quad ; \quad k = 2\pi / \lambda \qquad (5)$$

where θ' is the incident angle onto the bottom of prism which is related to the incident angle θ outside the prism through Snell's law as

$$n_c \sin(\theta - \alpha) = n_p \sin(\theta' - \alpha) \qquad (6)$$

where α denotes the toe angle of prism. If the propagation constant is in a range $n_c < \beta/k < n_f$ (guided mode), the guided wave is excited through the distributed coupling and penetrates as an evanescent wave into the film. In the waveguide film, β is expressed as

$$\beta = k n_{eff} \quad ; \quad n_{eff} = n_f \sin\theta'' \qquad (7)$$

where θ'' is the total reflection angle.

From the field equations for TE mode, the wave number in the x direction k_x can be derived in the form of *eigen value equation*,

$$k_x T = m\pi + \tan^{-1}(r_s/k_x) + \tan^{-1}(r_c/k_x) \qquad (8)$$

where $k_x = k(n_f^2 - n_{eff}^2)^{1/2}$, $r_s = k(n_{eff}^2 - n_s^2)^{1/2}$, $r_c = k(n_{eff}^2 - n_c^2)^{1/2}$, T is the thickness of the waveguide film and m=0, 1, 2,$\cdots$ denotes the mode number. The incident beam focused on the prism base in a synchronous direction can feed the optical energy into one of the waveguide modes of the film. Since the film scatters the optical energy in the excited mode into other modes, then the beam is coupled back to the outside medium by the same prism. This phenomenon gives a series of bright lines on the screen (m-lines). From the positions and the widths of m-lines, we can determine the mode spectra, the refractive index and the film thickness. Therefore, if we can observe m-lines by means of a prism-film coupler method, this is a good evidence of guided waves.

Waveguide Setup

For the direct observation of light propagation in the thin films, we should introduce the light into the film by either technique of a prism-coupling, a grating coupling or an end-fire coupling. In this study the spin coated polymer thin films on the fused silica substrate were used. The propagating light of CW YAG laser (1064 nm) in the film introduced by means of the prism coupling technique was monitored by IR sensitive CCD camera directly.

RESULTS AND DISCUSSION

THG: Maker Fringes

Each polymer sample spin-coated on a fused silica substrate was quite homogeneous in thickness of *ca.* 1.5 μm. Figure 6 is typical Maker fringe patterns (TH intensity as a function of the incident beam angle) for a fused silica substrate (a) and thermally crosslinked poly(1,9-decadiyne) on the substrate (b). The sinusoidally oscillating pattern of Figure 6(a) shows a good coincidence with the calcu-

Table 1. $\chi(3)$ Values of Samples (10-13 esu).

Fused Silica	0.28		
		PTh	3520
		PDTh	50
		PDD	1
		x-linked PDD	10
DEDBB	2.3	PDEDBB	120
H_2Pc	60	$TBH_2Pc/PMMA$	30
NiPc	46		
SnPc	790		
VOPc	1850	TBVOPc/PMMA	75

lated curve (solid line) of Equation (3). The minimum points are zero except the normal incidence (angle=0), which suggests the quality of substrate being quite good. From the periodicity a coherence length l_c can be estimated as 18.3 μm. In case of Figure 6(b), Maker fringes are the superposition of two components; one is a sinusoidally oscillating component and the other is an envelope of minimum points of the first one. The former is exactly the same as the fused silica substrate. The latter is due to the film itself, and its intensity follows Equation (4). In the calculation it is necessary to know the refractive indicies of the film at the fundamental (1907 nm) and TH (635 nm) wavelengths, respectively. We measured n_ω from the m-line at 1064 nm and $n_{3\omega}$ from both the ellipsometry and the m-lines at 632.8 nm for each sample. It should be noted that the values of $n_{3\omega}$ determined by both techniques are in good coincidence. These values are slightly changed to get a best fitting of Equation (3) to observed data points. The values of $\chi^{(3)}$ for various samples are estimated as listed in Table 1 by using that of fused silica [12]. The results for polythiophenes and polydiynes

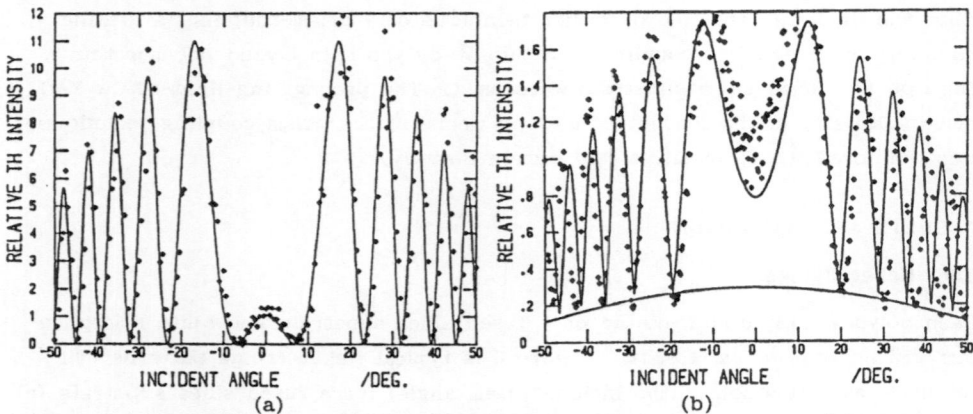

Figure 6. Maker fringe patterns of (a) fused silica substrate and (b) cross-linked poly(1,9-decadiyne) with fused silica substrate.

confirm that the formation of 1-D π-conjugation enhances the third order optical nonlinearity. The values of $\chi^{(3)}$ of TBH_2Pc and TBVOPc in PMMA are smaller than those of unsubstituted phthalocyanines. In case of VOPc's a large difference in $\chi^{(3)}$ may come from the stacking of Pc molecules, because VOPc film has an absorption peak in the longer wavelength region than Q-band (800-850 nm) which is unobservable in TBVOPc film. The difference in $\chi^{(3)}$ values of TBH_2Pc and TBVOPc indicates that the metal-to-ligand and/or ligand-to-metal charge transfer introduced by metal substitution contribute to the enhancement of third order susceptibility as discussed by Shirk et al [13].

M-lines, Nonlinear Refraction and Waveguides

The refractive indices of PDTh, heat treated PDD and TBVOPc/PMMA (38.6 wt%) films at 632.8 nm (He-Ne laser) are respectively 1.92, 1.58 and 1.78. Since those of a rutile prism and the fused silica substrate are respectively 2.874 and 1.457, it is possible to couple the incident beam (He-Ne and/or YAG laser) from the prism with polymer films via air gap as shown in Figure 5. In all the samples used we can clearly observe m-lines [6]. Here it should be noted that the m-line method is useful for the judgement of applicability of conjugated polymers and/or molecularly doped host/guest polymer systems to the optical waveguide. At the coupling conditions, for example, of polydoctylthiophene film (1.2 μm thick and n=1.64 for 1064 nm (Q-switched YAG)), the TE_0 mode line can be observed at the coupling angle of 66.40°. The output power P_{out} of guided wave light increases with an increase of the input power P_{in} in the linear fashion (Figure 7(a)). If the incident beam angle θ' is slightly detuned by ϕ, then P_{out} decreases and the relation between P_{out} and P_{in} becomes nonlinear as shown in Figure 7(b). That is, in case of ϕ=-0.8° the mode line almost disappears at lower P_{in} but P_{out} increases with increasing P_{in} nonlinearly because the refractive index increases as Equation (2). When ϕ is positive, on the other hand, the mode line is unobservable even at higher P_{in}. This result indicates that n_2 in Equation (2) is positive and the nonlinear refractive index change occurs in NLO active materials.

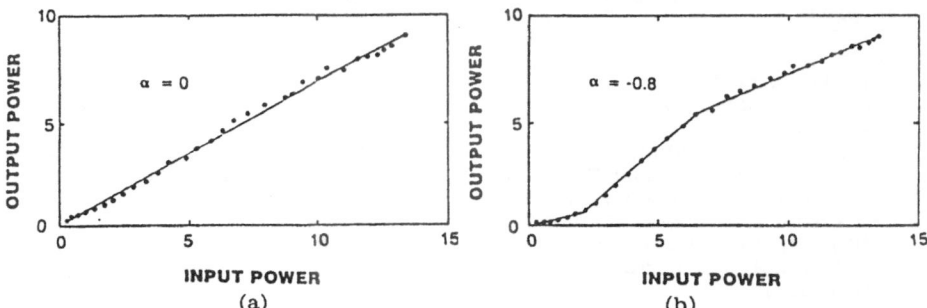

Figure 7.   Relation between input power of YAG laser and mode line intensity (output power) at (a) tuned angle and (b) detuned angle by -0.8°.

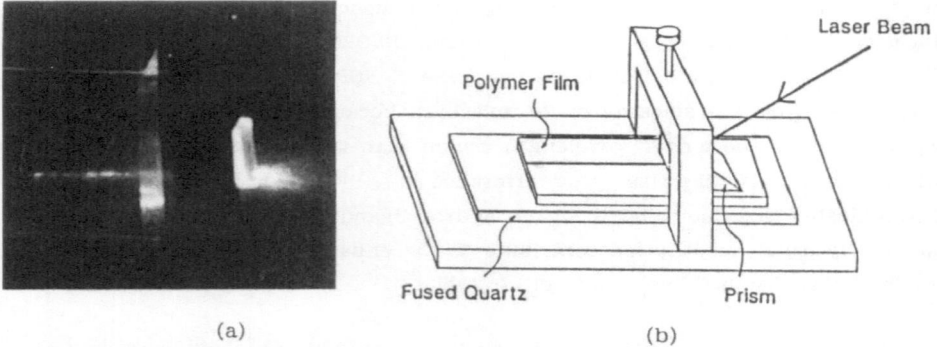

(a)   (b)

Figure 8. The trace of guided light in the polyoctylthiophene film (a) introduced by means of prism coupling technique (schematic representation in (b)).

Figure 8 shows an example of light propagation in the polyoctylthiophene film
introduced by means of a prism coupling technique. The trace of light path can
be detected by IR sensitive CCD camera due to the scattering of light in the film,
and the guided light stopped at the edge of the film. This is the direct indication of waveguide. The attenuation coefficient of light propagation will be calculated in the near future.

REFERENCES

1 Garito AF, Heflin JR, Wong KY, Zamani-Khamiri O (1988) In: Mat Res Soc Symp
Proc 109:91
2 Wu JW, Heflin JR, Norwood RA, Wong KY, Zamani-Khamiri O, Garito AF,
Kalyanaraman P, Sounik J (1989) J Opt Soc Am B6:707
3 Sauteret C, Hermann JP, Frey R, Pradere F, Ducuing J, Baughman RH, Chance RR
(1976) Phys Rev Lett 36:956
4 Smuelson L, Rahman AKM, Puglia GP, Clough S, Tripathy S, Inagaki T, Yang XQ,
Skotheim TA, Okamoto Y (1989) In: Proc Internatl Workshop on *Intelligent Materials*, Tsukuba, p139
5 Miyata S (1990) In: Proc SPIE 1147, San Diego, *in press*
6 Sasabe H, Wada T, Hosoda M, Ookawa H, Yamada A, Garito AF (1990) Mol Cryst
Liq Cryst, *in press*
7 Hosoda M, Wada T, Yamada A, Garito AF, Sasabe H (1990) In: Mat Res Soc Symp
Proc, *in press*
8 Wada T, Ojima J, Yamada A, Garito AF, Sasabe H (1990) In: Proc SPIE 1147, San
Diego, *in press*
9 Wada T, Matsuoka Y, Shigehara K, Yamada A, Garito AF, Sasabe H (1989) In:
Tazuke S (ed) *Photoresponsive Materials* Proc MRS Internatl Mtg on Advanced
Materials 12:75
10 Wada T, Yamada S, Matsuoka Y, Grossman CH, Shigehara K, Sasabe H, Yamada A,
Garito AF (1989) In: Kobayashi T (ed) *Nonlinear Optics of Organics and Semiconductors*, Springer Proc in Physics 36:292
11 Kajzar F, Messier J (1985) Thin Solid Films 132:11
12 Meredith GM, Buchalter B, Hanzlik C (1978) J Chem Phys 78:1533
13 Shirk JS, Lindle JR, Bartoli FJ, Hoffman CA, Kafafi ZH, Snow AW (1989) Appl
Phys Lett 13:1287

Free Radical Polymerizations in Compartmentalized Systems: The Role of Preformed Polymer

R.G. Gilbert, D.F. Sangster and D.H. Napper[*]

School of Chemistry, The University of Sydney, NSW 2006, Australia

Abstract: In emulsion polymerizations, the primary sites of polymerization are the monomer swollen latex particles. Polymerization thus takes place in the presence of preformed polymer. Such systems provide a convenient means by which the influence of polymer on the mechanistic steps involved in free radical polymerizations may be studied. This can be accomplished *inter alia* using polymerization kinetic studies, especially relaxation experiments on γ–initiated systems. The results demonstrate that the presence of preformed polymer can exert a profound influence on the magnitude of the rate coefficients for propagation and termination. The value of the rate coefficient for termination can be reduced by up to seven orders of magnitude; that for propagation can be diminished by several orders of magnitude if the polymer/monomer mixture passes through its glass transition point. An understanding of bimolecular termination processes may also require differentiation between free radicals attached to entangled and unentangled chains. The results obtained in compartmentalized systems imply that the observed failure of bulk systems that become glassy to polymerize to complete conversion is a consequence not of a dramatic reduction in the propagation process, as is commonly assumed, but rather coincides with a marked drop in the efficiency of initiation.

INTRODUCTION

In emulsion polymerizations, the latex particles once formed swell with monomer, capture free radicals from the aqueous phase and become the primary loci of polymerization. Polymerization thus takes place in the presence of preformed polymer, the weight fraction (w_p) of which may vary from, say, 0.3 to near 1.0. To understand the polymerization in such systems, in which the free radicals are compartmentalized within the latex particles, it is necessary to understand how the preformed polymer influences the course of any subsequent polymerization. Conversely, compartmentalized systems can be regarded as providing a convenient means by which the influence of preformed polymer on the subsequent course of a free radical polymerization may be investigated quantitatively.

Bulk and Solution Polymerizations: The Effects of Polymer

The mechanisms by which preformed polymer affects the kinetics of free radical polymerizations both in bulk and in solution have long been of interest (1). The classical, and by now well–established, mechanism by which such polymerizations proceed predicts a rate of polymerization that is given by (2,3)

$$-d[M]/dt = (fk^2_p\ k_d/k_t)^{\frac{1}{2}}\ [M]\ [I]^{\frac{1}{2}} \tag{1}$$

where [M] = monomer concentration, [I] = initiator concentration, f= capture efficiency of primary free radicals by monomer, k_p = propagation rate coefficient, k_d = initiator decomposition rate coefficient and k_t = bimolecular termination rate coefficient. The first order dependence of the rate upon the monomer concentration evident in equation (1) implies that a plot of the *percentage* conversion of monomer to polymer versus time should be

B. C. Anderson · Y. Imanishi (Eds.)
Progress in Pacific Polymer Science
© Springer-Verlag Berlin Heidelberg 1991

independent of the initial concentration of monomer (ie, there should be just one master curve). Figure 1 shows the results of Schulz and Harborth (4) for the solution polymerization of methyl methacrylate at 50˙C in benzene for various initial concentrations of monomer, as well as for its bulk polymerization. As Flory (1) has properly noted, it is readily apparent that the experimental results do not fall on a single master curve, except perhaps at low fractional conversions ($w_p \leqslant 0.1$, say). In as much as the classical mechanistic reaction scheme that gives rise to equation (1) should be operative over the entire course of the polymerization, so too should equation (1) be applicable over the entire polymerization. The failure noted above of the data points to lie on a single master curve is a consequence of the variation of some (or all) of the rate coefficients k_p, k_t and k_d, as well as the initiator efficiency factor f, with the fractional conversion. This is a consequence of polymer being generated in the system. It is for this reason that the term *"rate coefficient"* is preferred here over the alternative *"rate constant"* (2).

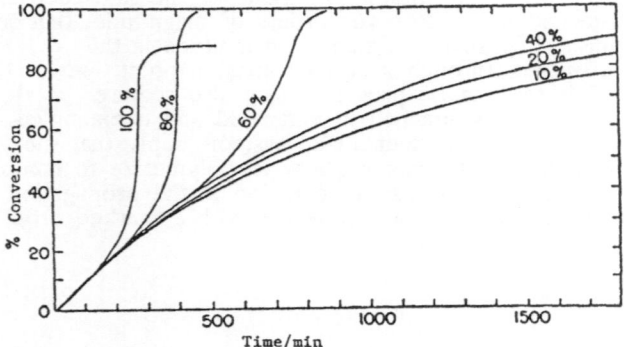

Fig 1   Conversion vs time curves for methyl methacrylate in benzene at 50˙C using different initial monomer concentrations (data of Schulz and Harborth (4)).

The kinetic results displayed in Figure 1 exhibit two other important features: first an autoacceleration (the Trommsdorff–Norrish or gel effect) is apparent if sufficient monomer is initially present; second, the bulk polymerization apparently proceeds to a limiting fractional conversion that is less that 1.00 (in this case *ca.* 0.85). It is generally agreed (1) that the Trommsdorff–Norrish effect is a consequence of a reduction is k_t with increasing w_p, but the physical origin(s) of this decrease has not yet been established definitively. Nor is it known whether the origin is the same for emulsion and bulk systems. The reason for the apparent limiting conversion in bulk polymerizations has long been attributed (1) to a catastrophic decrease in the propagation rate coefficient. This is postulated to result from the polymer/monomer mixture passing through its glass transition point as monomer is consumed. Propagation would then be expected to become diffusion controlled, rather than chemically controlled, since the diffusion of monomer through a glass is considered to be extremely slow; from this it has been concluded that propagation essentially ceases.

Experiments on seeded emulsion polymerizations allow the dependence of both k_p and k_t on w_p to be explored experimentally, at least over the accessible w_p range of 0.3 – 1.0. In what follows the results of such experiments are summarized, and the consequences of these results for the dependence of f on w_p for certain bulk systems are elaborated. It transpires that both k_p and k_t

may be profoundly influenced by the presence of preformed polymer, although the effect on k_t is usually much more dramatic and more widely observed.

PROPAGATION RATE COEFFICIENT k_p

Method of Measurement

One consequence of the compartmentalization of the free radicals within the latex particles in emulsion systems is that bimolecular termination between free radicals located in *different* particles is eliminated. In suitable circumstances, this compartmentalization may allow the concentration of propagating free radicals to build up to the point where it is detectable by ESR. The concentration of free radicals in the latex particles [M·] can then be determined quantitatively (7) and if a simultaneous measurement is also made (using, say, a dilatometer.) of the rate of polymerization, then the value of k_p can be calculated from

$$-d[M]/dt = k_p \, C_M \, [M\cdot] \qquad (2)$$

Here C_M is the concentration of monomer in the latex particles, the value of which may be readily estimated experimentally from studies of the partitioning of the monomer between the particles and the aqueous phase.

Exprimental Results

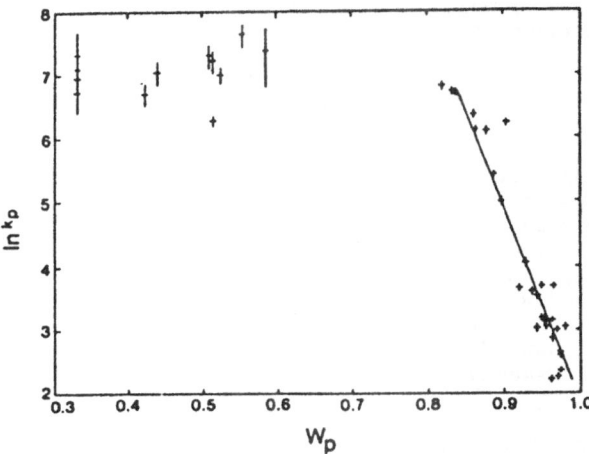

Fig 2   The dependence of k_p for methyl methacrylate upon the weight fraction of preformed polymer at 50°C.

The dependence of k_p on w_p for methyl methacrylate at 50°C, as measured by the ESR method (7), is shown in Figure 2. Although the precision of the data is poor when $w_p < 0.8$, the results suggest that k_p is relatively insensitive to the presence of preformed polymer at these lower values of w_p. (Unfortunately, it cannot be determined categorically from these data whether or not k_p is stricly independent of w_p.). Such insensitivity, nonetheless, is consistent with propagation being chemically controlled. For propagation to occur, monomer molecules must diffuse to the free radicals at the chain ends and, after due orientation and activation, be attached to the chains. At lower values of w_p, the latter chemical step would be expected to be rate determining and, unless a template effect were operative, the value of k_p should be insensitive to the presence of preformed polymer.

The bulk polymerization results mentioned above suggest that at 50°C, a polymer/monomer mixture with $w_p \geq 0.85$ is a glass (1). It is therefore not surprising that the measured values of k_p displayed in Figure 2 decrease (in this case exponentially) with w_p. It would be anticipated that at these high w_p values, the diffusion of monomer through the glassy matrix to the free radical sites would be the rate determining step rather than chemical activation. As the diffusion coefficients of most small species (including, presumably, monomers) decrease significantly with increasing w_p in this w_p range (8), the observed decrease in k_p *inter alia* reflects the variation of the monomer diffusion coefficient with w_p. Note, however, that the value of k_p decreases by only about two orders of magnitude over the w_p range of 0.8 – 1.0.

TERMINATION RATE COEFFICIENT k_t

Bulk and Solution Polymerizations: Theoretical Expectations

We begin this section with a brief discussion of the theoretically expected variation of k_t with w_p for bulk or solution systems. This is shown schematically in Figure 3 where two features merit attention: first, the presence of preformed polymer may decrease k_t quite drastically (by up to seven orders of magnitude); second, the dependence of k_t upon w_p displays considerable complexity. This complexity originates in changes in the rate determining step for bimolecular termination as the amount of preformed polymer increases. The common thread which unites mechanistically bimolecular termination events is that all are diffusion controlled. What differentiates the various domains evident in Figure 3 is the nature of the species whose diffusion is rate controlling and their respective diffusion mechanisms.

Benson and North (9) recognized that three sequential steps were necessary for the mutual annihilation of two free radicals attached to two different polymer chains. First, the centres–of–mass of the two annihilating chains must undergo translational diffusion so as to be in sufficiently close proximity to allow the second step, segmental diffusion, to bring the two attached free radicals into a contiguous relationship, presumably on the surface of the coils. The final step, the chemical reaction of the two free radicals by combination or disproportionation, is so rapid that it is rarely, if ever, rate limiting. Either centre–of–mass and/or segmental diffusion could be rate limiting according to this scheme.

There is evidence to suggest (9,10) that at very low values of w_p, centre–of–mass diffusion is relatively rapid and the rate determining step is segmental diffusion. The values of k_t reported in the literature (11) usually refer to low conversions and their order of magnitude (10^7-10^8 dm³ mol⁻¹s⁻¹) is consistent with this interpretation. At low w_p, the addition of preformed polymer presumably reduces the goodness of the solvency of the solvent for the polymer coils. This decreases the coil dimensions, thus allowing the annihilating free radicals to find one another more readily. Such a contraction accounts qualitatively for the *increase* in k_t with increasing w_p observed in certain systems (10).

As the value of w_p increases, so the growing polymer coils become increasingly entangled with preformed polymer chains. Engtanglements may so diminish their centre–of–mass translational diffusion that this becomes rate determining. The termination rate coefficient would then depend upon the molecular weight of the terminating chains. Furthermore, the value of k_t would be expected to drop dramatically (see Figure 3) with increasing w_p since translational diffusion now proceeds primarily by reptational motion. The drastic decrease in k_t with w_p would continue unabated were it not for the intervention of a new phenomenon, termed *reaction diffusion* or *residual* termination (12, 13), which is propagation–driven. It must not be forgotten that propagating chain ends possess spatial mobility as a consequence of their addition of monomer. It follows that free radical diffusion coefficients always contains a residual contribution arising from propagation events. This manifests itself as as irreducible minimum contribution to free radical annihilation while ever monomer is present. Of course, as mentioned above, propagation itself may become diffusion controlled if the monomer/polymer mixture passes through its glass transition point (see both Figures 2 and 3). Note that the residual termination rate coefficient would be expected to be independent of the molecular weight of the chains to which the free radicals are attached.

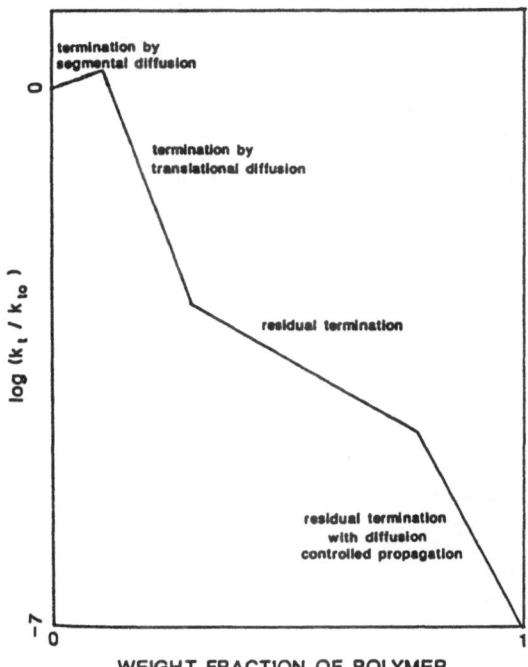

Fig 3   Schematic representation of the dependence of k_t upon the weight fraction of preformed polymer relative to its value at zero conversion.

Russell *et al.* (14) have developed theories for the upper and lower bounds for residual termination based upon the Smoluchowski approach to diffusion controlled chemical reactions. For the upper bound, termed the *flexible* limit, the value of k_t was calculated assuming that the chain end to which the free radicals are attached are capable of undergoing backbone rotational motions up to its nearest entanglement node on the time scale of termination events. The lower bound (the *rigid* limit) assumed that such motions were frozen, leaving just the terminal unit capable of rotational motion. These assumptions lead to

$$2\pi \; k_p \; C_M \; a^2 \; \sigma/3 \leqslant k_t \leqslant 4\pi \; k_p \; C_M \; a^3 \; j_c^{\frac{1}{2}}/3$$

where a^2 = mean square end-to-end chain displacement per monomer repeat unit (related to the characteristic ratio C_∞ of the chain), σ = Lennard-Jones diameter of the monomer and j_c = mean number of repeat units between entanglements. Note that there are no adjustable parameters in these formulae; all are subject to independent experimental determination.

Experimental Method

Seeded emulsion polymerizations provide convenient systems which allow their termination processes to be investigated directly via relaxation studies. Initiation in such systems can be readily accomplished by irradiation with γ-rays because exposure of water to high energy radiation generates $\cdot$OH, $\cdot$H and e$^-$(hyd). Free radical production in the aqueous phase can, however, be halted abruptly by the removal of the system from the radiation cavity. The polymerization relaxation kinetics, which are governed *inter alia* by termination events, can then be monitored (eg, dilatometrically), yielding in appropriate cases direct information about bimolecular termination. Successive insertion-relaxation cycles polymerize the monomer contained within the system and so allow the dependence of the termination events upon the weight fraction of polymer in the latex particles to be explored. Provided the average number of free radicals per particle ($<n>$) is sufficiently large (say $<n> \gtrsim 2$), the relaxation curves can often be analysed using the simple pseudo-bulk equation:

$$d<n>/dt \simeq \rho_o - 2c<n>^2 \tag{3}$$

Here ρ_o = entry rate coefficient for thermal initiation and $c = k_t/N_A V_S$, where V_S = swollen size of the latex particles and N_A = Avogadro constant.

Results

Methyl methacrylate: Figure 4 presents values obtained by relaxation experiments for k_t for methyl methacrylate at 50°C at w_p values in the range 0.3 - 0.95 (7). Also presented in this figure are the theoretical estimates for the upper and lower bounds for residual termination. It should first be noted that the absolute magnitude of the measured values of k_t (10^5–10^2 dm^3 mol^{-1}s^{-1}) is several orders of magnitude smaller than the literature value ($\simeq 10^8$ dm^3 mol^{-1}s^{-1}) reported for low conversions (11). In the w_p range studied, it seems reasonable to infer that residual termination is the major radical annihilation mechanism operative. The flexible limit for residual termination seems to be the more appropriate in these systems, at least at the lower values of w_p studied. This limit seems physically reasonable.

Styrene: It might be expected that the foregoing concepts developed for methyl methacrylate would be directly applicable to other monomers. This expectation was not, however, realized for styrene (15). Attempts to fit the relaxation curves in this case using the pseudo-bulk equation led to unphysical results. A more convincing interpretation of the relaxation data was obtained by incorporating a chain length dependence of the termination rate coefficient. This was accomplished by elaborating the proposal of Cardenas

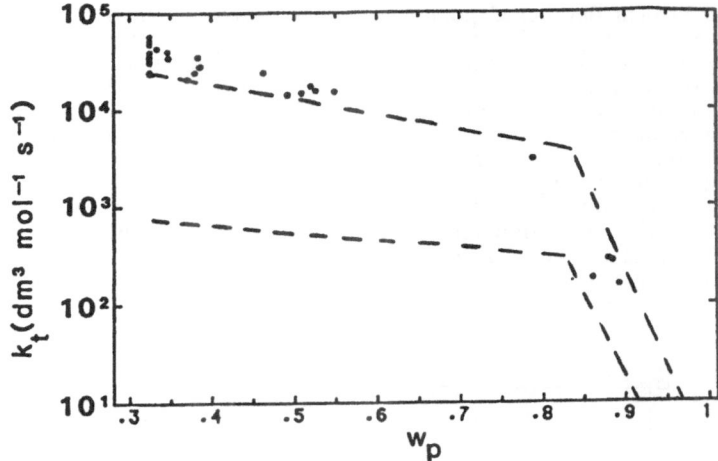

Fig 4   The dependence of k_t upon the weight fraction of polymer for methyl methacrylate at 50°C. The points represent the experimental results whereas the lines represent the theoretical upper and lower bounds for residual termination calculated for the flexible and rigid limits respectively.

and O'Driscoll (16) for high conversion bulk systems which required the population of radicals to be divided into two classes: those attached to chains whose centre–of–mass diffusional motion is reduced by entanglements with preformed polymer (so called *long* chains); and those whose mobility is substantially unaffected by entanglements (*short* chains). The pseudo–bulk equation must then be modified to read:

$$d\langle s\rangle/dt = \rho_A + k_{tr}\,C_M\langle L\rangle - k_pC_M\langle s\rangle/z - 2c_{ss}\langle s\rangle^2 - c_{SL}\langle S\rangle\langle L\rangle$$

$$d\langle L\rangle/dt = k_pC_M\langle s\rangle/z - k_{tr}C_M\langle L\rangle - c_{SL}\langle S\rangle\langle L\rangle - 2c_{LL}\langle L\rangle^2 \qquad (4)$$

where $\langle S\rangle, \langle L\rangle$ = average number of free radicals per particle attached to short and long chains respectively, ρ_A = entry rate coefficient, k_{tr} = rate coefficient for transfer to monomer, z = critical degree of polymerization at which short chains become long chains , and the subscripts S, L refer to termination events between species designated as short and long respectively.

Interpretation of the relaxation data for styrene using the short–long molecule resulted in the following picture for termination events. The rapidity of propagation causes short chains to be quickly transformed into long chains. As a result, the concentration of short chains, which are generated either by chain transfer to monomer or by entry, is relatively small (typically, eg, of order 1%). Despite this low concentration, the relatively high diffusional mobility of the short chains when coupled with the high concentration of long chains ensures that by far the most important contribution to termination events involves the mutual annihilation of short free radicals by long free radicals. Short–short events are relatively rare because of the low concentration of short free radicals; long–long events are also relatively unimportant because of the low centre–of–mass diffusional mobility of long, entangled chains. The rate coefficient for the dominant short–long termination events can be extracted from the relaxation data using eqns (4). This is an average rate coefficient spanning all short chain lengths from monomeric up to (but not including) the critical degree of polymerization for entanglement z. The results are shown in Figure 5. Also shown in this figure is the rate coefficient for the residual termination of long chains, although it must be admitted that the data are very insensitive to its actual value. The mean values of k_t obtained in this manner for short–long termination events are several orders of magnitude larger than those for residual termination, although comporable to the value ($\approx 10^8$ dm^3 mol^{-1}s^{-1}) reported (11) for styrene at zero conversion. It is interesting to note that the k_t results obtained for styrene from relaxation experiments predict the correct shape of chemically initiated seeded emulsion polymerization rate curves, including the occurrence of the Trommsdorff–Norrish effect. This suggests that in this case the form of the dependence of k_t upon w_p is responsible for the autoacceleration observed in the gel effect.

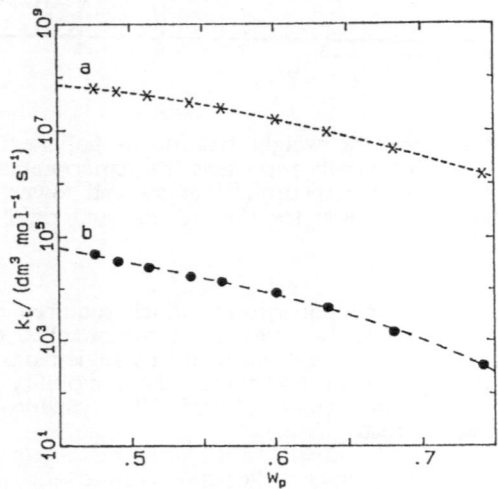

Fig 5   (a) The variation of k_t with weight fraction of polymer for short–long termination as a function of the weight fraction of polymer for styrene at 50°C (b) Also shown is the order of magnitude of the residual long–long termination rate coefficient.

EFFICIENCY OF RADICAL CAPTURE f

The experimentally measured decrease in k_p with w_p displayed in Figure 2 for methyl methacrylate at $w_p > 0.85$ appears to be exponential in character. This might be considered to lend credence to the conventional viewpoint that propagation must essentially cease in bulk systems at high conversions if the monomer/polymer mixture becomes glassy. However, close inspection of the data shows that k_p only decreases by *ca.* two orders of magnitude as w_p increases from $0.85 - 0.95$. This decrease is far too small to cause polymerization to stop abruptly, as in evident from the observation that emulsion polymerizations of methyl methacrylate proceed to essentially complete conversion over 24 hr.

An alternative, and indeed preferable, explanation for the rapid drop in the polymerization rate in bulk systems that have become glassy involves a drastic reduction in the efficiency with which initiation is effected, ie, the reason for the apparently limiting fractional conversion arises not from a marked decrease in k_p but rather from a sharp reduction in f. A significant decrease in f would be anticipated from the reduction in the diffusion coefficient of the primary free radicals in the glassy state. As a consequence, the primary free radicals display a strongly enhanced propensity to undergo geminate recombination. The measured values of k_p as a function of w_p discussed above, when coupled with the theoretically predicted values for k_t for residual

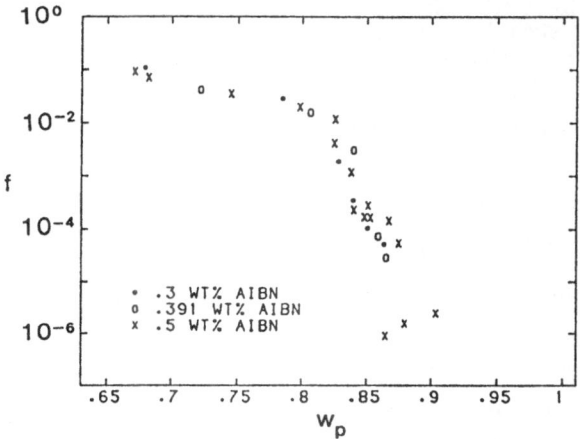

Fig 6   The dependence of the initiation efficiency factor f on the weight fraction of polymer for methyl methacrylate at 50°C using different concentrations of the initiator azo−di−isobutyronitrile.

termination, allow f for bulk systems to be calculated from appropriate experimental data (17). These are shown for methyl methacrylate in Figure 6. The values of f display a marked reduction by several orders of magnitude once the monomer/polymer mixture has become glassy. It is thus the failure of the initiation step rather than the propagation step that causes some bulk polymerizations to appear to cease prematurely.

ACKNOWLEDGEMENTS

We are grateful to the Australian Research Council and the Australian Institute of Nuclear Science and Engineering for financial support of these studies. The Electron Microscope Unit of this University is thanked for the provision of facilities.

REFERENCES

1   Flory PJ (1953) *Principles of polymer chemistry*, Cornell Univ Press, Ithaca; Jenkins, AD (1987) *Makromol Chem, Macromol Symp* 10/11:1

2   Schulz GV, Husemann E (1937) *Z physik Chem* B36:184; Flory PJ (1937) *J Am Chem Soc* 59:241

3   Price CC, Kell RW (1941) *J Am Chem Soc* 63:2798

4   Schulz GV, Harborth G (1947) *Makromol Chem* 1:106

5   Mills I (1988) *Quantities, units and symbols in physical chemistry,*
     Blackwell, Oxford

6   Norrish RGW, Smith RR (1942) *Nature* 150:336;
     Trommsdorff E, Köhle H, Lagally P (1947) *Makromol Chem* 1:169

7   Ballard MJ, Gilbert RG, Napper DH, Pomery PJ, O'Sullivan PW,
     O'Donnell JH (1986) *Macromolecules* 24:1027;
     Ballard MJ, Napper DH, Gilbert RG, Sangster DF (1986) *J Polym Sci A:
     Polymer Chem* 24:1027

8   Lee JA, Frick TS, Huang WJ, Lodge TP, Tirrell M (1987) *Polym Prepr*
     28:369

9   Benson SW, North AM (1959) *J Am Chem Soc* 81:1339; 84:935

10   Mahabadi HK (1987) *Makromol Chem, Macromol Symp* 10/11:127

11   Brandrup J, Immergut, EH (eds) (1975) *Polymer Handbook*. Wiley, New
     York

12   Schulz GV (1956) *Z Phys Chem (Frankfurt am Main)* 8:290

13   Gardon JL (1968) *J Polym Sci A−1* 6:2851

14   Russell GT, Napper DH, Gilbert RG (1988) *Macromolecules* 21:2133

15   Adams ME, Russell GT, Casey BS, Gilbert RG, Napper DH (1990)
     Macromolecules, accepted for publication

16   Cardenas JN, O'Driscoll KF (1976) *J Polym Sci Polym Chem Ed*
     14:883

17   Russell GT, Napper DH, Gilbert RG (1988) *Macromolecules* 21:2141

Miscibility Behavior in Polyethersulfone/Polyimide Blends with and without Solvents

K. Liang, L. Wu, J. Grebowicz, F.E. Karasz and W.J. MacKnight

Department of Polymer Science and Engineering, University of Massachusetts
Amherst, Massachusetts 01003 USA

Abstract: The miscibility behavior in blends of polyethersulfone (Victrex PES) with the polyimide PI 2080, (the condensation product of 3,3',4,4'-benzophenone tetracarboxylic dianhydride [BDTA] and a 4:1 mixture of 2,4-toluene diisocyanate and 4,4'-diphenylmethane diisocyanate) or with the polyimide XU 218 (the condensation product of BDTA and 5(6)-amino-1-(4'-aminophenyl)-1,3,3'-trimethylindane) was investigated using differential scanning calorimetry, dynamic mechanical analysis and thermogravimetric analysis. The effects of solvents (dimethylacetamide, tetramethylene sulfone, dimethyl sulfoxide and 1-methyl-2-pyrrolidone) on miscibility were studied and one solvent, tetramethylene sulfone was found to have a plasticizing effect. In the absence of solvent, the equilibrium phase boundary for these blends was in the experimentally inaccessable region below the T_g-composition line. The phase boundary at zero solvent concentration was obtained by extrapolation using data collected in the presence of the plasticizer.

INTRODUCTION

In the last few years, high performance miscible polymer blends have attracted attention in the search for new materials at lower costs. Recently, two new families of high performance blends consisting of an aromatic polybenzimidazole and aromatic polyimides [1-3] and aromatic polyimides and polyethersulfone [4] have been reported. However, even though these blends appear to be miscible over the entire range of compositions, phase separation on heating above T_g is irreversible. This contribution describes the miscibility behavior in polyether-sulfone/polyimide PI 2080 and polyethersulfone/polyimide XU 218 blends with and without solvents.

Polyethersulfone (PES) is a high performance engineering thermoplastic with the repeat unit:

B. C. Anderson · Y. Imanishi (Eds.)
Progress in Pacific Polymer Science
© Springer-Verlag Berlin Heidelberg 1991

The polyimide 2080 (PI 2080) is also a thermally stable polymer; its repeat unit is:

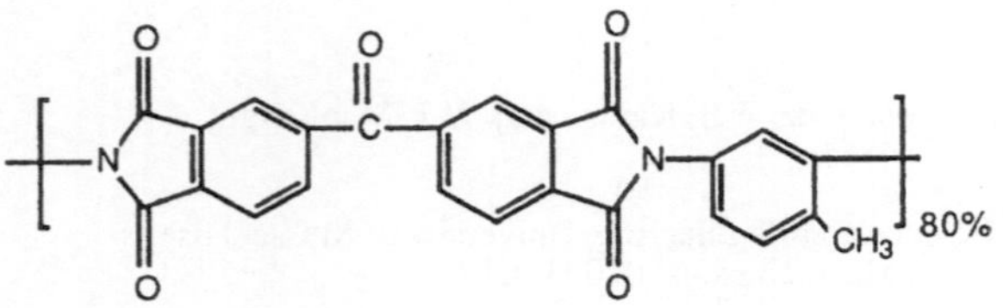

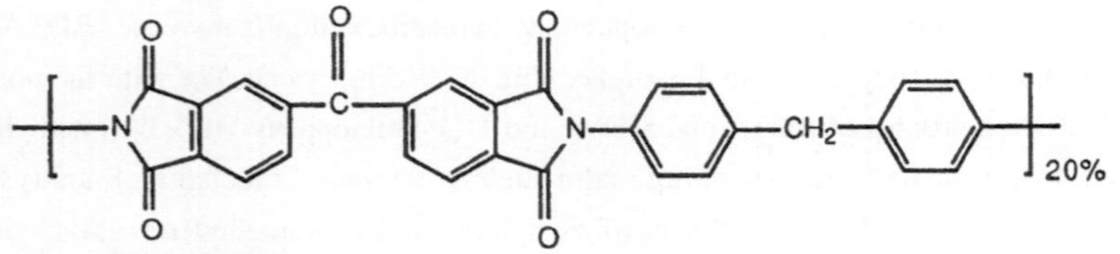

80%

20%

PI XU 218 has the repeat unit:

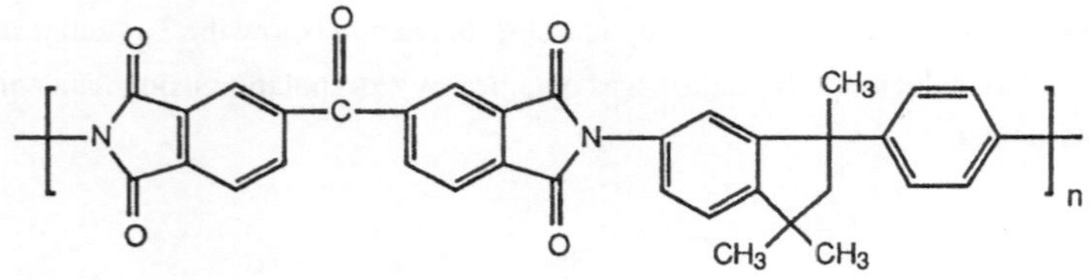

PES/polyimide blends are interesting for several reasons. First, little information about these blends is available in the literature although PES has been reported to be miscible with poly(ethylene oxide) [5,6] and with phenoxy resin [7]. Second, PES costs less than polyimide and has somewhat similar mechanical properties. Third, films of the blends have indicated that the brittle nature of pure PI is decreased by mixing with PES. Thus PES/PI blends offer a potential improvement over the pure materials.

Miscibility in polymer blends is controlled by thermodynamic factors such as the polymer-polymer interaction parameter [8,9], the combinatorial entropy [10,11], polymer-solvent interactions [12,13] and the "free volume" effect [14,15] in addition to kinetic factors such as the blending protocol, including the evaporation rate of the solvent and the drying conditions of the samples. If the blends appear to be miscible under the given preparation conditions, as is the case for the blends descibed here, it is important to investigate the reversibility of phase separation since the apparent one-phase state may be only metastable. To obtain reliable information about miscibility in these blends, the miscibility behavior was studied in the presence and absence of solvents under conditions which included a reversibility of phase separation. An equilibrium phase boundary was then obtained for the binary blend systems by extrapolating to zero solvent concentration.

EXPERIMENTAL

Materials

The polyethersulfone, Victrex PES, was supplied by ICI Americas, Inc. Its reported grade and molecular weight were 4800p and M_n = 22,000, respectively; its glass transition temperature was 230°C. PI 2080 (T_g = 310°C) was supplied by Dow Chemical Co. PI XU 218 (M_w = 80,000, M_n = 10,000-11,000, T_g = 320°C) was obtained as a yellow powder from Ciba-Geigy.

The solvents, N,N,-dimethylacetamide (bp = 165-166°C), tetramethylene sulfone (bp = 285°C), 1-methyl-2-pyrrolidone (bp = 202°C) and dimethyl sulfoxide (bp = 189°C) were all obtained from Aldrich Chemical Co.

Preparation

PES/PI 2080 Blends PES was dried for more than four hours at 150°C to remove residual water before use. Solutions (3 wt%) of the blends were prepared by dissolving the two pure polymer constituents in the desired ratio. Films of the blends were cast from these solutions on Pyrex glass dishes at 80-90°C under vacuum. To obtain solvent-free samples, the films were gradually heated to 240-250°C over a period of three days. To obtain samples of varying solvent contents, the films were dried at lower temperatures for shorter lengths of time. The solvent content of the samples was measured by thermogravimetric analysis.

PES/PI XU 218 Blends PES was dried as described above and pure PES and PI XU 218 were dissolved in the desired proportions in methylene chloride. Tetramethylene sulfone (TMS) was the only other solvent used in the blends containing PI XU 218. PES, PI XU 218 and TMS ternary mixtures were prepared by dissolving all three components in the desired proportions in methylene chloride at room temperature. Because TMS is a solid below 27°C, it was first liquified by heating and was then added dropwise to mixtures of PES and PI XU 218. Films were cast from solution on Pyrex glass dishes. Methylene chloride was evaporated by placing the dishes is a dessicator with a hole in the lid for several days and then by transferring the dishes to a fume hood. Final traces of solvent were removed under vacuum, annealing at 60°C for three days.

Measurements

Thermogravimetric analyses (TGA) were carried out using a Perkin-Elmer TGA-7 instrument. The temperature range was from 50 to 700°C and the heating rate was 20°C/min in a nitrogen atmosphere.

The glass transition temperatures of the pure polymers and the blends were measured with a Perkin-Elmer DSC-7 differential scanning calorimeter at a heating rate of 20°C/min for those samples not containing TMS and 10°C/min for samples containing TMS. All measurements were carried out under nitrogen. All samples except those containing TMS were encapsulated in the standard aluminum pans.

Dynamic mechanical thermal analyses were carried out using a Polymer Laboratories DMTA equipped with a high temperature (500°C) head operating at a frequency of 1 Hz and a constant strain, x 1, in the single cantilever mode. The thickness of the PES/PI 2080 films was in the range of 0.1-0.2 mm and that of the PES/PI XU 218 films was 0.2-0.3 mm.

RESULTS AND DISCUSSION

A. Physical Appearance of the Blends

All polymer solutions were transparent and films cast from these solutions also appeared to be transparent; those containing PI 2080 were slightly yellow to light brown in color. The brittle nature of PI 2080 was improved by mixing with PES in that films of these blends qualitatively were more flexible than films of pure PI 2080.

B. Thermogravimetric Analysis

Since solvents present in blend samples can act as efficient plasticizers, the most important step in sample preparation is to eliminate residual solvent. Figure 1 and Table 1 show TGA results for several "solvent-free" PES/PI 2080 blends prepared in N,N-dimethyl acetamide (DMAc) and for PES/PI XU 218 blends prepared in methylene chloride. It is clear that for PES/PI 2080 blends the residual solvent contents at all blend compositions were less than 1 wt% after drying. For PES/PI XU 218 blends prepared without tetramethylene sulfone (TMS) there was no obvious decomposition above 300°C. Figure 1 also shows that with increasing PI 2080 content, the thermal stability of the blends increases as indicated by the gradual decrease in weight loss from 100°C to 700°C.

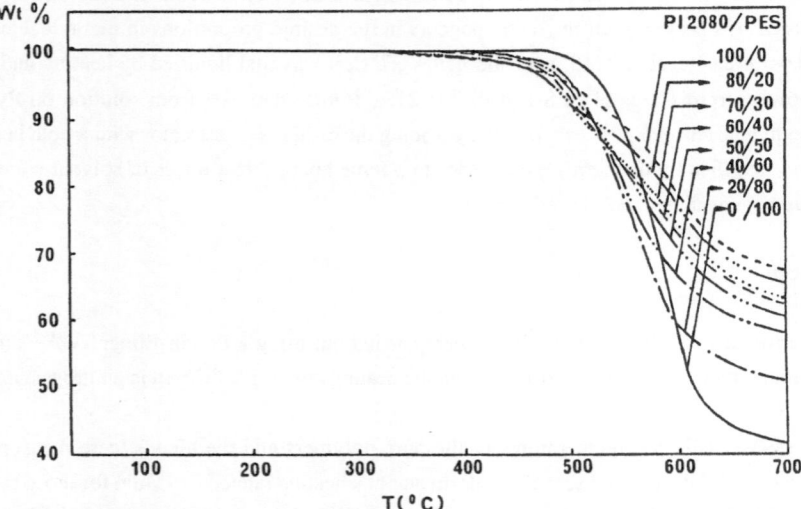

Figure 1. TGA curves for films of pure PES and PI 2080 and blends obtained by casting from DMAc solution. The films were previously dried at 245-250°C for 15 hr under vacuum.

Table 1

TGA results for PES/PI 2080 and PES/PI XU 218 blends

blend composition wt% PES/PI 2080	weight loss (%) 100-400°C	weight loss (%) 100-700°C	onset of decomposition (°C)
100/0	0.90	35.7	463
90/10	0.76	36.0	446
80/20	0.61	39.5	442
70/30	0.79	38.6	439
60/40	0.79	41.7	442
50/50	0.54	43.8	450
40/60	0.42	47.4	448
30/70	0.19	47.3	456
20/80	0.14	50.7	450
10/90	0.19	55.8	467
0/100	0.11	58.8	483
PES/PI XU 218 wt%	weight loss (%) below 350°C	weight loss (%) at 300°C, 2 hr	
100/0	0.90	-	
80/20	0.60	0.20	
70/30	0.40	0.30	
60/40	0.90	0.40	
50/50	0.50	0.30	
40/60	0.40	0.30	
30/70	0.70	0.30	
20/80	0.50	0.20	
10/90	0.40		
0/100	0.50		

For PES/PI XU 218 blends prepared with TMS in methylene chloride, TGA results and elemental analyses indicated that TMS was not lost from the samples during drying (Figure 2). After several

218

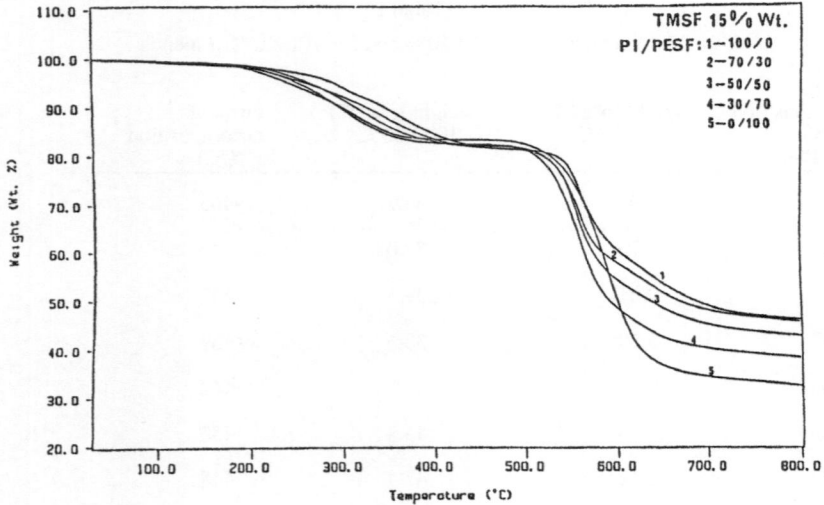

Figure 2. TGA curves of PES/PI XU 218 blends containing 15 wt% TMS.

C. Enthalpy Relaxation Phenomena

Since the thermal behavior of polymers is known to be sensitive to thermal history and traces of residual solvent and moisture, all samples were annealed at temperatures sufficiently high to remove solvent and moisture but low enough to prevent phase separation. In this way reproducible glass transition temperatures were obtained; all data presented are average values of at least three measurements.

Initial differential scanning calorimetry (DSC) scans of PES/PI 2080 blends always showed one or more enthalpy relaxation peaks while those of the PES/PI XU 218 blends showed one broad relaxation peak for each composition investigated. The relaxations observed for PES/PI XU 218 blends increased smoothly with increasing PI content.

The presence of enthalpy relaxation peaks may affect the determination of glass transition values. To erase a relaxation peak within the glass transition, the samples were heated to 10°C over T_g and then quenched at a rate of 320°C/min to 50°C. DSC curves were then recorded again with a heating rate of 20°C/min.

Figure 3 shows typical DSC curves for 80/20 wt% PES/PI 2080 blends. It can be seen that curves A, C and D display relaxation peaks while curves B and E correspond to scans obtained after the previous thermal history has been "erased". In addition, curves A, B and C indicate that the blends are

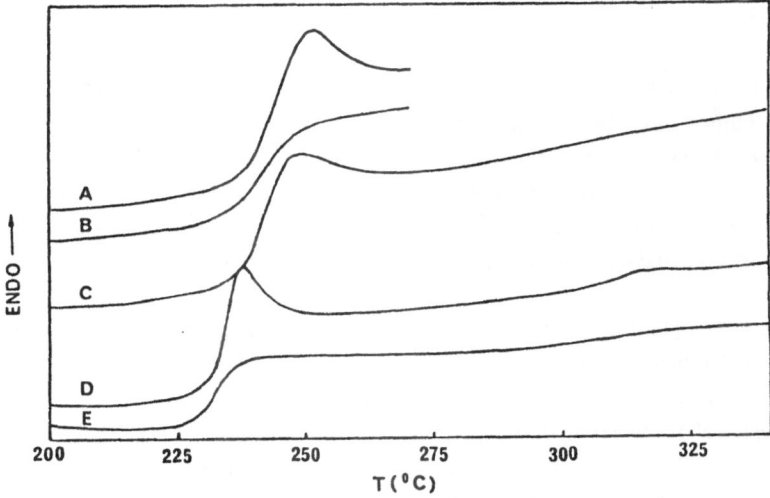

Figure 3. DSC curves for an 80/20 wt% PES/PI 2080 blend. One film was subjected to the following process in sequence: A: initial scan; B: after quenching from 270°C; C: after slow cooling from 270°C (1°C/min); D: after slow cooling from 350°C (1°C/min); E: after quenching from 350°C.

PES/PI XU 218 blends containing TMS exhibited lower T_g values as expected and always showed a single T_g, independent of TMS content. Figure 4 shows the glass transition temperatures (measured by DSC at 10°C/min) of the ternary mixtures as a function of TMS content and weight percent PI XU 218. The most notable effect on the reduction of T_g in the ternary blends (as compared to blends prepared without TMS) was observed with the initial addition of TMS.

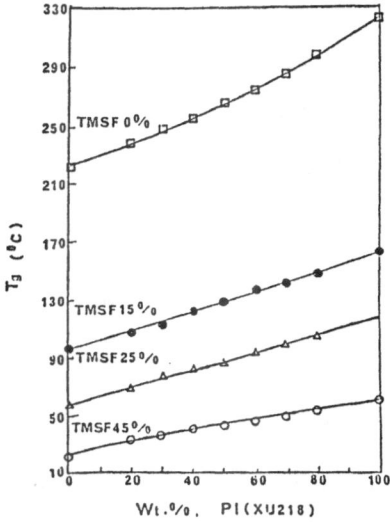

Figure 4. T_g-composition curves for PES/PI XU 218 blends containing different amounts of TMS.

220

D. Phase Behavior

The DSC and dynamic mechanical thermal analysis (DMTA) results obtained for solvent-free PES/PI 2080 samples after the thermal history had been "erased" and before phase separation had occurred show only one glass transition indicating that under these conditions PES is miscible with PI 2080 over all blend compositions. However, when solvent-free samples were first heated to 350°C (330°C for PES/PI XU 218 samples), then cooled and subjected to DSC and DMTA measurements, two T_g's were always observed indicating that phase separation had taken place.

It is expected [16-18] that thermally induced phase separation can be reversed when polymer blends are annealed below their respective LCST's if the system is in equilibrium. However, in the present PES/PI systems, as in the PBI/PI systems [19], reversibility is not observed. It is obvious that the data presented above represent only apparent phase boundary curves. Because of the ridigity of the PES and PI chains, the mobility of the segments is limited and the system is highly viscous. The observed one-phase system corresponds to a homogeneous "frozen" structure formed from a solution of the constituents as the solvent evaporates. When relaxation times are sufficiently reduced, the chains have sufficient mobility to form a stable two-phase state. The location of the true phase boundary curve in the present case for PES/PI blends may lie below the T_g-composition line.

In Figure 5 it can be seen that the "window" between T_g and the apparent phase boundary for PES/PI XU 218 blends is narrow, especially when the PI content is greater than 60 wt%.

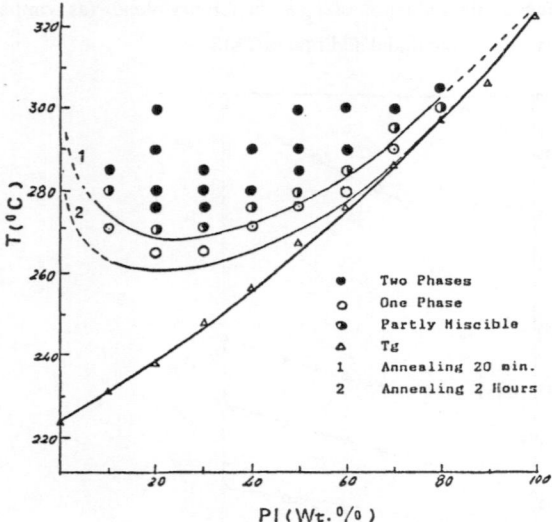

Figure 5. Apparent boundary curve for PES/PI XU 218 blends (without TMS).

E. Effects of Solvent on Phase Behavior

The addition of a compatible low molecular weight solvent to the PES/PI systems would have the effect of reducing the viscosity of the system and thus reducing the relaxation time of the segments. Therefore, a series of films of pure polymers and blends prepared in different solvents was subjected to DSC investigation. The results showed that the T_g values decreased with increasing solvent concentration. Figure 6 (pure PES) shows typical results.

Since TMS has a relatively high boiling point (bp = 285°C) and showed a low extent of evaporation from the samples, it was used in the preparation of samples for further study.

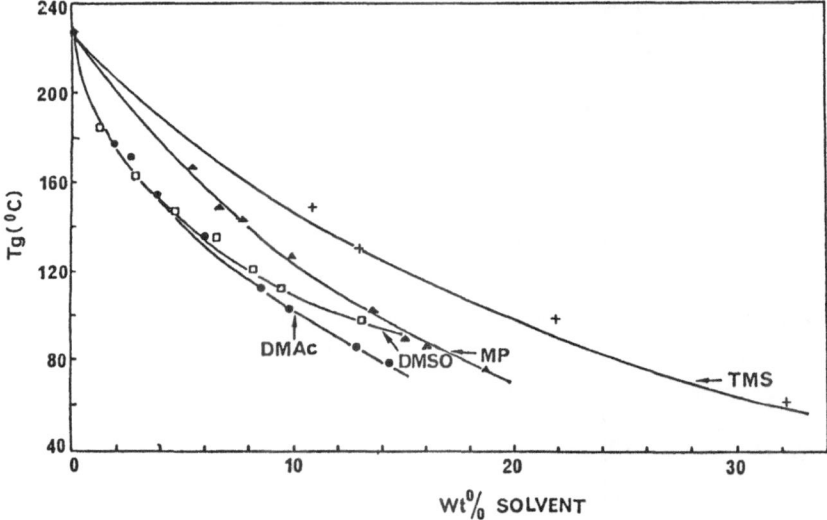

Figure 6. T_g as a function of solvent content for pure PES.

To determine the apparent phase boundaries in PES/PI blends with and without TMS a series of annealing experiments was performed. Each blend was annealed at a different temperature for 20 min for the system without TMS and for 10 min for the PES/PI/TMS ternary system and then analyzed by DSC. The lowest temperature for which two T_g's were detected was taken as the location of the apparent phase separation boundary.

The effects of the addition of TMS are shown in Figures 7 and 8 for PES/PI XU 218 and PES/PI 2080 blends, respectively. The glass transitions and phase separation occur at lower and higher temperatures, respectively than in the absence of TMS and the "window" between the T_g-composition line and the apparent boundary curve is expanded.

As discussed above it is clear that the phase boundary curves obtained under these conditions are true equilibrium curves. Reversible phase separation was observed for PES/PI XU 218 blends at a TMS content of 25 wt% or greater and for PES/PI 2080 blends at a TMS content of 30 wt% or greater.

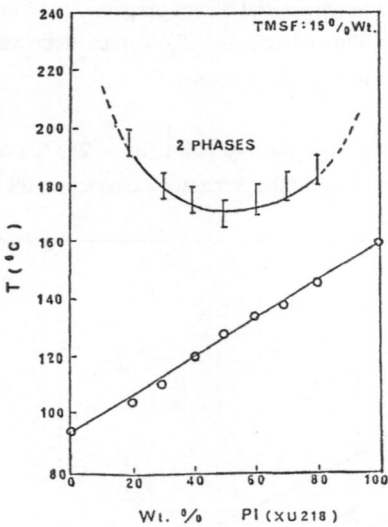

Figure 7. T_g-composition line and phase boundary curve for PES/PI XU 218 blends containing 15 wt% TMS.

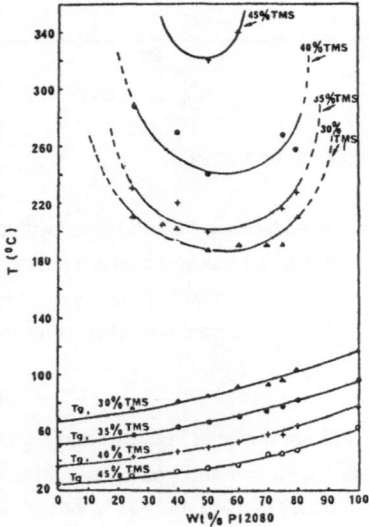

Figure 8. T_g-composition lines and phase boundary for PES/PI 2080 blends containing different amounts of TMS.

To obtain a true equilibrium phase boundary curve for solvent-free PES/PI blends, the data collected for samples containing different amounts of TMS were extrapolated to zero solvent content. For a 50/50 wt% PES/PI XU 218 blend a phase separation temperature of 140°C is obtained, a temperature lower than the blend T_g (267°C). Similar results are observed for PES/PI 2080 blends, Figure 9.

We note some similarities between the present PES/PI/TMS blends and the PMMA/SAN/DMP system reported by Bernstein et al. [20]. These authors found that phase separation in the PMMA/SAN system is not reversible but addition of 15 wt% or more of DMP induced reversibility. These effects were explained by the proposal that "equation of state" and "entropy of mixing" factors are working in opposition. In this system it is our belief that the reversible and irreversible processes merely reflect thermodynamic and kinetic considerations. The displacement of the phase boundary curve for the PES/PI/TMS system to higher temperatures is due to contributions to the free volume from TMS [21-23].

Crosslinking was not observed in annealing studies with PES/PI XU 218 blends; all samples with and without TMS could be dissolved in methylene chloride to form transparent solutions after thermally induced phase separation and annealing. However, PES/PI 2080 samples heated to 350°C several times could not be dissolved in DMAc suggesting that a network structure had formed.

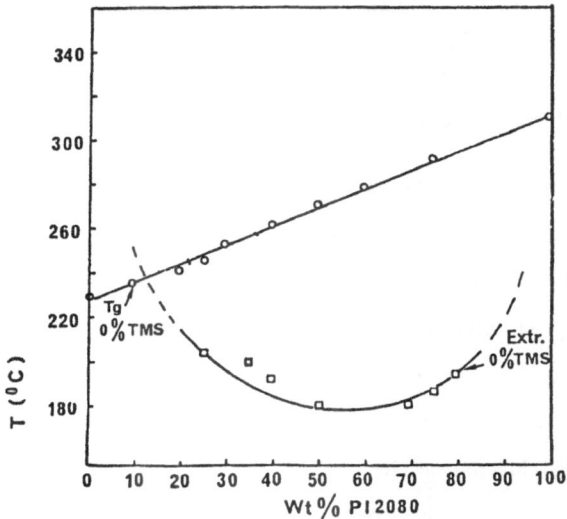

Figure 9. T_g-composition line and phase boundary for PES/PI 2080 blends extrapolated to zero TMS content.

CONCLUSIONS

The apparent miscibility in the PES/PI blends is due to the "frozen" structure of the mixture produced during the film preparation protocol. After the system is phase separated at a temperature above T_g it cannot revert to a one-phase mixture. Thus the structure of the blend samples is not thermodynamically stable; the equilibrium phase boundary for these mixtures must be below the respective blend T_g's.

When tetramethylene sulfone was added to PES/PI blends the blend T_g's were sufficiently depressed and the window between the phase boundary curve and the T_g-composition line was expanded so that equilibrium data could be obtained. At TMS contents greater than 25 or 30 wt% an equilibrium LCST could be detected which occurred at higher temperatures as the TMS content increased. Concurrently, the phase separation process became experimentally reversible.

ACKNOWLEDGEMENT

This work was supported by the Air Force Office of Scientific Research through grant # 88-0011.

REFERENCES

1 Leung, L, Williams, DJ, Karasz, FE, MacKnight, WJ (1986) Polym Bull 16:1457

2 Guerra,G, Choe, S, Williams, DJ, Karasz, FE, MacKnight, WJ (1988) Macromolecules 21:231

3 Guerra,G, Williams, DJ, Karasz, FE, MacKnight, WJ (1988) J Polym Sci Polym Phys Edn 26:301

4 Liang, K, Karasz, FE, MacKnight, WJ (1988) Bull Am Phys Soc 33:326

5 Walsh, DJ, Singh, VB (1984) Makromol Chem 185:1979

6 Walsh, DJ, Zoller, P (1987) Makromol Chem 188:2193

7 Singh, VB, Walsh, DJ (1986) J Macromol Sci Phys B25:65

8 Zeman, L, Patterson, D (1972) Macromolecules 5:513

9 Hsu, CC, Prausnitz, JM (1974) Macromolecules 7:320

10 Roe, RJ, Zin, WC (1980) Macromolecules 13:1221

11 ten Brinke, G, Karasz, FE (1984)Macromolecules 17:815

12 Silverberg, A (1968) J Chem Phys 48:2835

13 Huggins ML (1971) Macromolecules 4:274

14 Robard, A, Patterson, D (1975) Macromolecules 10:1021

15 Bank, M, Leffingwell, J, Thies, C (1971) Macromolecules 4:43

16 Fried, JR, Karasz, FE, MacKnight, WJ (1978) Macromolecules 11:15

17 Alexandrovich, P, Karasz, FE, MacKnight, WJ (1977) Polymer 18:1022

18 Vukovic´, R, Karasz, FE, MacKnight, WJ (1983) Polymer 24:529

19 Choe, S, Karasz, FE, MacKnight, WJ (1990) in "Multiphase Macromolecular Systems", ed Culbertson, WM, Plenun, New York

20 Bernstein, RE, Cruz, CA, Paul, DR, Barlow, JW (1977) Macromolecules 10:681

21 Flory, PJ, Orwall, RA, Vrij, A (1964) J Am Chem Soc 86:3507, 3515

22 McMaster, LP (1973) Macromolecules 6:760

23 Lacombe, RH, Sanchez, IC (1976) J Phys Chem 80:2568

SIMS Depth Profiling of Polymeric Materials Having Concentration Gradient

Riichirô Chûjô

Department of Biomolecular Engineering,
Tokyo Institute of Technology,
12-1 Ookayama 2-chome, Meguro-ku,
Tokyo 152, Japan

Abstract: Concentration gradient was found in the blends of poly-
(vinylidene fluoride) and poly(methyl methaccrylate).   It was
analyzed with terms of surface tension and the existence of oxydized
compounds.   Similar gradients were found in the polymer optical
fibers, video tapes, and surface-fluorinated polyolefin films.

Concentration gradient in polymeric material is becoming one of the
most important factors in the molecular design of them.   SIMS
(Secondary Ion Mass Spectrometry) is applicable to the depth profiling of polymeric materials; this technique is quite promising to
obtain the information on the concentration gradient (1-3).   SIMS
is more advantageous than the other methods such as XPS(X-ray Photoelectron Spectroscopy) in two points; 1) the maximum of detectable
depth in XPS is limited to 5 nm due to the finite escape depth of
photoelectron, while there is no limit in SIMS, and 2) the information is obtained for any individual depth in SIMS, while it is
obtained in the integral form from the surface to the desired depth
in XPS.   There is, of course, disadvantage in SIMS than in XPS;
XPS depth profiling can be done nondestructively, while SIMS one
is destructive.   Nevertheless, SIMS is, totally, more advantageous
than XPS.   In this paper, results from our studies on SIMS depth
profiling will be surveyed; they are applications to polymer blends,
optical fibers, video tapes, and surface-fluorinated polyolefin
films.

SIMS was observed with a model A-DIDA-3000 equipped by Atomika
Technische Physik GmbH.   The primary Ar^+ ion beam was accelerated
with a voltage of 3kV and its beam current was 50nA.   Such a low
current is effective to prevent from any change (thermal or chemical)
of the sample surface.   Furthermore, the sample was covered by an

B. C. Anderson · Y. Imanishi (Eds.)
Progress in Pacific Polymer Science
© Springer-Verlag Berlin Heidelberg 1991

228

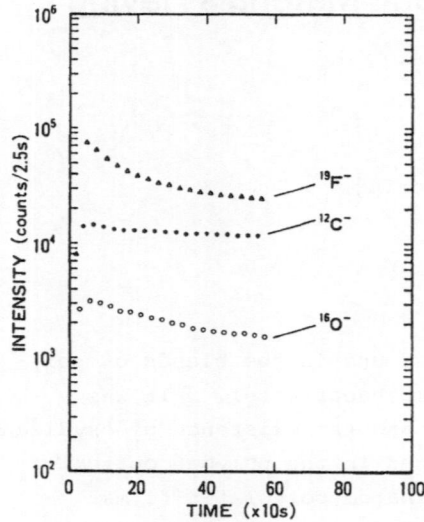

Figure 1.   Depth profiling
of the polymer blend between
PVDF and PMMA (50/50 v/v).

electron shower compensating for
the charge-up.   During measurements the pressure was kept at
3×10^{-9} torr.

POLYMER BLENDS
Poly(vinylidene fluoride) (PVDF)
is miscible with poly(methyl methacrylate) (PMMA).   SIMS depth
profiling was done for the blends
of PVDF and PMMA (50/50 v/v) making use of the time dependence of
the intensities of secondary $^{12}C^-$,
$^{16}O^-$, and $^{19}F^-$ ions.   "Time"
means the time of measurement from
the start.   By the bombardment of
primary ions, crator was produced
on the sample surface, the depth
of which increases with increase
of the time.   If the depth is,
therefore, proportional to the time, the time dependence should be
equivalent to the depth dependence.   In the plasma etching of polymeric materials including PVDF and copolymer of MMA and methacrylic
acid linear relationships were found between the etch depth and the
time except for induction period (4).   This is one support to the
above equivalency.   The other support is the constant intensity of
$^{12}C^-$ irrespective of time.   In Figure 1 is shown the SIMS depth
profiling of a blend.   Intensity of $^{12}C^-$ is almost kept at constant
except for the region close to the surface.
Intensity of $^{19}F^-$ decreases repidly from the surface to the inside.
It means PVDF is more abundantly distributed in the surface than in
the inside.   Intensity of $^{16}O^-$ also decreases from the surface to
the inside.   However, this is not due to abundant distribution of
PMMA in the surface, but to an existence of oxidized compounds in the
surface.   This statement will be confirmed in the next section.
In the thermodynamics of polymer blends the solubility parameter is
the most important parameter describing the miscibility between
component polymers.   However, this parameter is insufficient to
explain the above finding.   The samples were prepared in the form
of film by casting from dimethyl formamide (DMF) solution.   The
values of solubility parameters are 14.7, 18.93, and $24,8 \times 10^3$

$J^{-1/2}$m$^{-3/2}$ for PVDF, PMMA, and DMF, respectively(5). Therefore, if the solubility parameter is the good parameter describing the concentration gradient, PMMA should be much abundantly distributed in the surface in which the concentration of PMMA becomes larger due to the evaporation of solvent. Instead, the surface tension is the best parameter describing the gradient. The values of surface tension are 25 and 39 x 10^{-3}N m^{-1} for PVDF and PMMA, respectively (6). Besides the solubility parameter, a new parameter, surface tension must be introduced as a parameter describing the concentration gradient in polymer blends (1). Similar finding has been done in the XPS depth profiling for the diblock polymers between polystyrene and poly(ethylene oxide) (7). Furthermore, structure factor was also studied for the blends of PVDF and PMMA(8). In this study the structure factor is always larger in the blends than in homopolymers. This means there are concentration gradients in the blends. The origin of these gradients was clarified by our depth profiling.

OPTICAL FIBERS

Optical fibers are, at least, composed of two components in order to accomplish total reflection of optical rays. SIMS depth profiling is promising to analyze the two-component structures. The profiling was done for the optical fibers composed of PVDF (clad) and PMMA (core) (2). Similar to the last section, the intensities of $^{12}C^-$, $^{16}O^-$, and $^{19}F^-$ ions were used for the profiling. Results were shown in Figure 2 (2). In the former half in the time domain, the intensity of $^{19}F^-$ was almost constant and very large; this means the clad is actually composed of PVDF. In the latter half, the intensity of $^{16}O^-$ behaved similarly suggesting the existence of PMMA in the core. Furthermore, in the very early former part $^{16}O^-$ was intense. This reflects an existence of oxidized compounds. C-F bonds are easily degraded compared with other kinds of bond. It implies the possibility of the observation of carbonization

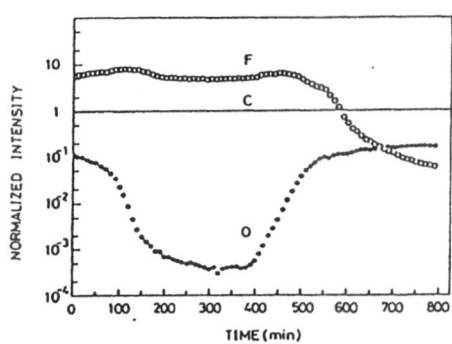

Figure 2. Depth profiling of the optical fiber composed of PVDF (clad) and PMMA (core) normalized with the intensity of $^{12}C^-$.

process instead of depth profiling in the last section. The
constancy of $^{19}F^-$ intensity in Figure 2 liberates us from such an
implication. The large intensities in the very early stage both
in Figures 1 and 2 support the statement in the last section: the
existence of oxidized compounds in the surface.

VIDEO TAPES

Video tapes are composite materials composed of polymeric base films
and magnetics. In order to know the concentration gradients, four kinds
of video tapes of the grade of T-120 were profiled making use of the
intensities of $^{12}C^+$, $^{16}O^+$, $^{27}Al^+$, $^{35}Cl^+$, $^{52}Cr^+$, $^{56}Fe^+$, and $^{59}Co^+$ ions.
In Figure 3 is shown the profiling of the sample A (3). In order

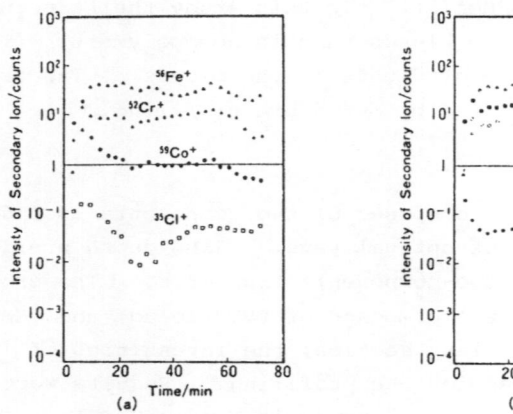

Figure 3. Depth profiling of the video tepe A
normalized with intensity of $^{16}O^+$.

to avoid the confusion due to the crowdness of experimental points,
figure was divided into (a) and (b), in which the intensities of
$^{35}Cl^+$, $^{52}Cr^+$, $^{56}Fe^+$, and $^{59}Co^+$ were compiled in (a), while $^{12}C^+$,
$^{14}N^+$, $^{27}Al^+$, and $^{56}Fe^+$ were in (b). For the convenience of comparison, intensities of $^{56}Fe^+$ were compiled both in (a) and in (b).
Intensities of secondary ions are much enhanced by the existence
of oxygen (9). This phenomenon is known as oxygen effect. In
order to remove this oxygen effect, all intensities of the secondary
ions were normalized by the intensities of $^{16}O^+$ in Figure 3.
Comparing with the profiling patterns between four samples, we find
some differences between them (differences due to the differences
of manufacturing companies). Generally speaking, the samples A and

D (the profiling patterns are not shown here for the samples B, C, and D) are close with each other, and B and C are.   More precisely, 1) Cl is more abundantly distributed in the region close to the magnetics in the sample A.   This trend is explainable with a term of the distribution of antistatic agent which includes Cl.
2) Cl is more abundantly distributed in the opposite side (base film) in the samples B and C, while Cl is in both sides in the sample D.
3) The time required for the depth profiling over total thickness is the shortest in the sample A in spite of almost equal thickness in these four samples.   This finding is explainable with a term of the effectiveness of the antistatic agent.
4) Metallic elements, Al, Fe, Cr, and Co are localized in the area close to the surface (magnetics) in the samples A and D, while they are almost equally distributed in the whole range of the samples B and C.   A small amount of these elements is observed even in the inside of base film in the samples A and D.   It is still unclear whether these traces are due to actual migration to inside or poor resolution in the measurements.
5) An increase of N is observed corresponding to the decrease of the above metallic elements in the inside of the whole samples.
6) Among physical properties, both of static and dynamic friction constants are not correlated with any finding from the SIMS depth profiling.
7) Dynamic resistance is well correlated with the intensity of Cl in the region close to the surface.

SURFACE-FLUORINATED POLYOLEFIN FILMS

Surface fluorination of polyethylene accomplished by the contact with fluorine gas was proposed in 1954 (10). Enthalpy balance of this reaction is given by (11)

$$F_2 \rightleftarrows 2F \cdot \qquad\qquad \Delta H = + 154 \ kJ \ mol^{-1}$$
$$F \cdot + - CH_2 - CH_2 - \rightarrow \ - \overset{\bullet}{C}H - CH_2 - \ + HF \quad \Delta H = - 142 \ kJ \ mol^{-1} \qquad (1)$$
$$F_2 + - \overset{\bullet}{C}H - CH_2 - \rightarrow - CHF - CH_2 - \ + F \cdot \quad \Delta H = - 285 \ kJ \ mol^{-1}$$

Therefore, if a fluorine molecule is dissociated into two fluorine radicals, succeeding reactions proceeds automatically.   Nevertheless, this simple technique is considerably new in practical application (12).   For instance, the effectiveness of fluorination was found in high density polyethylene (HDPE) gasoline container; gasoline packaged in an untreated HDPE container and kept at 50°C rapidly diffused and evaporated through the container wall, losing

over half its weight in less than 4 weeks.   When packaged in fluorinated HDPE containers, gasoline permeation was virtually negligible.   In order to characterize the surface-fluorinated polyethyl-
enes, SIMS depth profiling was done making use of the intensities
of intense secondary anions whose mass numbers are 12, 13, 16, 19,
31, 47, 50, and 69 (13).   The anion of m = 16 was observed for the
consideration of oxygen effect.   Remaining anions are assigned to
C^-, CH^-, F^-, CF^-, COF^-, $CF_2{}^-$, and $CF_3{}^-$ in the order of increasing
mass numbers.   Anions of m = 28 and 29 were also intense.   However,
they are not assigned to $C_2H_4{}^-$ nor $C_2H_5{}^-$, but $^{28}Si^-$ and $^{29}Si^-$ due to
the traces of contamination.   In Figure 4 is shown the profiling

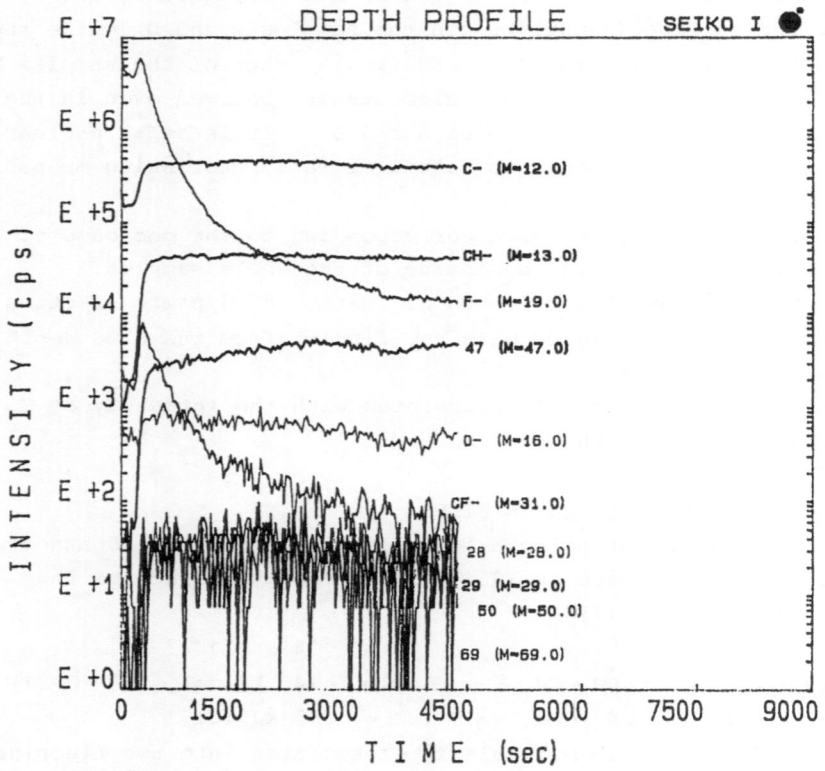

Figure 4.   Depth profiling of surface-fluorinated
polyethylene which contains fluorine atoms of 9.0 µg cm^{-2}.

of a sample which contains fluorine atoms of 9.0 µg cm^{-2}.
The diffusion equation in one-dimensional system is given by

$$\frac{\partial C}{\partial t} = D \frac{\partial^2 C}{\partial x^2} \qquad (2)$$

where x, t, and C stand coordinate, time, and the concentration of substance, respectively. D is the diffusion coefficient which is assumed to be independent of the concentration. If the diffusion is accompanied by chemical reaction, Eq (2) is modified as follows:

$$\frac{\partial C}{\partial t} = D \frac{\partial^2 C}{\partial x^2} - \frac{\partial S}{\partial t} \qquad (3)$$

where S is the concentration of immobilized solute. In the reaction (1) the concentration of -CHF- is given by

$$\frac{d(-CHF-)}{at} = k_2 (F_2) (-\overset{\bullet}{C}H-) \qquad (4)$$

where k_2 is the rate constant in the third stage of the reaction (1). In stationary state, $(-\overset{\bullet}{C}H-)$ should be constant. Eq. (4) is, therefore, simplified into

$$\frac{d(-CHF-)}{at} = k (F_2) \qquad (5)$$

where k is equal to the product of k_2 and (F_2). The quantities $(-CHF-)$ and (F_2) in Eq (5) correspond to S and C, respectively, in Eq (3). Eq (3) is, therefore, simplified into

$$\frac{\partial C}{\partial t} = D \frac{\partial^2 C}{\partial x^2} - k C \qquad (6)$$

This equation has been used to describe the dyeing process by reactive dyestaffs, and this can be used to describe the reaction (1) without any modification. In our case, the initial condition

$$t = 0, \quad x > 0. \quad C = 0 \qquad (7)$$

and the boundary conditions

$$t > 0, \quad x = 0, \quad C = C_o$$
$$t > 0, \quad x \to \infty, \quad C = 0 \qquad (8)$$

are practical, because the reaction was done for polyethylene film by the contact with fluorine-containing gas. The analytical solution of Eq (6) under the conditions of Eq (7) and (8) is obtained as follows (14):

$$C = \frac{C_o}{2} \{ \exp(-x \sqrt{k/D}) \, \mathrm{erfc}((x/2\sqrt{Dt}) - \sqrt{kt}) $$
$$+ \exp(x \sqrt{k/D}) \, \mathrm{erfc}((x/2\sqrt{Dt}) + \sqrt{kt}) \} \qquad (9)$$

where erfc(z) is the coerror function of error function, erf(z) and
is given by

$$erfc(z) = 1-erf(z) = (2/\sqrt{\pi}) \int_z^\infty exp\,(-\xi^2)\,d\xi \qquad (10)$$

The quantity, S(t), is, therefore, given by

$$S = \int_0^t kC\,dt$$

$$= (kCo/2\{(t-(x/2\sqrt{Dk}))\,exp\,(-x\sqrt{k/D})\,erfc\,((x/2\sqrt{Dt})-\sqrt{kt})$$
$$+ (t+(x/2\sqrt{Dk}))\,exp\,(x\sqrt{k/D})\,erfc\,((x/\sqrt{2Dt}+\sqrt{kt})\} \qquad (11)$$

In depth profilings, the ordinate is the intensity, I, of secondary
ion, and the abscissa is the time, τ, for measurement. Therefore,
if proportionalities

$$I = \lambda S$$
$$\tau = \mu x \qquad (12)$$

hold, Eq (11) can be written as

$$I = (k\lambda Co/2)\,\{(t-(\tau/2\mu\sqrt{Dk}))\,exp\,(-\tau\sqrt{kt}/\mu\sqrt{Dt})\,erfc\,((\tau/2\mu\sqrt{Dt})-\sqrt{kt})$$
$$+ (t+(\tau/2\mu\sqrt{Dk}))\,exp\,(\tau\sqrt{kt}/\mu\sqrt{Dt})\,erfc\,((\tau/2\mu\sqrt{Dt})+\sqrt{kt})\} \qquad (13)$$

In each depth profiling, the behavior in very small τ cannot be
analyzed by Eq (13) due to rather complicated phenomena in surface
region, and the behavior in very large τ cannot also due to insufficient memory size of computer in less intense region. Eq (13)
is, therefore, used to analyze the depth profiling pattern in intermediate τ values. Results are tabulated in Table I. Contents

Table I. Parameters for surface-fluorinated polyethylene

Sample	Fluorine Contents/$\mu g\,cm^{-2}$	$k\lambda C_0 t/10^6$	$\mu\sqrt{Dt}/10^3$	$\sqrt{kt}$
E	4.6	1.03	2.12	1.02
F	61.3	3.97	1.73	1.04
G	64.2	2.57	0.45	2.27
H	88.2	2.77	1.66	1.30
I	150.2	2.08	1.56	0.99

of fluorine atoms determined from fluorescent X-ray analysis are
slso tabulated in the Table.

From the above-mentioned four examples, readers can find SIMS depth
profiling is a promising technique in the characterizetion of
polymeric materials having concentration gradients.

1 Chûjô R, Nishi T, Sumi Y, Naitoh H, Frenzel H (1983) Polym Lett 21:487

2 Chûjô R, Nishi T, Sumi Y, Adachi T, Naitoh H, Frenzel H (1984) Secondary Ion Mass Spectrometry, SIMS IV, Springer-Verlag, Berlin

3 Chûjô R, Nishi T, Naitoh H, Adachi T (1988) Bull Inst Chem Res Kyoto Univ 66:312

4 Pedersen LA (1982) J Electrochem Soc 129:205

5 Small PA (1953) J Appl Chem 3:71

6 Zisman WA (1962) Adhesion and Cohesion, Elsevier, Amsterdam

7 Thomas HR, O'Malley JJ (1979) Macromolecules 12:323

8 Wendorff JH (1980) Polym Lett 18:439

9 Benninghoven A (1975) Surf Sci 53:596

10 Rudge AJ (1954) Brit Pat 710523

11 Clark DT, Feast WJ, Musgrave WKR, Ritchie I (1975) J Polym Sci Chem 13:857

12 Buck DM, Marsh PD, Milcetich FA, Kallish KJ (1986) Plast Eng No. 4:33

13 Chûjô R, Wada M (1989) unpublished work

14 Danckwerts PV (1950) Trans Faraday Soc 46:300

NMR Studies of Copolymerizations of Acrylonitrile

J.D. Borbely, D. Graham, D.J.T. Hill,
A.P. Lang, P. Munro and J.H. O'Donnell

Polymer Materials and Radiation Group
The University of Queensland
Brisbane, Australia, 4067

INTRODUCTION

The study of the mechanisms of copolymerization reactions has attracted the attention of polymer scientists over many years. Copolymerization reactions have been described in terms of various mathematical models, terminal, penultimate, complex, etc., which are characterized by a particular set of reactivity ratios. These mathematical models can be used to make predictions about the variation of the copolymer composition or the copolymer microstructure (for example, the triad fractions, number average sequence lengths, etc.) with comonomer feed composition or monomer conversion (1).

The early studies of copolymerization mechanisms were based mainly upon analysis of the variation of the copolymer composition with feed composition, from which a set of reactivity ratios could be calculated. However, over the last decade, advances in the techniques of NMR spectroscopy have allowed much more detail about copolymer microstructure, such as triad fractions or stereosequence arrangements to be determined (2). This extra information has allowed the mechanisms of copolymerization reactions to be evaluated more precisely.

From the point of view of providing information about copolymer microstructure, ^{13}C NMR has proved to be more informative than ^{1}H NMR, particularly because of the absence of $^{13}C-^{13}C$ spin-spin coupling. In addition, the development of the various multipulse techniques (e.g. DEPT) and 2D-NMR techniques (e.g. heteronuclear chemical shift correlation spectroscopy) have allowed the resonance peaks associated with the various types of carbon atoms in the polymer to be identified (2).

In this paper we review our studies of some copolymerizations of acrylonitrile, in which we have made extensive use of ^{1}H and ^{13}C NMR to obtain microstructural information, in order to evaluate which of the various possible copolymerization models best describes each system.

STYRENE-ACRYLONITRILE COPOLYMERIZATIONS

There have been many studies of the copolymerization of styrene (S) and acrylonitrile (AN), both in bulk and in solution. Frequently the terminal model has been used to describe the polymerization, for example studies conducted by Pichot and Pham (3), and Arita et al. (4). However, other workers have reported deviations from the terminal model, and these have been explained in terms of the penultimate model, for example Ham (5) and Guyot and Guillot (6), or the complex model, Sandner et al. (7) or Kucher et al. (8).

We have studied this copolymerization in bulk and in several solvents at 60°C in order to identify which of the proposed mechanisms best represents the copolymerization data (9)(10)(11). In this we have used a variety of NMR techniques to provide information about copolymer microstructure.

The ^{1}H NMR solution spectrum of SAN copolymers, Figure 1, shows that the aromatic and backbone protons are well separated and the peak areas are easily quantified to provide the composition of the copolymers. However, it is not possible to obtain sequence distribution information from the ^{1}H NMR spectrum of these polymers.

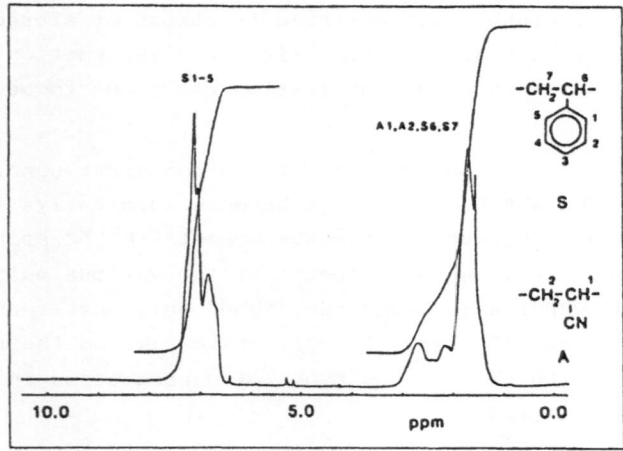

Figure 1: ^{1}H NMR spectrum of poly(styrene-co-acrylonitrile).

The ^{13}C NMR solution spectra provide a wealth of information about the microstructure of the copolymers. A typical spectrum is shown in Figure 2. The copolymer composition can be obtained from the spectrum in several ways, by comparisons of the areas of the side chain or main chain carbon resonances. The copolymer compositions obtained by ^{1}H NMR, ^{13}C NMR and elemental analysis are in excellent agreement (9). The variation on the copolymer composition obtained by ^{13}C NMR with monomer feed composition is shown in Figure 3 for the bulk polymerization of the monomers at 60°C.

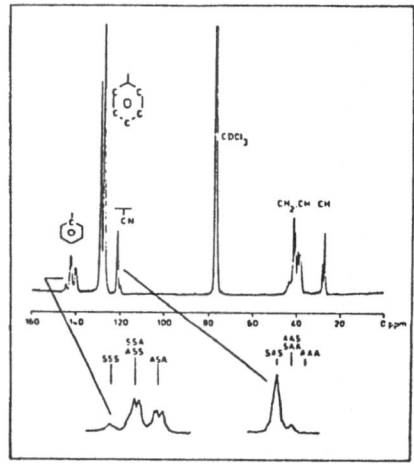

Figure 2: ^{13}C NMR spectrum of poly(styrene-co-acrylonitrile).

In addition to the copolymer composition, the ^{13}C NMR spectrum can provide information about the triad fractions of both the acrylonitrile and the styrene units in the copolymer. The acrylonitrile triad fractions can be obtained from the splittings in the cyano carbon resonances, and the styrene triad fractions can be obtained from the splittings in the resonance of the quaternary carbon in the styrene side chain, as shown in Figure 2. The experimentally determined triad fractions are given in Figure 4 for all triads over the range of monomer feed compositions, for bulk copolymerization of the monomers at 60°C.

We have examined the copolymer composition and triad fraction data in terms of the various polymerization models. Mathematical analysis of the composition and triad fraction data using a non-linear least squares method, followed by application of the statistical F-test to discriminate between the fit of each of the models to the data (1), indicates that the penultimate model provides the best description of the bulk copolymerization (9,11).

The penultimate model reactivity ratios obtained from the mathematical analyses of the compositions and the triad fractions are given in Table 1, together with the standard error, S_y, in the fitted data parameter. The two sets of reactivity ratios are in excellent agreement. The fit of the model to the experimental data is demonstrated in Figures 3 and 4.

Table 1

Penultimate model reactivity ratios at 60°C for AN and S polymerized in bulk. The standard error in the experimental data (S_y) is calculated from the curve fit.

Ratio	Composition	Triad
r_{AA}	0.04	0.06
r_{SA}	0.10	0.09
r_{SS}	0.23	0.24
r_{AS}	0.66	0.58
S_y	0.005	0.02

The aliphatic region of the ^{13}C NMR spectrum of a SAN copolymer (Figure 2) shows that the acrylonitrile methine carbon resonance at 25-30 ppm is well separated from the other aliphatic carbon resonances at 30-45 ppm, which are overlapping. The latter resonances consist of the methine carbon for the styrene units and the methylene carbons for both the styrene and acrylonitrile units. Application of the DEPT multipulse technique (12) allows these overlapping resonances to be resolved. The methine and methylene carbon subspectra can be generated separately, as shown in Figure 5.

The methine carbon subspectrum contains two major peaks (one for the AN and one for the S centred sequences) both of which have unresolved fine structure. This fine structure results from the various sequence and configurational arrangements around these carbon atoms.

The methylene carbon subspectrum contains three major peaks, each of which has unresolved fine structure. The major splittings in these spectra are believed to arise as a consequence of the various diad arrangements in the copolymers, as displayed in Figure 6.

Integration of these spectra allow the diad fractions to be calculated. In Table 2 these experimental diad fractions are

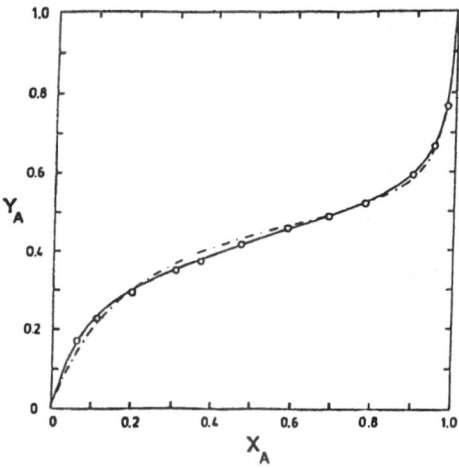

Figure 3: Copolymer composition curve for bulk polymerization of styrene and acrylonitrile at 60°C. Y_A = mole fraction AN in the polymer; X_A = mole fraction AN in the monomer; experimental data o ; terminal model —·—, penultimate model —→ .

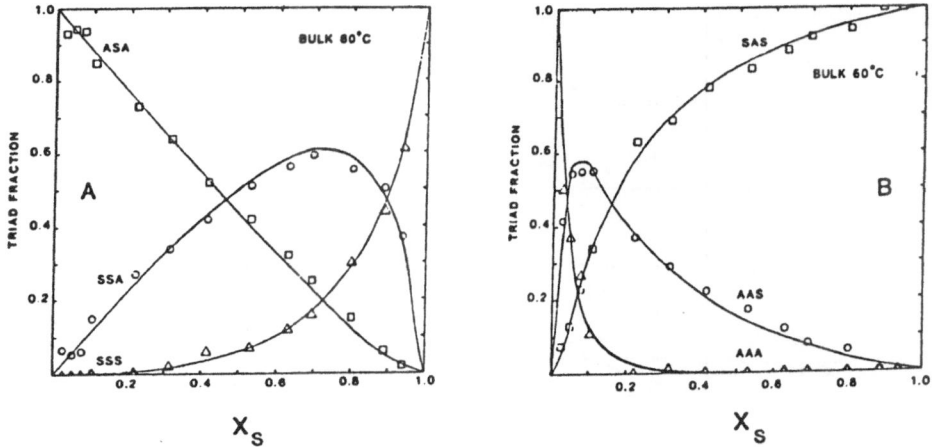

Figure 4: Triad fractions for bulk copolymerization of styrene and acrylonitrile at 60°C. X_S = mole fraction S in the monomer; A = styrene centred triads, B = acrylonitrile centred triads; penultimate model —— .

compared with the diad fractions calculated from the reactivity ratios for the penultimate model. The agreement between the experimental and predicted diad fractions is excellent.

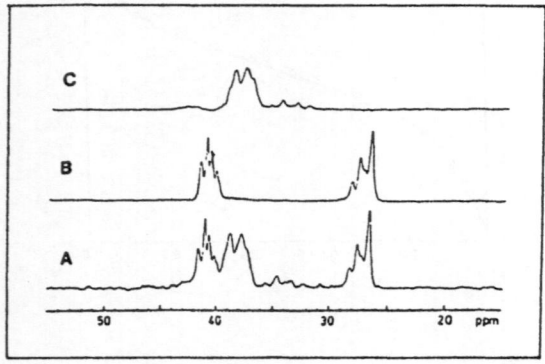

Figure 5: DEPT subspectra of poly(styrene-co-acrylonitrile). A = normal ^{13}C NMR spectrum; B = methine subspectrum; C = methylene subspectrum.

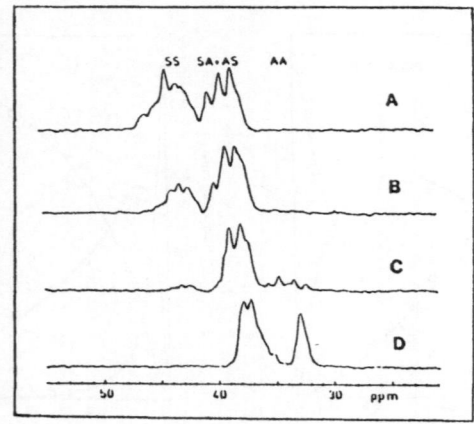

Figure 6: DEPT methylene subspectra of poly(styrene-co-acrylonitrile). A-Y_A = 0.23; B-Y_A = 0.35; C-Y_A = 0.52; D-Y_A = 0.59.

We have recently extended our study of the SAN copolymerization to examine polymerizations in solution. In particular, we have examined copolymerizations in toluene, a styrene-like solvent, and acetonitrile, an acrylonitrile-like solvent. Again we have determined copolymer compositions and triad fractions using ^{1}H and ^{13}C NMR. The composition data are shown in Figure 7.

Table 2

Comparison of the diad fractions obtained from ^{13}C NMR methylene subspectra of SAN copolymers and from calculations based upon the penultimate copolymerization model, at various comonomer mole fractions, x_A, for bulk copolymerization at 60°C.

x_A	F_{AA}		F_{AS+SA}		F_{SS}	
	Exp	Calc	Exp	Calc	Exp	Calc
0.594	0.27	0.21	0.73	0.76	0.00	0.03
0.524	0.14	0.12	0.80	0.82	0.06	0.06
0.350	0.04	0.01	0.68	0.69	0.28	0.30
0.228	0.00	0.00	0.48	0.48	0.52	0.52

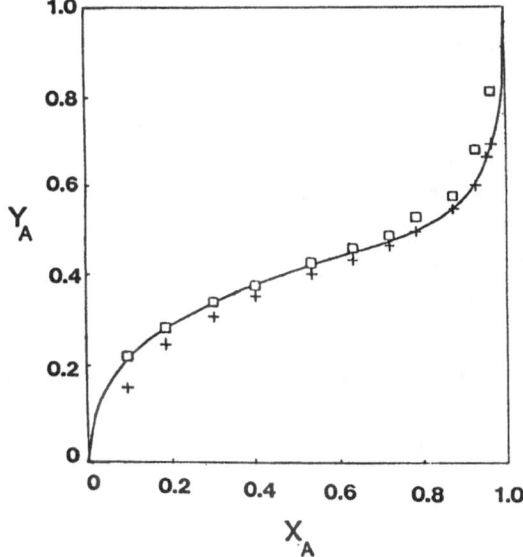

Figure 7: Copolymer composition curves for copolymerization of styrene and acrylonitrile at 60°C. Acetonitrile + ; toluene ☐ ; bulk —— .

Mathematical analysis of the copolymer composition data and the triad fraction data indicates that the copolymerizations are best described by a penultimate model. The penultimate model reactivity ratios obtained from the analysis of the copolymer compositions are given in Table 3 for each solvent.

Table 3

Penultimate model reactivity ratios calculated from analysis of copolymer compositions for copolymerizations at 60°C.

Ratio	Toluene	Acetonitrile
r_{AA}	0.28	0.06
r_{SA}	0.05	0.07
r_{SS}	0.25	0.42
r_{AS}	0.43	0.60
S_y	0.007	0.005

Although the data analyses indicate that a penultimate model best represents the polymerizations in bulk, toluene and acetonitrile, the values of the reactivity ratios vary with the nature of the solvent. This type of variation was also observed by Sandner et al. (7) for a range of different solvents. They correlated the variation with various solvent parameters and reported that the effect followed the donor number of the solvent.

The data in Figure 7 clearly show that, at low mole fractions of AN, the bulk polymerization follows closely the polymerization in toluene, whereas at high mole fractions of AN, the bulk polymerization follows closely that in acetonitrile. This observation suggests that the solvation sheath around the growing chain end plays an important role during the monomer addition step.

ACRYLONITRILE-METHACRYLIC ACID COPOLYMERIZATIONS

The ^{1}H NMR spectrum of acrylonitrile (AN) – methacrylic acid (MA) copolymers, Figure 8, is characterized by three regions: the methyl protons 0.8 – 1.5 ppm, the methylene protons 1.5 – 2.4 ppm and the methine protons 2.5 – 3.2 ppm. The copolymer composition can be obtained from the ratio of the area of the methyl (MA) or methine (AN) protons to that of the methylene protons. The methine proton resonance is split into three peaks on the basis of the (AN) triad sequences, as shown in Figure 8, thus allowing the acrylonitrile triad fractions to be calculated.

The ^{13}C NMR spectrum of an ANMA copolymer is shown in Figure 9. The acrylonitrile cyano carbons at 118 – 124 ppm are split on the basis of the AN triad sequences, while the carbonyl carbons of the

methyacrylic acid units at 174 - 180 ppm are split on the basis of
the MA triad sequences. The triad fractions for the AN and MA
sequences can be calculated from these resonance splittings.

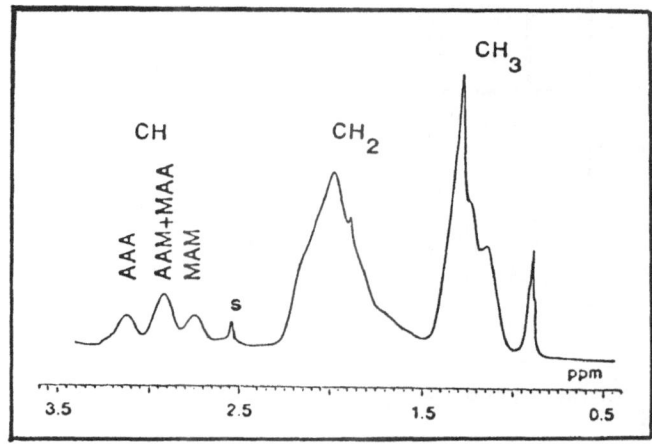

Figure 8: ¹H NMR spectrum of poly(methacrylic acid-co-acrylonitrile).

The aliphatic carbon region of the ¹³C NMR spectrum of ANMA
copolymers is characterized by a very large number of splittings of
the methyl, methylene, methine and quaternary carbon resonances.
Generation of the DEPT subspectra, Figure 10, greatly simplifies
assignment of these resonances, but incomplete cancellation of some
resonance peaks in the development of the subspectra makes a
definitive assignment impossible. This difficulty can be overcome by
the application of 2D heteronuclear chemical shift correlation
spectroscopy.

The 2D-HETCOR spectra of the methyl-methine region, Figure 11, and of
the methylene-quaternary region, Figure 12, clearly show the extent
to which these carbon resonances overlap. This allows the various
carbon resonances to be assigned, as shown in Figures 11 and 12. The
methyl carbon resonances in Figure 11 are split on the basis of the
MA triad sequences and the methine carbon resonances are split on the
basis of the AN triad sequences. The methylene carbon resonances
shown in Figure 12 are split on the basis of the tetrad sequences for
the AA, AM and MM diads.

246

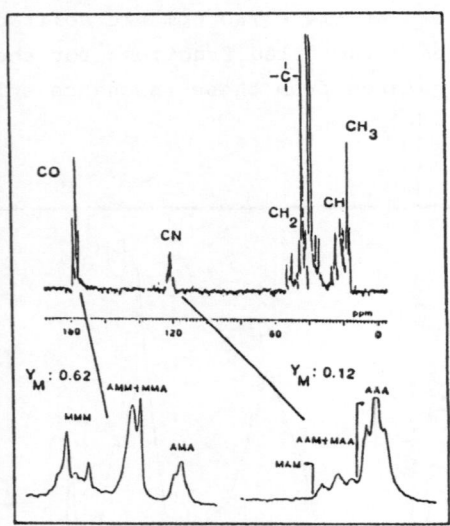

Figure 9: ^{13}C NMR spectrum of poly(methacrylic acid-co-acrylonitrile).

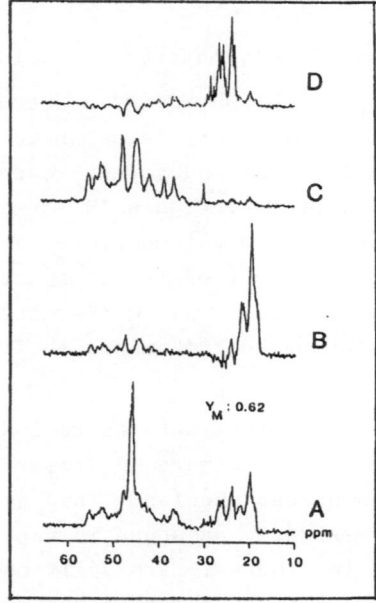

Figure 10: DEPT subspectra of poly(methacrylic acid-co-acrylonitrile). A = normal ^{13}C NMR spectrum; B = methyl subspectrum; C = methylene subspectrum; D = methine subspectrum.

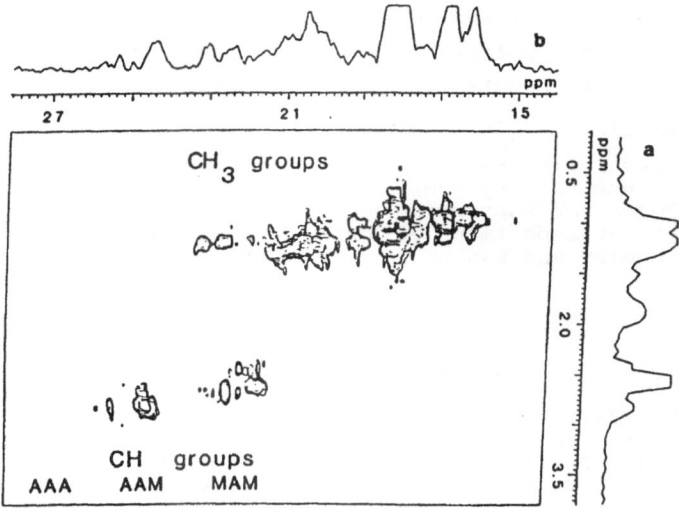

Figure 11: 2D-NMR HETCOR spectrum of poly(methacrylic acid-co-acrylonitrile) for the methyl and methine region. a = ^{1}H NMR; b = ^{13}C NMR.

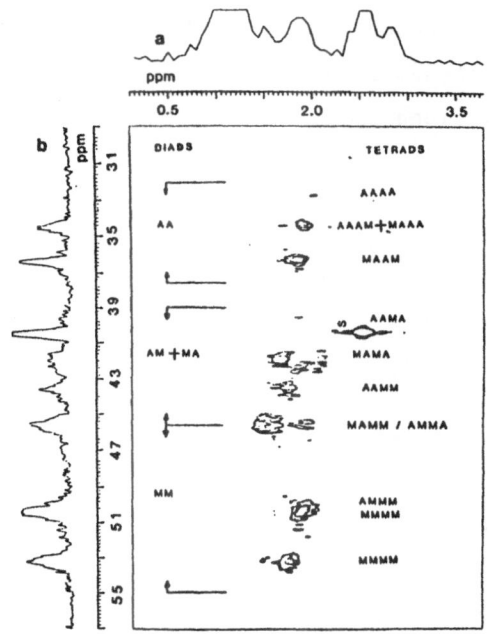

Figure 12: 2-D NMR HETCOR spectrum of poly(methacrylic acid-co-acrylonitrile) for the methylene region. a = ^{1}H NMR; b = ^{13}C NMR.

Mathematical analysis of the composition or triad fraction data
indicates that this copolymerization can be adequately described by a
terminal model, with reactivity ratios $r_A = 0.19$ and $r_M = 3.7$.

Acknowledgements:

We thank the Australian Research Committee, the Australian Institute
of Nuclear Science and Engineering and the Industrial Research and
Development Board for supporting our research. Dr. J.D. Borbely was
a visitor from the Department of Applied Chemistry, Kossuth L.
University, Hungary and was supported by the University of
Queensland.

References:

1 Hill DJT, O'Donnell JH (1987) Macromol Chem, Macromol Symp
   10/11:375
2 Tonelli AE (1989) NMR Spectroscopy and Polymer Microstructure,
   VCH Publishers, New York
3 Pichot C, Pham QT (1979) Makromol Chem 180:2359
4 Arita K, Ohtomo T, Tsurumi Y (1981) J Polym Sci, Polym Lett Ed
   19:211
5 Ham GE (1954) J Polym Sci 14:87
6 Guyot A, Guillot J (1968) J Macromol Sci Chem A2:889
7 Sandner B, Kraemer S, An BT (1979) Acta Polym 30:265
8 Kucher RV, Zaitsev YuS, Bondarenko AV, Zaitseva V (1979) Theor
   Exp Chem 14:631
9 Hill DJT, O'Donnell JH, O'Sullivan PW (1982) Macromolecules
   15:960
10 Barron PF, Hill DJT, O'Donnell JH, O'Sullivan PW (1984)
   Macromolecules 17:1967
11 Hill DJT, Lang AP, O'Donnell JH, O'Sullivan PW (1989) Eur
   Polym J 25:911
12 Doddrell DM, Pegg DT (1980) J Am Chem Soc 102:6388

Methacrylate Star Polymers via Group Transfer Polymerization

Harry J. Spinelli

Polymer Products Department
E. I. DuPont de Nemours

Star polymers are highly branched polymers with linear arms radiating out from a central core. Methacrylate stars can be easily made by using Group Transfer Polymerization technology. They are made by synthesizing blocks of monofunctional (e.g. MMA) and polyfunctional methacrylates (e.g. ethyleneglycol dimethacrylate EGDM). The monofunctional monomer forms the arm of the star. The polyfunctional monomer forms a highly crosslinked core. The polyfunctional monomer is added at a higher level than the moles of initiator or living ends. A variety of compositions and sizes can be obtained by controlling both the process and the raw materials used to make them.

BACKGROUND

Star polymers may be considered to be highly branched polymers that have linear chains radiating out from a central area. This area may be one atom, a small molecule, or a "core". The "core" is a quasi-spherical structure as opposed to a linear structure that would be present in a conventional comb or branched polymer. An early example of stars made from small molecules is the star polymer of Schaefgren and Flory (1) who polymerized E-caprolactam in the presence of a tetrafunctional or octafunctional carboxylic acid to produce polymers that have 4 or 8 arms radiating out from a central molecule. Other examples use the coupling of "living" anionically polymerized polystyrene with silicone tetrachloride (2) or chloromethyl-benzene (3). Recent work in this area includes that of Fetters (4) who has made 12 and 18 arm stars with this general technique.

A second general process to making stars involves the addition of polyfunctional monomers to active or living polymers. This has been done by adding divinyl benzene to living polystyrene (5). However this process limits one to a maximum of twenty arms (6). The reverse process, the addition of styrene to a living core of divinyl benzene, allows one to put more arms on the star. However it is necessary to run the core forming step under dilute conditions (2.5%) in order to avoid gelation.

Both of these general anionic processes have been tried with methacrylate monomers. Andrews and Sharkey (7) prepared acrylic star polymers having three or four arms by coupling anionic prepared living polymethyl methacrylate with bromomethyl benzenes. This procedure is limited by the number of possible branch points present in the coupling agent. A multiarmed star was prepared by the addition of ethyleneglycol dimethacrylate to "living" PMMA (8). However a broad dispersity of star sizes and number of arms resulted. In addition it was unclear how many arms really attached to the star.

The problem with anionic polymerizations of methacrylates is the low temperatures (-78°C) required to maintain some level of livingness. Even at these low temperatures

B. C. Anderson · Y. Imanishi (Eds.)
Progress in Pacific Polymer Science
© Springer-Verlag Berlin Heidelberg 1991

methacrylates have a limited lifetime and large amounts of unattached arms result. Webster (9) suggested that, thru Group Transfer Polymerization, three and four arm stars can be made by either reacting living GTP methacrylates with coupling agents or by using a 3 or 4 functional initiator.

STAR FORMING PROCESSES

A conventional branched polymer results when a dimethacrylate is copolymerized in a random fashion with monofunctional monomers. However it is necessary to keep the dimethacrylate/initiator ratio less than 1/1 in order to prevent gelation. When the dimethacrylate/initiator ratio equals or exceeds a 1/1 ratio, there should be an infinite number of branches with each chain connected to at least two other chains, resulting in a gel. Table 1 shows that gels result when a 2/1 dimethacrylate/initiator ratio is used and the dimethacrylate is randomly copolymerized with a monofunctional monomer. This is true whether the polymerization is free radical or GTP.

Table 1. Effect of Dimethacrylate on Polymer Structure

DIMETH./INIT.	PROCESS	COMPOSITION	RESULTS
2/1	FREE RADICAL	MMA/EGDM COPOLYMER	GEL
2/1	GTP	MMA//EGDM	STAR PARTICLES
2/1	GTP	EGDM//MMA	STAR PARTICLES
8/1	GTP	EGDM//MMA	STAR PARTICLES

If however the same 2/1 molar ratio is used but all of the dimethacrylate is located at one end of the polymer chain, the reaction does not gel. Instead very small particles (stars) form. This is true whether the dimethacrylate is polymerized first or last. By polymerizing the dimethacrylate in a block, difunctional to initiator ratios that exceed 4/1 can be used and no gel will form.

The reason why stars do not gel when the arms are formed first is that as the dimethacrylate is added to the living linear polymers (the arms), it starts to react with the living ends. Since there are two functional sites on the dimethacrylate, arms can start to couple. The addition of more dimethacrylate joins coupled arms and the structure starts to grow in size as more and more coupled arms are bound together. However at some point there will soon be a condition where it is sterically difficulty to interact one growing branched system with another. Therefore excess dimethacrylate will only be able to enter and react into the "core" of the star. Cores will be sterically prevented from readily reacting with each other, thus preventing gelation.

There are several processes used to make stars. These include: the arm first process, the core first process, and the arm/core/arm process. The process used controls both the size of the star and the number of arms per star, Table 2.

In the arm first process, initiator is reacted with monofunctional monomers, such as MMA, forming the arms of the stars. Polyfunctional monomers, such as ethyleneglycol dimethacrylate, are then added to react with the living ends of the arms. The dimethacrylate joins the arms together and forms the core of the star, Figure 1. Stars made from this process can be from 100,000 to several million in molecular weight.

Table 2. Star Sizes

PROCESS	MOL. WT.	D(Å)	NO. OF ARMS
ARM FIRST	4.2×10^5	321	35
CORE FIRST	5.6×10^8	2057	4400
ARM/CORE/ARM	2.3×10^6	--	200

Figure 1. Arm First Process for Star Formation

CHARGE INITIATOR

I - X

FEED IN MONOFUNCTIONAL MONOMER

I-X + MMA $\longrightarrow$ I ∿∿∿—X

FEED IN DIFUNCTIONAL MONOMER

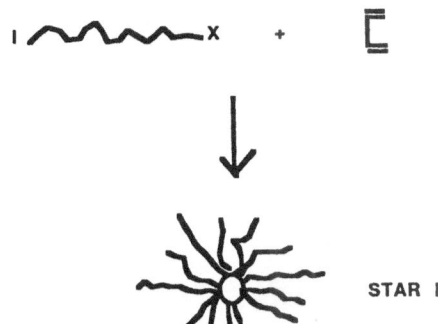

STAR POLYMER

In the core first process, initiator is first reacted with the polyfunctional methacrylate to make an active core. Monofunctional monomers are then added to grow arms out from the active core. This process, in general, makes stars that are larger than those from the arm first process and are more difficult to control. Stars made from this process can range from several hundred thousand to several billion in molecular weight.

In the arm/core/arm process, monofunctional monomer is first polymerized to make an arm, then polyfunctional monomer is added to attach the arms together and make the core, and finally more monofunctional monomer is added to grow a second group of arms out from the core. This process can be used to put two different types of arms onto the same star. The molecular weights of these stars are in between those of the other two processes.

The size and molecular weight of the star, the number of arms attached to the star, and the size of the arms can be determined by using a combination of analytical techniques. These include Quasi-Elastic Light Scattering QUELS, static light scattering, viscosity measurements, and gel permeation chromatography. With these techniques, stars were found to range in size from 200 to 7,000 Angstroms in diameter, in molecular weight from 50,000 to several billion, and to have from 10 to 10,000 arms per star, Table 2. In general, arm first stars are smaller in size and molecular weight than core first stars. Arm/core/arm stars are intermediate between the other two main processes.

ARMS

Size

With the arm first process, the size of the arms of a star can be controlled and this influences the overall size of the star. Arms of almost any size can be made before adding the core-forming polyfunctional methacrylate. The arm size influences the size, molecular weight, and number of branches of the star. With the arm first process, stars made with shorter arms tend to be larger, Table 3. A star with arms of Mw = 8,000 will have a molecular weight 100 times greater than a star with an arm Mw = 27,900. Short arm stars tend to have more arms per star, be larger in size, and have a greater molecular weight than long arm stars. A possible reason for this is the longer arms provide more steric hindrance and minimize the coupling of active, growing clusters, thus minimizing the size of the star.

Table 3. Effect of Arm Length on Star Size for an Arm First Process

ARM Mw	STAR MOL. WT.	STAR D (Å)	NO. OF ARMS
8,000	3.3×10^7	1000	4120
16,600	3.0×10^6	387	185
27,900	4.3×10^5	155	15

With the core first process, the theoretical size of the arm does not have a large influence on star size and weight, Table 4. The size of these stars seem to be controlled by core size, composition and solids.

Table 4. Effect of Arm Size on Star Size for both Arm First and Core First Processes

ARM THEO. Mn	PROCESS	STAR DIAM. (A)	STAR MOL. WT. (QUELS)	NO. OF ARMS
5,000	CORE	4380	$5.8 * 10^9$	730,000
	ARM	1000	$3.3 * 10^7$	4,100
10,000	CORE	3600	$1.8 * 10^9$	109,000
	ARM	387	$3.0 * 10^6$	185
20,000	CORE	2900	$1.6 * 10^9$	55,600
	ARM	155	$4.3 * 10^5$	15

Composition

A variety of different types of acrylic and methacrylic star polymers have been made. The composition and Tg of the arms have ranged widely from methyl to butyl to lauryl methacrylate and acrylate, Table 5. The arms of the stars can be homopolymers, copolymers, or block polymers. Stars with two different types of arms have been made by using the arm/core/arm process in which the second monomer addition differs in composition.

Table 5. Star Compositions

Homopolymers	Tg
MMA	105
EMA	65
BMA	20
EHMA	-10
EA	-22
LMA	-65

Stars have been made with various functional groups. These include hydroxyl, acid, amine, epoxy, and reactive double bonds. The functional group can be incorporated randomly in the arm, in blocks, in the core, or at the end of the arms. The easiest way to functionalize the ends of the arms is to use a functional initiator. Of course functional groups that contain active hydrogens, such as hydroxyl and acid, need to be protected during the GTP reaction. The blocking groups can be removed after polymerization is complete.

CORES

Size

The amount of dimethacrylate used in the core forming step influences the formation of the star, Table 6. Higher levels of dimethacrylate lead to larger stars. However there is a limit to the amount that can be added at a specific solids. At 40% solids with 10,000 Mn arms, a Dp 8 core can be made but a Dp 12 core star will gel. Stars with very large cores can be made by running the star process under dilute conditions. A Dp 20 core star can be made at 10% solids. This stars has a core that is almost 30% of the total weight.

Table 6. Effect of EGDM Level on Star Size for P-MMA Arm First Star

DP EGDM*	STAR SOLIDS	STAR MN (GPC)	D
1	40	14,400	2.22
2	40	12,800	2.32
4	40	296,000	1.67
8	40	400,000	5.52
12	40	GEL	
10	30	266,000	2.05
10	10	~ 400,000	
20	30	GEL	
20	10	~ 500,000	

*DP EGDM is the ratio of ethylene glycol dimethacrylate to initiator

255

Composition

Other polyfunctional monomers can be used in the core and these alter the crosslink densities of the cores as well as the size and properties of the stars. Some examples of these are given in Table 7. Even trifunctional methacrylates can be used.

Table 7. Comparison of Polyfunctional Monomers

Core*	Process	DH	Star Mol. Wt.	No. of Arms	Mol. Wt. Per Arm
EGDM	Core	2057	5.6×10^8	4400	125,000
	Arm	321	2.5×10^6	264	9,000
BDDM	Core	348	2.7×10^6	153	17,000
	Arm	265	1.2×10^6	167	7,000
HDDM	Core	230	7.3×10^5	124	6,000
	Arm	321	2.4×10^6	259	9,000
TEGDM	Core	244	6.7×10^5	99	7,000
	Arm	479	9.2×10^6	600	15,000
TMPTM	Arm	200	5.0×10^5	106	10,000

* EGDM = Ethyleneglycol dimethacrylate
  BDDM = Butanediol dimethacrylate
  HDDM = Hexanediol dimethacrylate
  TEGDM = Tetraethyleneglycol dimethacrylate
  TMPTM = Trimethylolpropane trimethacrylate

As the spacer in the dimethacrylate increase in size, from ethyl (EGDM) to hexyl (hexanediol dimethacrylate HDDM), the size of the star decreases in size and molecular weight, especially for the core first process. A possible reason for this is that the longer bridging section may allow for some cyclization of the two functional groups. This cyclization reduces the apparent amount of difunctional material. The cyclization process changes a difunctional material into a monofunctional monomer. It is known that lower levels of difunctional core material lead to smaller stars, Table 6.

EXPERIMENTAL

Standard Group Transfer Polymerization techniques were followed (9). All solvents and monomers were purified by running them through a column of neutral alumina and storing them over sieves. Initiators and catalyst (9,10) were used as is. The procedures to make many types of GTP star polymers are published (10,11,12). The following is representative of process to make a standard star.

A one liter flask is charged with 300 gms of THF, 3.48 gms of 1-trimethylsiloxy-1-methoxy-2-methylpropene, and 0.20 ml of a 1.0 M tetrabutylammonium chlorobenzoate. Feed I, 200 gms of methyl methacrylate, is added over 60 minutes. Twenty minutes after completion of Feed I, Feed II is started and added over 10 minutes. Feed II is 15.8 gms of ethyleneglycol dimethacrylate. Sixty minutes after completion of Feed II, the reaction is quenched with 10 gms of methanol to make a p-MMA star polymer of about 500,000 molecular weight.

SUMMARY

Star polymers are highly branched polymers with linear arms radiating out from a central core. They can be made using GTP technology by making blocks of monofunctional (e.g. MMA) and polyfunctional methacrylates (e.g. ethyleneglycol dimethacrylate EGDM). The monofunctional monomer forms the arm of the star. The polyfunctional monomer forms a highly crosslinked core.

The formation of star polymers can be done using a variety of processes. The three main processes for making all acrylic star polymer are arm first, core first, and arm-core-arm.
Factors that influence the size of a star include.

• Process - In general arm first makes small stars, arm/core/arm makes bigger stars, and core first makes the largest stars.

• Arm size - Shorter arms tend to give larger stars

• Dimethacrylate levels - Higher dimethacrylate levels give larger stars and fewer free arms

• Type of dimethacrylate - The type of dimethacrylate used affects star size and free arm content. Long chain dimethacrylates may be undergoing an internal cyclization that reduces their efficiency.

REFERENCES

1 Schaegren JR, Flory PJ (1948) J Am Chem Soc 70, 2709

2 Morton M, Helminiak T E, Gadkary S D, Bueche F (1962) J Polym Sci 57, 471

3 Wenger F, Yen, S P S, (1962) Polymer Reprints ACS 3 (1), 162

4 Fetters L J, et al (1986) Macromolecules 19, 215

5 Decker D, Rempe P, (1965) C R Acad Sci 261, 1977

6 Eschley H, Hallensleben M L, Burchand W, (1973) Makromolekulare Chemie, 173, 235

7 Andrews G W, Sharkey W H, US Patent 4,351,924

8 Zilliox J G, Rempp P, Parrol J (1968) J Polymer Sci, Part C No 22, 45

9 Webster, US Patent 4,711,942

10 Spinelli, US Patent 4,659,782

11 Spinelli HJ, US Patent 4,659,783

12 Spinelli HJ, US Patent 4,810,756

REFERENCES

Hydrogen Bonding in Aromatic Polyamides

R. A. Gaudiana* and R. F. Sinta

Polaroid Corporation, 750M-5C
Cambridge, MA 02139

Abstract : Rigid, rodlike polyamides, polyesters and polyesteramides
characteristically exhibit a high degree of intermolecular association resulting
in crystalline and liquid crystalline materials with high thermal transitions
and moderate to poor solubility. Certain structural modifications can diminish
intermolecular attractive forces which leads to nearly complete elimination of
highly-correlated structures. Many of these materials exhibit unique spectral,
thermal, optical and morphological properties, and several are very soluble in
common solvents but do not form lyotropic solutions. The specific structural
modification primarily responsible for this unusual combination of properties
are the 2,2'-disubstituted-4,4'-biphenylene diacid, diol or diamine
comonomers. This substitution pattern forces non-coplanarity between the
phenyl rings while maintaining the rodlike conformation of the backbone. This
paper describes an investigation of the effect of orientation on hydrogen
bonding in polyamides containing a non-coplanar biphenyl diamine in the
backbone. The results help substantiate the notion that weak intermolecular
associations are primarily responsible for many of the properties exhibited by
these polymers.

INTRODUCTION

Our interest in amorphous polymers exhibiting a unique set of optical
properties including high refractive index and birefringence was originally
described in a series of patents on the subject of non-absorbing polarizers and
filters [1-3]. The utility of these devices was limited by the availability of
polymeric materials exhibiting the high birefringence ($\Delta n \geq 0.4$) required to
maximize polarization efficiency.

The development of processable, highly oriented polyamides [4-6], polyesters
[7-10] and esteramides which began in the early seventies has continued over
the last twenty years. The combination of high inherent polarizability, rodlike
or semi-rodlike geometry, and high orientability of these materials generated
unusually high birefringence ($\Delta n \geq 0.7$ [11]). Unfortunately, the crystallinity,
biaxial optics, and strawlike color (polyamides) of these polymers preclude
their use for most optical applications. The goals of our investigation were the

B. C. Anderson · Y. Imanishi (Eds.)
Progress in Pacific Polymer Science
© Springer-Verlag Berlin Heidelberg 1991

design and synthesis of modified, rodlike polymers that could be processed into highly oriented, optically uniaxial, colorless, transparent (noncrystalline) films and fibers suitable for optical applications. Most of these goals were realized by the incorporation of 2,2'-disubstituted biphenyls into the polymer repeat units.

Many optical applications require optical elements which are completely non-absorbing over a large wavelength range. Furthermore, in devices where precise control of the refractive index is improtant over a broad range of wavelengths, it is essential that the absorption band of each optical element, including the polymer, occur at a wavelength as remote as possible from the wavelength being utilized, thereby minimizing dispersion effects. Most aromatic polyamides, such as poly(p-phenylene)- terephthalamide are yellow in solution and generate pale-yellow films and fibers [4-6]. This absorption is due to the very strong double character of the amide C-N bond (rotational barrier ≈20Kcal/mol [12]) which extends the conjugation between adjacent phenyl/amide groups over several repeat units. We have domonstrated that the incorporation of sterically bulky groups in the 2,2'-positions of biphenylenes generates completely colorless polymers [13]. The size and electronic nature of the substituents directly affect the rotational barriers around the 1,1'-bond of the biphenyl and diminish the extent of multi-repeat unit conjugation by limiting p-orbital overlap of the 1,1'-carbon atoms. The non-coplanar biphenyls effectively limit the conjugation to a phenyl/amide/phenyl chromophore which has no absorption beyond 400nm.

In order to realize the inherently high birefringence of aromatic polyamides, they must be oriented by solution processing, in many instances from corrosive or expensive solvents, because they do not exhibit thermal transitions which allow for thermal processing. Although polyesters and several examples of polyesteramides have limited solubility in less corrosive solvents, their major advantage derives from their thermotropic behavior which allows them to be melt processed. Unfortunately, para-linked, completely aromatic polyesters, such as poly(p-hydroxy-benzoate), require processing temperatures in excess of 400°C [14]. Processing temperatures in this range are particularly inconvenient because they require significant modifications to standard processing equipment, and the polymers partially degrade when held at elevated temperatures for an extended period.

Subsequent development of the thermotropic polyester field centered around molecular structural modifications which would allow lower processing temperatures while maintaining rigidity. While several very successful strategies were devised by various research groups, the common theme of all of them is the disruption of chain packing. One of the most successful of these stratagies is the incorporation of a 2,6-dicarboxy or hydroxy naphthalene moeity. Since the these positions are colinear, the monomer does not disrupt the parallelism of connecting bonds but merely offsets the chain axis of the bonds relative to one another thereby disrupting chain packing [9]. The crystalline and liquid crystalline transition temperatures of the resulting polymers are lowered by as much as 175°C. Another approach is the incorporation of aromatic diols and diacids substituted with alkyl or alkoxy groups, halogens or phenyl groups [15-24]. Some of these examples exhibit crystalline/nematic transition temperatures as low as 170-200°C. Unfortunately, these polymers remain highly crystalline and therefore highly scattering. However, we have synthesized several high molecular weight polyesters containing disubstituted biphenyls which are noncrystalline and have glass to liquid crystal transitions in the 150-200°C range [25, 26]; the trifluoromethyl group is a particularly effective biphenyl substituent. This combination has been shown to be effective in the aromatic polyesteramides as well. Typically, thermotropic transitions in this class of polymers is usually in the 300-400°C, but in biphenylene-containing polymers the transitions drop to 250-350°C [27].

Polymers comprising the disubstituted biphenyls also exhibit unusual solution and solubility properties. For example, highly concentrated solutions of high molecular weight samples of all three classes of polymer show no evidence of lyotropism. As the concentration is increased the viscosity increases up to the limit of solubility at which point the polymer precipitates instead of forming a biphasic solution. The solution properties of several of these polymers have been examined in great detail, and in every example the criteria for lytropism, e. g., persistence length, solubility, etc., has been met or exceeded.

High solubility is one of the unusual features of nearly all of these 2,2'-disubstituted polymers. For example, unlike typical wholly-aromatic polyamides, most of these polymers dissolve rapidly in LiCl-containing amide solvents, and most are soluble in these solvents without salt. Even more surprisingly, several of these polymers are soluble in THF, glyme, and/or acetone up to 40 - 50% (wt/v) [28]. Similarly wholly-aromatic polyesters are insoluble or exhibit very limited solubility in chlorinated phenols and cresols. However, the 2,2'-disubstituted biphenyl polyesters are very soluble in a variety of solvents such as methylene chloride, chloroform, thtrachloroethane, and THF as well as phenols. Several of these polyesters are soluble to as much as 50% (w/v) in THF [26] and yet we have not been able to detect any lyotropic phase. The polyesteramides also exhibit excellent solubility relative to their non-biphenyl-containing analogues [27].

In regard to morphology, the great majority of non-biphenyl-containing polymers in these classes of polymers are highly crystalline, and as a result films and fibers are highly scattering. The biphenyl-containing polymers on the other hand exhibit unique morphology. For example, wide-angle X-ray analysis of oriented and unoriented samples of polyamide films indicates that crystallinity is at most 10% [29]. X-ray diffraction of highly oriented films of all three classes of polymers generates broad, diffuse arcs and broad, nondescript densitometer peaks typical of nematic liquid crystals [30]. These films are colorless and almost completely non-scattering. Light scattering indicates that microdomains are on the order of 1000Å [31].

There are several factors which affect morphology and solubility in these polymers : permanent and induced dipoles and polarizability (van der Waala forces); geometric regularity such as positional isomerism, comonomer length, and coaxiality; the size and electronic nature of substituents. Furthermore, we have demonstrated that the almost completely ignored asymmetry of the non-coplanar biphenyls, i. e., atropism, introduces another very significant and effective level of irregularity into the polymers. We believe that the degree of hydrogen bonding in the biphenyl amides and esteramides also plays a key role in determining the unique properties of these materials and is itself directly affected by the presence of these groups.

ORIENTATIONAL EFFECTS

In general, one would expect to observe little or no change in the relative amounts of free and NH bonded species in amide-containing polymers on orientation unless alignment induced crystallization. Manipulations such as extrusion, stretching and annealing frequently increase crystallinity and enhance crystalline perfection and as a result generate a high degree of hydrogen bonding. Aromatic polyamides such as poly(p-phenylene)tere-phthalamide are highly hydrogen bonded in the crystalline state [32] and aliphatic polyamides owe much of their strength to interchain hydrogen bonding. A recent report convincingly argues that thermotropic liquid crystallinity in a series of polyesteramides is strongly dependent on hydrogen bonding [33]. On the other hand, analysis of the NH and CO stretching regions of polyether urethanes, by Seymour, et al. [34], clearly showed that orientation, up to 300% elongation, had very little effect on the relative amounts of hydrogen bonded species.

Although our interest was primarily focused on the dichroism of the carbonyl region, it was observed that IR spectra of oriented films in the NH stretching region of all of the polymers studied to date gave unique and very unexpected results.

The polyamide used for this study exhibits well seperated free and bonded NH peaks which makes it a particularly good candidate for studying this anomolous behavior. Figure 1 shows the NH stretching region of this polymer at two

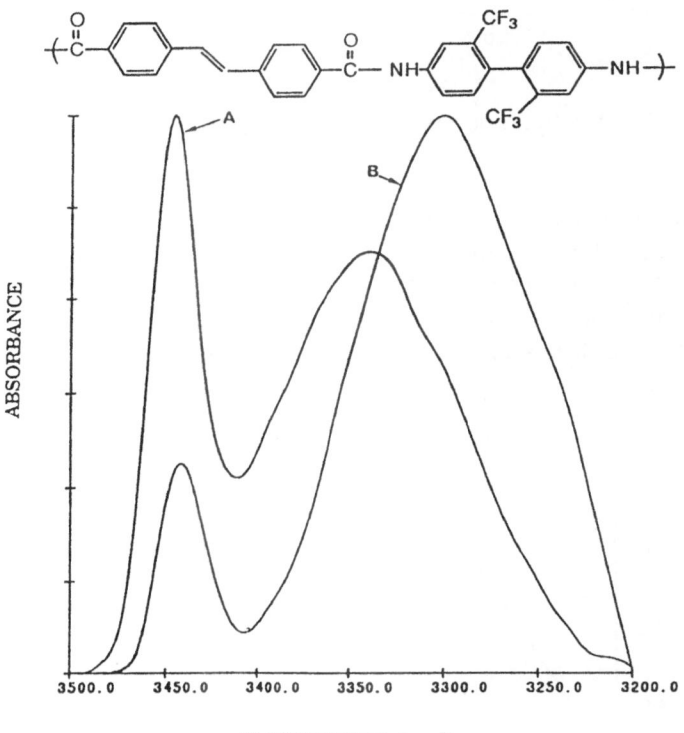

Figure 1. The effect of orientation on the NH stretching vibration (curve A - 77% oriented; curve B - unoriented).

different orientations, 77.0% for curve A and 0% for B. The most striking feature about these spectra is the significant increase in the size of the free NH peak and a concomittant decrease in the bonded peak on orientation. It is also interesting to note that the NH bonded peak shifts to higher frequency on orientation. A plot of peak position versus orientation (Figure 2) begins to deviate significantly from linearity at approximately 40%.

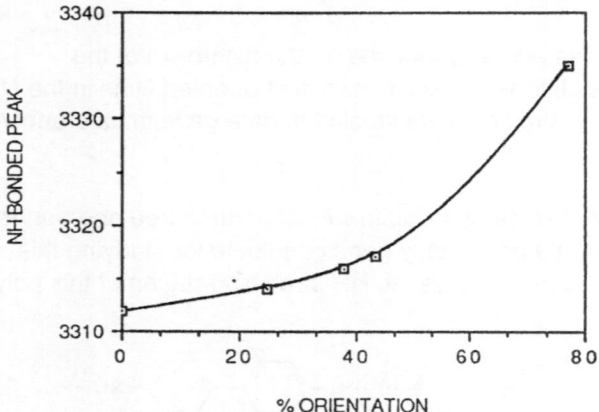

Figure 2. The shift in the hydrogen bonded stretching vibration as a function of orientation.

Calculation of the percent of free amide species via deconvolution of the amide carbonyl band for five film samples from 0% to 77.0% orientation is shown in Figure 3. The data clearly shows that 42% of the amide groups are not hydrogen bonded in the unoriented sample, and as orientation increases to 43%, the percent of free amide carbonyls increases almost linearly to 80.6% free.

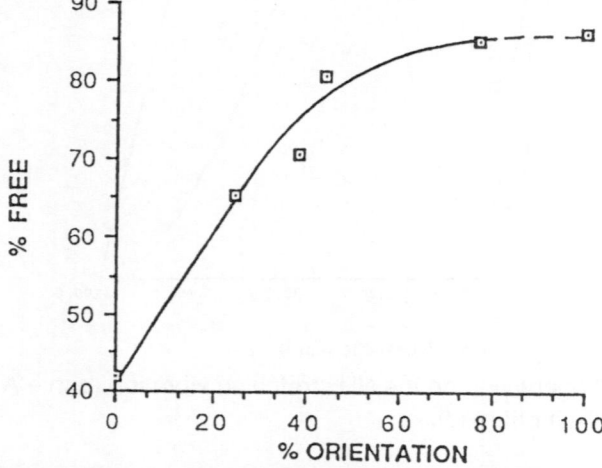

Figure 3. The population of free amide groups as a function of orientation.

Beyond 43% orientation, the population of free species levels off and reaches a maximum extrapolated value of 86%. The hypsochromic shift in the position of the NH bonded peak as strain increases (Figures 1 and 2) indicates that orientation breaks weaker hydrogen bonds preferentially.

Our interpretation of these observations is shown schematically in Figures 4 and 5. Before stretching, when the rodlike molecules are randomly oriented, hydrogen bonding may fortuitously occur at points of intersection (Figure 4), and therefore the polymers should contain a significant percentage of hydrogen

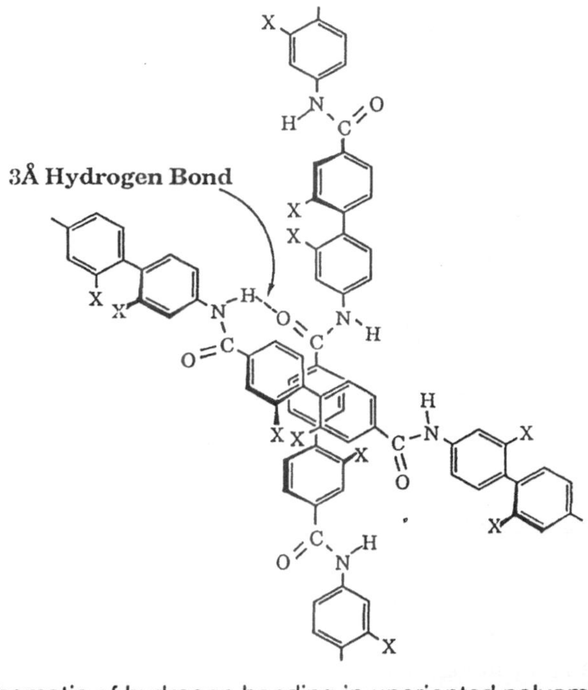

Figure 4. A schematic of hydrogen bonding in unoriented polyamides.

bonded NH species with a lesser amount of free NH species. When the sample is stretched, the rodlike molecules reorient by rotating and slipping past each other. Even at low levels of orientation interchain hydrogen bonds are irreversibly broken and converted to free NH species, although a few may slip into register with new amide groups. However, as orientation proceeds to higher levels and the rodlike molecules approach parallelism, the population of intersecting points significantly decreases. Because the molecular diameter of the rod is so large (8.2Å from spacing filling models), as parallelism is attained the interchain distance between a carbonyl oxygen atom and an amide hydrogen reaches a minimum of 7Å (Figure 5a) which is much too large for hydrogen bonding to occur (preferred distance = 3Å). X-ray analysis of highly crystalline fibers of poly(p-phenylene)-terephthalamide indicates that the phenyl rings can rotate out of the plane of the amides by ~30° which allows the chains to attain the optimum distance for hydrogen bonding (Figure 5b) [33]. In the biphenylene-containing polymers, when one of the phenyl rings becomes noncoplanar with an adjacent amide group, the second phenyl is forced to rotate in unison; the result is that interchain steric effects cannot be alleviated by rotation, and hydrogen bonding is adversely affected.

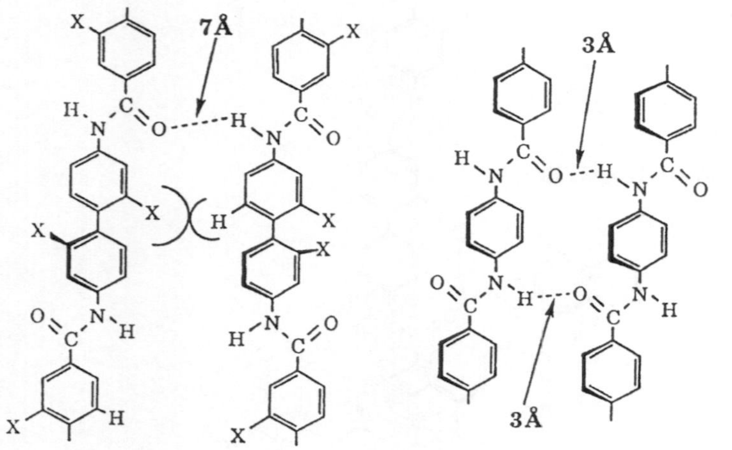

Weak Hydrogen Bonding   Strong Hydrogen Bonding

(a)   (b)

Figure 5. A schematic of hydrogen bonding in completely oriented polyamides : a) 2,2'-disubstituted polyamide; b) crystalline poly(p-phenylene)tere-phthalamide.

CONCLUSION

The incorporation of the 2'2'-disubstituted biphenyl moiety into the polymer backbone of wholly-aromatic polyamides, polyesters, and polyesteramides significantly affects their optical, morphological, and solution properties. Experimental evidence strongly suggests that intermolecular attractive forces, such as hydrogen bonding in polyamides, are significantly weakened in these materials.

References:

1   Rogers HG (1965) US Patent 3,213,753
2   Land EH (1970) US Patent 3,506,333
3   Rogers HG (1971) US Patent 3,610,729
4   Kwolek SL (1972) US Patent 3,671,542
5   Morgan P (1974) US Patent 3,801,528
6   Blades H (1975) US Patent 3,869,429
7   Roviello A, Sirigu A (1975) J Polym Sci, Polym Letters 13:455
8   (a) Kuhfuss HF, Jackson WJ (1973) US Patent 3,778,410
     (b) ibid (1974) US Patent 3,804,805
9   (a) Calundan GW (1978) US Patent 4,067,852
     (b) Calundan GW, Jaffe M (1982) Anistropic polymers, their synthesis and
      properties, The Robert A. Welch Conferences on Chemical Research XXVI,
      Synthetic Polymers, pp 247-291
10 (a) Jackson WJ (1982) Contemporary topics in polymer science, Plenum
      Press, New York NY, pp 177-208
     (b) Lenz RW (1985) Polymer J 17:105
     (c) Krigbaum WR, Hakemi H, Kotek R (1985) Macromolecules 18:965
     (d) Ballauff M (1986) Makromol Chem Rapid Commun 7:407
     (e) Hutchings DA, Sieloff GM, Lee DM, Willard GF (1986) US Patent
      4,614,790
     (f) Jin J-I, Choi E-J, Jo B-W (1987) Macromolecules 20:934
11 (a) Hamza AA, Sikorski J (1978) J microsc, Oxford 113:15
     (b) Hamza AA, El-Farahaty KA (1986) Textile Res J 56:580
12 Robin MB, Bovey FA, Basch H (1970) The chemistry of amides, Zabichy
     J (ed) Wiley-Interscience, New York pp 7, 27, 34, 46
13 Rogers HG, Gaudiana RA, McGowan C (1985) J Polym Sci, Polym Chem
     23:2669
14 Economy J, Storm RS, Matkovich SG, Cottis SG, Nowak BE (1976) J
     Polym Sci, Polym Chem 14:2207
15 Blackwell J, Biswas A (1986) Macromol Chem Macromol Symp 2:21
16 Kwolek SL (1971) US Patent 3,600,350 A polymer composed of 60 mol %
     HBA, 20 mol % TA and 20 mol % HQ exhibits at $T_m \approx 425°C$
17 (a) Jackson WJ Jr, Kufuss HF (1980) US Patent 4,238,600
     (b) Jackson WJ Jr (1984) Contemp Top Polym Sci 5:117
18 Jackson WJ Jr (1980) Br. Polym J 12:153
19 Kwolek SL, Luise RR (1975) Macromolecules 19:1789
20 Kyotani M, Kanetsuna H (1986) Kobunshe Robunshu 43:43
21 Majnusz J, Catala JM, Lenz RW (1983) Eur Polym J 19:1043
22 Dicke HR, Lenz RW, (1983) J Polym Polym Sci Chem 21:2581
23 Hutchings DA, Sieloff GM, Lee DM, Willard GF, (1986) US Patents
     4,614,790 and 4,614,791

24 See reference 12 (d) and Ballauff M and Schmidt GF (1987) Liq Cryst 147:163

25 Sinta R, Minns RA, Gaudiana RA, Rogers HG (1987) J Polym Sci, Polym Lett 25:11

26 Sinta R, Gaudiana RA, Minns RA, Rogers HG (1987) Macromolecules 20:2374

27 Sinta R, Gaudiana RA, Rogers HG, (1989) J Macromol Sci, Chem A26:773

28 Gaudiana RA, Minns RA, Rogers HG, Sinta R, Taylor LD, Kalyanaraman P, McGowan C (1987) J Polym Sci, Polym Chem 25:1249

29 Rogers HG, Gaudiana RA, Hollinsed WC, Kalyamaraman PS, Manello JS, McGowan C, Minns RA, Sahatjian R (1985) 18:1058

30 Sinta RF, Gaudiana RA, (1988) Proc Mat's Res Soc 14.3:282

31 Hsiao BS, Rojstaczer S, Stein RS, Weeks N, Gaudiana RA, (1988) Proc Mat's Res Soc 14.3:281

32 Herlinger H, Knoell H, Manzel H, Schleafer J (1973) J Appl Polym Sci, Polym Sympos 21:215

33 Aharoni SM, (1988) Macromolecules 21:1941

34 Seymour RW, Estes GM, Cooper SL (1970) Macromolecules 3:579

The reason why I take up these polymers in this review is that (1) CB-BR is a cure type solid rubber carrying OH groups and Cl atoms as active species besides double bonds, (2) viologen elastomers are thermoplastic elastomers (TPE) derived from liquid rubber, indicating some characteristic behaviors, i.e., high tensile properties and special functions, and (3) ABA type liquid crystalline elastomers are also new TPE at room temperature, but behave as liquid rubber at high temperature.

2. 1-CHLORO-1,3-BUTADIENE—1,3-BUTADIENE RUBBER (CB-BR)

1-Chloro-1,3-butadiene (CB) obtained as a by-product in the course of chloroprene production is copolymerizable with 1,3-butadiene by emulsion process and affords a novel rubber, CB-BR, having chlorine atoms attached to 1,4-butadiene units.[1,2]

In Table 1 is shown its representative structure. These moieties in CB-BR enable us to perform a variety of chemical reactions with vulcanizers of new types, reactive softeners, reactive antioxidants, and reactive reinforcing fillers like silica or lignin.[3-5]

Here is concerned with development of lignin-reinforced CB-BR and moisture curing of CB-BR.

Table 1   Structure of CB-BR

$-(Bd)_k(CH-CH=CH-CH_2)_l(CH-CH=CH-CH_2)_m(CH_2-CH)_n-$
$\quad\quad\quad\quad Cl \quad\quad\quad\quad\quad OH \quad\quad\quad\quad\quad\quad\quad CH=CHCl$
k=97.6 l=0.6 m=0.9 n=0.9

[OH]×10⁴ (mol/g)	1.64
[Cl]active ×10⁴ (mol/g)	1.06
[Cl]total×10⁴ (mol/g)	2.86
Mooney Viscosity (ML₁₊₄,100°C)	46

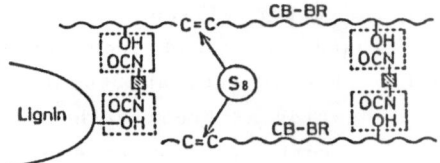

Fig. 1   Schematic curing mechanism of novel rubber compound.

2.1   MODIFIED LIGNIN-REINFORCED CB-BR[6,7]

A large amount of fuel oil and a great quantity of electric energy have to be consumed when carbon black is manufactured[8]. From the points of saving energy and an effective use of natural resource, much attention has been attracted to utilizing lignin from the pulp wastes as a substitute for carbon black[9-11].

A rubber/lignin compound, which is usually given by coprecipitated from a rubber latex, shows a large strength and a high heat resistance in its vulcanizate. However, the rubber/lignin compound has a big problem for a practical use, that is, an extremely high viscosity or a low processability in the unvulcanized state.

Synthesis and Properties of Some Characteristic Elastomers

S. Yamashita

Fuculty of Engineering and Design,
Kyoto Instisute of Technology,
Matsugasaki, Sakyo,
Kyoto 606, JAPAN

Abstract: Along our studies, three sorts of characteristic elastomers are described in this review focusing on their synthesis, structure and properties.
(1) A novel moisture-curable rubber and a high performance lignin-reinforced rubber with good processability were obtained by using newly developed 1-chloro-1,3-butadiene—1,3-butadiene rubber (CB-BR) derived from the copolymerization of 1-chloro-1,3-butadiene (CB) and 1,3-butadiene by emulsion process.
(2) The ionene-type viologen elastomers having poly(butadiene-co-acrylonitrile) (NBV) were obtained by the reaction of α,α'-dichloro-p-xylene (DCX), 4,4'-bipyridine (BP) and dimethylamine-terminated liquid nitrile rubber (AT-NBR). The tensile properties of these elastomers were found to depend upon the number of cation-site in a structural unit of the ionene, and to attain 46 MPa in tensile strength and 340% in elongation at break without any reinforcement and vulcanization.
(3) Two sorts of ABA type liquid crystalline elastomers terminated with mesogenic groups were synthesized. The polymeric segment of B is polyolefin with molecular weight of about 2500. Mesogenic groups of A carring two or three phenylene groups combined by ester linkage, were prepared by the reaction of terephthalic acid and p-hydroxybenzoic acid. These polymers were liquid crystalline and could be formed into elastic films. Their thermal and mechanical properties were clarified.

1. INTRODUCTION

High performance and novel functionality in diene or olefin rubber could be obtained by various chemical reactions through suitable functional groups attached to their polymeric chains. Along our studies, this review describes 1-chloro-1,3-butadiene—1,3-butadiene rubber (CB-BR), viologen elastomers and ABA type liquid crystalline elastomers.

B. C. Anderson · Y. Imanishi (Eds.)
Progress in Pacific Polymer Science
© Springer-Verlag Berlin Heidelberg 1991

To improve the processability of the rubber/lignin compound without losing the good mechanical and heat-resistant properties, we used CB-BR, which has hydroxyl groups to make chemical bond with lignin (Fig.1), and a chemically modified lignin whose phenolic hydroxyl groups were converted into aliphatic hydroxides (Fig.2).

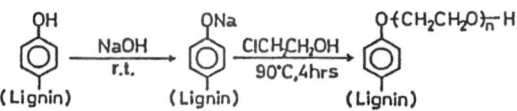

Fig. 2 Reaction scheme of lignin and ethylene chlorohydrin.

The preparation method of the modified lignin-reinforced CB-BR vulcanizate is schematically shown in Fig.3. The compounding recipes illustrated in Table 2 were used for preparing five sorts of CB-BR vulcanizates, i.e., the modified or unmodified lignin (ML or L)-reinforced CB-BR cured by sulfur with or without a blocked triisocyanate (designated as IS or S, respectively) and HAF black-loaded CB-BR cured by sulfur (sample C). This sample C was used as a control one.

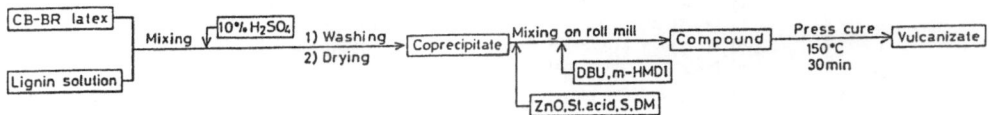

Fig. 3 Preparation of CB-BR/lignin vulcanizate.

Table 3 shows the Wallace plasticities of these compounds and the tensile properties of the vulcanizates cured at the optimum conditions. From this table, the sample IS-ML showed the best tensile

Table 2	Compounding	recipes(phr)	
No.	IS	S	C
CB-BR[a]	100	100	100
Lignin[b]	60	60	——
Carbon black	——	——	45
ZnO	5.0	5.0	5.0
Sulfur	2.2	2.2	2.2
Stearic acid	1.0	1.0	1.0
DM[c]	1.7	1.7	1.7
m-HMDI[c]	26.3	——	——
DBU[d]	0.3	——	——
Process oil	——	——	5.0

a) [OH]=1.64×10⁻⁴ mol/g

a) $[OH]=1.64 \times 10^{-4}$ mol/g
b) Phenolic $[OH]=9.27 \times 10^{-4}$ mol/g
c) $[NCO]/([OH]_{rubber} \cdot phenolic [OH])=1$
c) m-HMDI (Blocked Isocyanate)

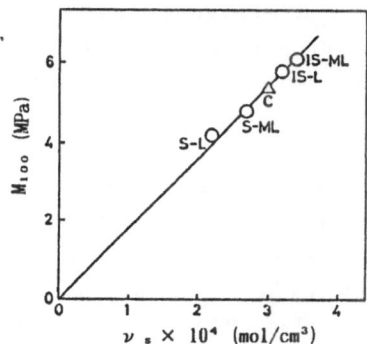

d) 1,8-Diazabicyclo   e) Dibenzothiazyl disulfide
   [5,4,0] undecene-7

Fig. 4 Relationship between M_{100} and ν_s of lignin-or carbon black-loaded CB-BRs.
M_{100}: Tensile modulus at 100% elongation.
ν_s: Network-chain density by swelling in benzene.

strength at break, T_B, and modulus at 100% elongation, M_{100}, in all lignin-loaded CB-BR vulcanizates, and its M_{100} is higher than that of HAF black-loaded stock, sample C, in spite of the lowest Wallace plasticity, i.e., the lowest viscosity, of the uncured stock. These results indicate that the chemical modification of lignin by ethylene chlorohydrin resulted in recognizable good effects for improving processability and reinforcing CB-BR. High moduli of IS samples, especially IS-ML, were concluded to be due to the increase of chemical crosslinks as shown in Fig.4.

Table 3   Physical properties of lignin- or carbon black-loaded CB-BRs

Cure	Filler	W.P.[a]	M_{100}[b] (MPa)	T_B[c] (MPa)	E_B[d] (%)
IS[e]	L[g]	64	5.8	9.5	160
	ML[h]	50	6.1	12.3	220
S[i]	L	74	4.2	10.0	290
	ML	61	4.8	9.0	250
S	C[j]	86	5.4	12.9	190

a) Wallace plasticity.
b) Tensile modulus at 100% elongation.
c) Tensile strength.
d) Elongation at break.
e) Isocyanate-sulfur cure.

f) Sulfur cure.
g) Unmodified lignin.
h) Modified lignin.
i) Carbon black.

As other characteristic properties of samples IS-L and IS-ML, it should be pointed out that these samples have good heat-resistance. This result seems to be due to high heat stability of urethane crosslinks.

In conclusion, the results of the study of the modified lignin-reinforced CB-BR on characteristic properties are of interest in three points, i.e., good processability, high tensile properties and good aging properties. Of course, the present method for producing the lignin-reinforced elastomer is valuable from the point of useful utilization of natural resources and low energy processing.

2.2 Moisture-Cure of CB-BR

We reported that CB-BR having active chlorine atoms attached to 1,4-butadiene units, is moisture-curable by using 3-aminopropyltriethoxy-silane (APS)[12] as shown in Fig.5 (with only one condensation for simplicity, which does not mean that only one ethoxy group is reactive).

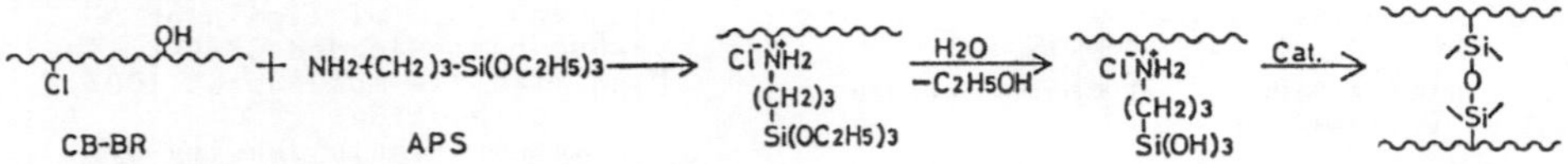

Fig. 5   Reaction mechanism of moisture cure.   CB-BR/APS reaction system.

It has already been reported that halogenated butyl rubbers are moisture-curable with APS[13,14]. The details of the cure system, however, have not been studied. Therefore, we studied on the moisture-cure of halogenated rubbers[15-17] other than CB-BR. However, the moisture-cure of halogenated rubber by APS has a shortcoming of relatively lower reactivity of APS toward the halogen atoms attached to rubber molecules compared with the rate of condensation reaction between the triethoxysilane groups in APSs under usual vulcanization conditions.

Thus we developed a novel moisture cure of CB-BR by using 4-trimethoxysilyl-1,2,3,6-tetrahydrophthalic anhydride (MSTP)[18]. in Fig. 6 is shown the reaction mechanism of the moisture cure of CB-BR by MSTP. The moisture-cured film was obtained by curing the CB-BR cast film at 50°C. Dibutyltin dilaurate (DBTDL) was used as a catalyst for silanol condensation reaction.

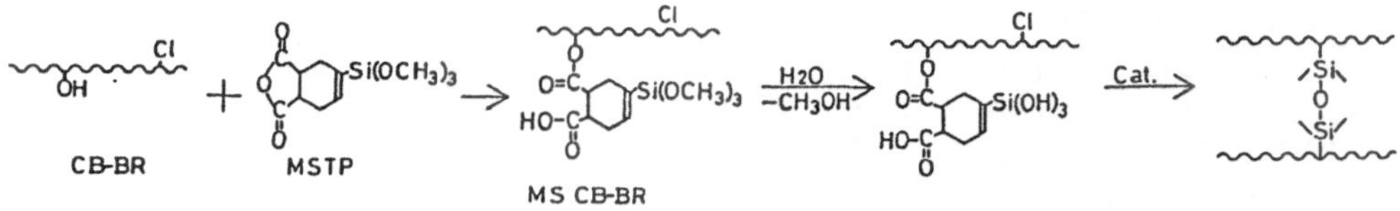

Fig. 6   Reaction mechanism of moisture cure.   CB-BR/MSTP reaction system.

As the result it was found that the consumed amount of hydroxyl groups in CB-BR attains about 70mol-% at 50°C for several hours. Table 4 shows the kinetic data including our previous experiments carried out by using APS, indicating that the addition reaction of MSTP with CB-BR is faster than that of APS toward chlorinated rubbers including CB-BR.

The CB-BR films obtained at 60°C for 24h by the method described above were soaked in hot water at prescribed conditions. Fig.7

Table 4   Kinetic data of polymer/silane coupling agent reaction system.

Polymer	Temp. (°C)	Final conv. (%)	$k \times 10^4$ (l/mol·sec)	Ref.
B-IIR[a]	90	69	13	(1)
C-IIR[a]	90	40	0.44	(1)
CR[a]	90	37	1.4	(2)
CB-BR[a]	90	—	31.0	(3)
CB-BR[b]	50	70	195	this paper

a) APS reaction system
b) MSTP reaction system

(1) S.Yamashita et al.,Makromol.Chem.186,1373 ('85).
(2) S.Yamashita et al.,ibid.,188,2553('87).
(3) C.S.Yoo et al.,34th Ann.Meeting of Polym. Sci.,Jpn, Kobe,July 8,1988.

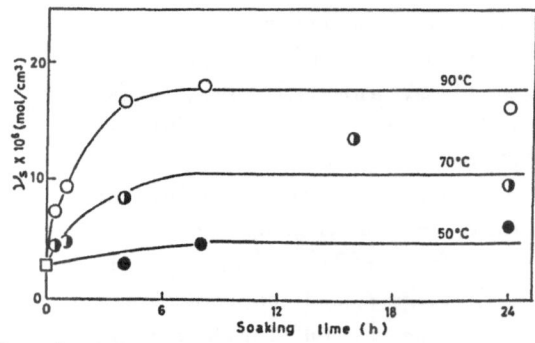

Fig. 7   Relationship between network-chain density of MS CB-BR and soaking time. (Sample size: 11.36mm in diameter, 3.07mm in thickness.)

indicates the relationship between the network-chain density calculated from the swelling data and the soaking time in hot water.

The moisture-curable CB-BR is also made by usual rubber processing. Table 5 gives a typical compounding recipe with or without hydrated silica. The moisture-cured films were obtained by soaking the samples in water at prescribed temprature for 24h. These samples were prepared by using a two-roll mill, followed by pressing at 120°C for 20min.

Table 6 includes R value which is the ratio of observed network-chain density, $\nu_{s,obs}$, to the calculated one, $\nu_{s,cal}$, i.e., $\nu_{s,obs}/\nu_{s,cal}$. From the data given in Table 6, it was found that compounding of the silica enables us to increase R value.

Table 5 Compouding recipe of moisture-curable CB-BR(phr)

	CB-BR	MSTP[a]	DBTDL[b]	Silica
MSF	100	4.4	1.02	40
MS	100	4.4	1.02	—

a) molar ratio [MSTP]/[OH] = 1.0
b) molar ratio [DBTDL]/[MSTP] = 0.1

Table 6 Calculated and observed network-chain densities

Sample	Soaking temp(°C)	Soaking time(h)	$\nu_{s\,cal}$[a] x 10^5 (mol/cm³)	$\nu_{s\,obs}$ x10^5 (mol/cm³)	R[b]
MS	50	24	14.83	2.09	0.14
MS	70	24	14.83	2.22	0.15
MS	90	24	14.83	3.13	0.21
MSF	50	24	14.83	10.85	0.73
MSF	70	24	14.83	10.42	0.70
MSF	90	24	14.83	13.96	0.94

a) $\nu_{s\,cal} = \rho \times c$ b) R = $\nu_{s\,obs}/\nu_{s\,cal}$
c: MSTP (mol/g rubber)
ρ: density of CB-BR (0.917g/cm³)

From these facts, we may conclude that the moisture-cure of CB-BR by MSTP is more favorable than that by APS from the view point of a higher reactivity of MSTP toward CB-BR.

3. VIOLOGEN ELASTOMERS

Viologen has been known as a reduction-oxidation-type organic dye. Recently, viologen derivatives attract interests among reseachers owing to their special functions such as photochromism[19,20], electrochromism[21-23] and thermochromism[24,25]. In order to use in practical applications, it would be much more convenient if the material be obtained in thin flexible form, e.g., elastomeric polymer film.

We have reported the synthesis and properties of an ionene-type elastomer consisting of poly(oxytetramethylene) units and viologen units (PTV)[26,27], and clarified that PTV behaves as a typical photochromic and photomechanical material[28]. Lately, there appeared some reports on synthesis and characterization of such viologen elastomers[29-31].

Lately, we obtained the ionene-type viologen elastomer having poly-(butadiene-co-acrylonitrile) by the reaction of α,α'-dichloro-p-xylene (DCX), 4,4'-bipyridine (BP) and dimethylamine-terminated liquid nitrile rubber (AT-NBR)[32,33]. We synthesized three sorts of NBV (m=1, 2 and 3) by using variable quantities of AT-NBR, DCX and BP as well as NBR ionene, NBI (m=0 in NBV).

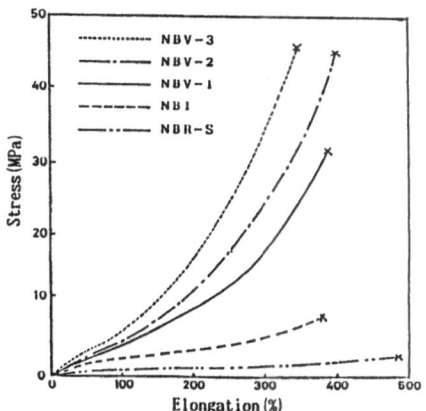

NBI (m=0), NBV:m (m=1, 2 and 3)

poly(butadiene-co-acrylonitrile)/viologen moiety

The synthetic method is as follows: The theoretical amounts of DCX and BP were dissolved in 10% THF solution of AT-NBR. The solution was cast on a glass plate and was heated at 100°C for 1hr after removing THF. AT-NBR was obtained by the reaction of carboxy-terminated liquid nitrile rubber (CT-NBR) and N,N-dimethyl-1,3-propanediamine (DMPDA).

The stress-strain curves of sulfur-cured nitrile rubber having 29wt% acrylonitrile (NBR-S), NBI, NBV-1, NBV-2 and NBV-3 are shown in Fig.8. From Fig.9 it is to be noted that tensile strength, T_B, and tensile modulus at 100% elongation, M_{100}, depend strongly on the number of BP in a structural unit of the ionene, m.

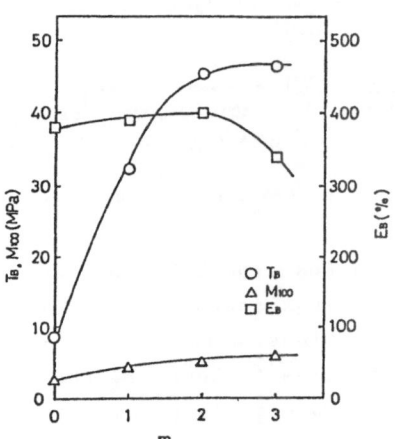

Fig. 8 Stress-strain behaviors of ionenes.

Fig. 9 Relationship between m and T_B, M_{100} or E_B.

In conclusion, viologen elastomers derived from the reaction of telechelic liquid rubber with viologen moiety have high tensile properties and thermoplastic elastomeric properties. Further experiments on functionalizations of NBV & are being carried out.

4. ABA TYPE LIQUID CRYSTALLINE ELASTOMERS[34,35]

Teleblock copolymers of ABA type (A: hard block, B: soft block) are widely used as thermoplastic elastomers (TPEs), i.e., SBS, SIS, SEBS etc. The average molecular weights of A block and B block are usually higher than 5,000 and 50,000, respectively. On the other hand, ABA type teleblock copolymers, carrying mesogenic groups to each end of soft blocks, are expected to be a new type of TPE whose molecular weight in hard block is relatively low compared with usual TPE (Fig.10).

Thus, it is of interest to pursue the possibility of the realization of new elastomer having a characteristic property of reversible phenomenon, i.e.,

$$\text{Elastomer} \xrightleftharpoons[\text{low temp.}]{\text{high temp.}} \text{liquid rubber.}$$

Lenz and his coworkers[36] have prepared some thermotropic liquid crystalline elastomers (TLCE) having mesogenic groups at terminal chain ends in oligomers having average molecular weight of several thousand. However, no experiment has been made to elucidate their phase structure and mechanical properties.

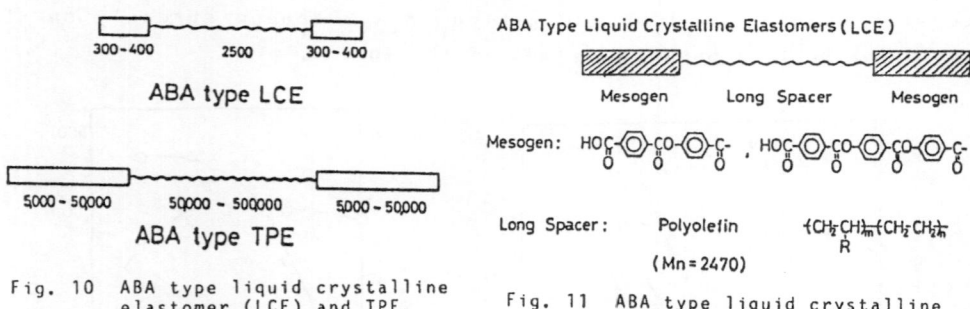

Fig. 10 ABA type liquid crystalline elastomer (LCE) and TPE.

Fig. 11 ABA type liquid crystalline elastomers (LCE).

An ABA type TLCE derived from telechelic polytetrahydrofuran and 4-[(4'-alkoxybenzoyl)oxy]benzoyl compound was synthesized, and its phase transitions were determined by measurements of linear viscoelastic properties[37]. However, the mesogen-induced cross-linking showed little resistance to large amplitude shearing.

Here, an attempt was made to elucidate the structure-properties relationship of ABA type TLCE in order to develop high performance TLCE.

4.1 PREPARATION OF ABA TYPE TLCE

We synthesized two ABA type liquid crystalline elastomers

terminated with two kinds of mesogenic groups (A segments) (Fig.11). One was a polyolefin terminated with p-carboxybenzoyl-p-oxybenzoyl esters (LCPO-1), and the other was terminated with p-carboxybenzoyl-p-oxybenzoyl-p-oxybenzoyl esters (LCPO-2). B segments in both LCPO-1 and LCPO-2 were polyolefin (HT-PO) having molecular weight of 2,470. Mesogenic groups were synthesized from p-hydroxybenzoic acid and terephthalic acid.

The synthetic routes of LCPO-1 and LCPO-2 are shown in Fig.12. The properties of HT-PO, LCPO-1 and LCPO-2 are given in Table 7.

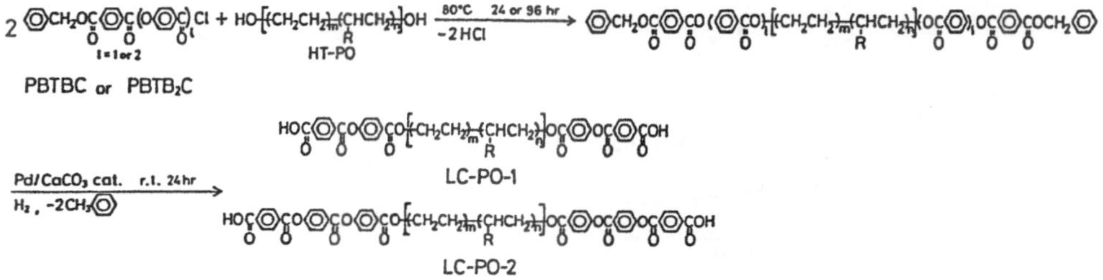

Fig. 12 Syntheses of LC-PO-1 and LC-PO-2.

Table 7   Properties of HT-PO, LC-PO-1 and LC-PO-2

		Yield (%)	$Mn^{a)}$	Molar ratio$^{b)}$ of LC block
HT-PO		—	2470	—
LC-PO-1	c)	57.5	3560	1.98
LC-PO-2		20.8	3650	2.10

a) By VPO. b) Calulated from Mn and liquid crystalline block concentration. c) Y.Nakae et al.

Table 8   Phase transition temperatures of HT-PO, LC-PO-1 and LC-PO-2

		$Tg^{a)}$(°C)	$Tm^{b)}$(°C)	$Tc^{c)}$(°C)
HT-PO		-40	—	—
LC-PO-1	d)	-32	65(68^{e})	90(87^{e})
LC-PO-2		-32	78(84^{e})	150(157^{e})

a) Glass transition temperature. b) Melting or softening temperature. c) Clearing temperature.
d) Y.Nakae et al. e) Visual observation.

4.2 PHYSICAL PROPERTIES OF ABA TYPE TLCE

Both LCPO-1 and LCPO-2 were found to be liquid crystalline and could be formed into elastic films. Phase transition temperatures of these polymers illustrated in Table 8 show that mesomorphic temperature range (M.R.) of LCPO-2 (78-150°C) is higher and broader than that of LCPO-1 (65-90°C) due to the increase of aromatic groups in the mesogenic groups.

Temperature dispersions of mechanical damping index, Δ, and relative rigidity, Gr, for LCPO-1 and LCPO-2 measured by torsional braid analysis (TBA) method showed two peaks corresponding to the glass transition and melting temperatures as shown in Fig.13 and table 9, suggesting the presence of microphase separation in both polymers. Fig.14 shows the result of tensile test, indicating that mechanical properties, especially tensile strength, T_B, increased with the increase of the rigidity of mesogenic groups (LCPO-1, T_B = 0.77MPa : LCPO-2, T_B = 2.10MPa).

Table 9   Results of TBA and DSC

	Tg(°C)	Tm(°C)	Tc(°C)
LC-PO-1	-28(-32[a])	42(65[a])	95(90[a])
LC-PO-2	-25(-32[a])	98(84[a])	155(157[a])

a) By DSC.

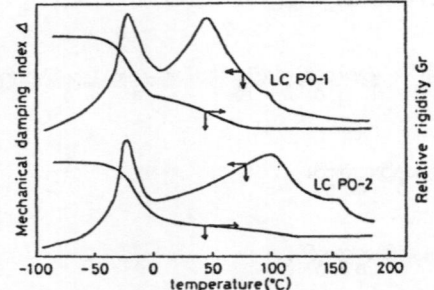

Fig. 13   Effect of temperature on mechanical damping index   and relative rigidity Gr for LC-PO-1 and LC-PO-2.

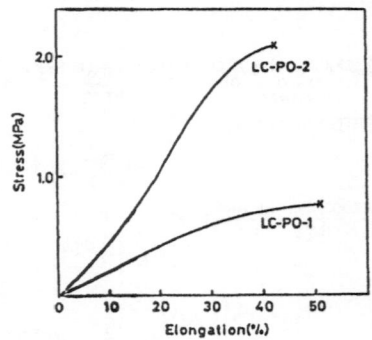

Fig. 14   Stress-elongation curves of LC-PO-1 and LC-PO-2.

SAXS measurements provided us with the informations on phase structure: Bragg spacings were 73 Å for LCPO-1 and 83 Å for LCPO-2, which confirmed the microphase separation in these polymers. The SAXS pattern of LCPO-2 at 100°C (liquid crystalline state) shows three well-defined peaks which appeared at q(scattering vector)=0.09, 0.16, and 0.18 as shown in Fig.15. The results of the SAXS measurement of

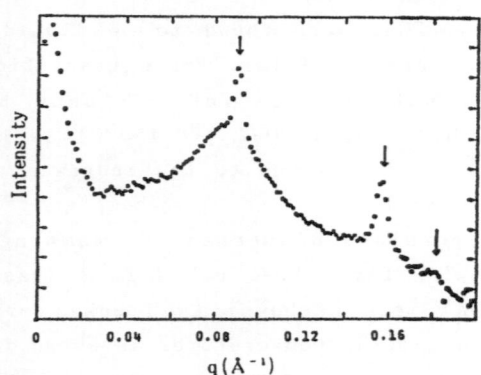

Fig. 15   Small angle X-ray scattering curve of LC-PO-2 at 100°C.

LCPO-2 are listed in Table 10. From the ratio of q (1: 3:2), a cylindrical domain structure based on the aggregation of the mesogenic groups in non-polar polymer matrix was proposed as illustrated in Fig.16 (A).

Table 10 Results of SAXS mesurement of LC-PO-2

temp.(°C)	q(Å⁻¹)ᵃ⁾ first peak	second peak	third peak	ratio of q	d(Å)ᵇ⁾
25	0.076	—	—	—	82.7
100	0.090	0.156	0.180	$1:\sqrt{3}:2$	69.8
145	0.093	0.161	0.186	$1:\sqrt{3}:2$	67.6
160	0.095	0.163	—	$1:\sqrt{3}:-$	66.1

a) $q=4\pi \sin \theta /2$
b) $d=2\pi /q$

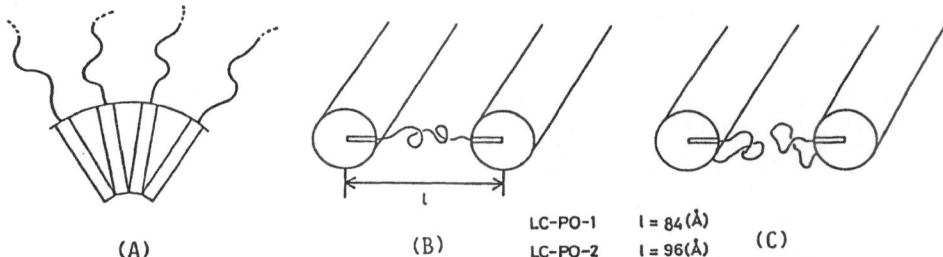

(A) (B) LC-PO-1 l = 84(Å) LC-PO-2 l = 96(Å) (C)

Fig. 16 Structure in LC-PO representing aggregation of mesogenic groups (A), tie molecule (B), and reentry molecules (C).

The tensile properties of LCPO-2 are not so good as shown in Fig.14 in spite of its good phase separation as indicated in Fig.13 and Table 10. The reason is not clear, but it seems that the mesogens are held together in domains by relatively weak forces, i.e., the cross-linking due to clustering of mesogenic groups did not show so large resistance to tensile deformation, and or that there are some polymers whose hard segments enter the same cylinder domain as shown in Fig.16 (C). Further experiments on clarification of these presumptions are being carried out.

Reference

1 Yamashita S, Tamura M, Terada J, Kohjiya S (1977) Rubber Chem Technol 50:364
2 Kohjiya S, Takeuchi H, Kawamoto K, Yamashita S (1981) Bull Chem Soc Jpn 54:3245

3 Yamashita S (1982) Polym Seminar 1982, Nov 18-20, Kuala Lumpur
4 Yamashita S, Kohjiya S (1988) Int Seminar on Elastomers, Oct 26-28 1988, Akron Ohio, Prepr p8-18
5 Yamashita S (1989) In: Saegusa T, Higashimura T, Abe A (eds) Frontiers of Macromolecular Science p201-206. Blackwell, London Boston Melbourne
6 Yamashita S (1987) In: Res on Effective Use of Energy in Chemical Process, Reports of Special Project Res on Energy p109-110 and p117-122
7 Yamashita S, Hiramatsu K, Miura K, Kidera A (1988) Int Rubber Conf, Oct 10-14 1988, Sydney,
8 Anon (1980) Europ Rubber J, Oct 29
9 Keilen J J, Pollak A (1947) Ind Eng Chem 29:480
10 Tibenham F J, Grace N S (1954) Ind Eng Chem 46:824
11 Nando G B, De S K (1980) J Appl Polym Sci 25:1249
12 Yamashita S, Shigaraki M, Orita M, Nishimura J, Sato K (1980) Rev Gen Caoutch Plast 606:126; Chem Abstr (1981)94:104636u
13 US 3366612 (1968) Esso Res & Eng Co, invs: Baldwin F P, Malatesta A (1968) Chem Abstr 68:60386u
14 Yamashita S, Matsumoto T, Kakehi M, Ikeda S, Technical Polym Res Group (1981) 4th Meeting, Kobe
15 Yamashita S, Yamada A, Ohata M, Kohjiya S (1985) Makromol Chem 186: 1373
16 Yamashita S, Yamada A, Ohata M, Kohjiya S (1985) Makromol Chem 186: 2269
17 Yamashita S, Nakawaki Y, Kidera A (1987) Makromol Chem 188:2553
18 Yamashita S, Maeda K, Tokuyama K, Kohjiya S (1989) Ann Meeting Soc Rubber Ind Jpn, May 18 1989, Osaka
19 Brown G H (1971) Photochromism, in Tech Chem Vol 3. Wiley-Interscience New York
20 Iwabuchi S, Kamogawa H (1976) Sanka-kangen Jushi p181. Kodansha Tokyo
21 Elofson R M, Edsberg R L (1987) Can J Chem 35:646
22 Osa T, Kuwana T (1969) J Electroanal Chem 22:389
23 Yasuda A, Mori H, Mizuguchi J (1987) J Appl Phys Jpn 26:1352
24 Moore J S, Stupp S I (1986) Macromolecules 19:1815
25 Kitagawa M, Ono A, Kohjiya S, Yamashita S (1987) Polym Prepr Jpn 36:797
26 Kohjiya S, Ohtsuki T, Yamashita S (1981) Makromol Chem Rapid Commun 2:417
27 Kohjiya S, Ikeda Y, Moriya N, Hashimoto T, Yamashita S, Shibata Y (1988) MRS Int Meeting on Advance Material, May 30-June 3 1988 Tokyo, Lecture No R-14
28 Kohjiya S, Yamashita S (1988) Int Seminar on Elastomers, Oct 26-28 1988, Akron Ohio, Prepr p20-26
29 Lee B, Wilkes G L, McGrath J E (1988) Polym Prepr 29(1) 136
30 Huynh-Tran C, Riffle J S, McGrath J E (1988) Polym Prepr 29(1) 138
31 Kohjiya S, Hashimoto T, Yamashita S (1989) Makromol Chem Rapid Commun 10:9
32 Yamashita S, Tanaka H, Inaki M, Kohjiya S (1989) Ann Meeting Soc Rubber Ind Jpn, May 17 1989, Osaka
33 Yamashita S, Kohjiya S, Ikeda Y (1989) Int Rubber Conf, Aug 29-Sep 1 1989, Prague, Book of Summaries I p41-42
34 Inaki M, Ikeda Y, Yamashita S (1989) 58th Spring Meeting of the Chem Soc Jpn, April 2 1989, Kyoto, Prepr 2IK-28 p1132
35 Inaki M, Ikeda Y, Yamashita S, Kidera A, Hayashi H (1989) 38th Ann Meeting Soc Polym Sci Jpn, May 14 1989, Yokohama, Polym Prepr Jpn 14R 27 p E453
36 Hoshino H, Jin J-I, Lenz R W (1984) J Appl Polym Sci 29:547
37 Lin Y G, Zhou R, Chien J C W, Winter H H (1988) Macromolecules 21: 2014

Modifications of Natural Rubber

A. Kadir
Rubber Research Institute of Malaysia
P.O. Box 10150, 50908 Kuala Lumpur, Malaysia

Natural rubber (NR) is an excellent elastomer with a well-balanced set of properties. For this reason, it is used in a wide range of applications, including very demanding engineering components such as structural bearings. Many applications of rubber, however, require special properties for which NR is deficient. Others require that the rubber be of a different physical form than is normally obtained with conventional NR. Thus, modifications of NR either physically or chemically have been an important subject of research carried out by the Malaysian Rubber Research and Development Board.

Physical Modification

NR, as is usually available, is in the form of bales. The increasing importance of powder or particulate technology, especially, in plastics processing, has prompted the development of specialised granulated and powdered synthetic rubbers. Equivalent materials are also available with NR.

Production of particulate forms of NR is fraught with difficulty because its high tack will cause the particles to coalesce if nothing is done to keep them separated. A process of surface clorination has been found to be effective in ensuring that a granulated form of NR remains in this state. An effective way of producing NR in powdered form is to make it in the form of a carbon black masterbatch powder. The carbon black not only decreases the tackiness of the rubber but also acts as a partitioning agent to prevent agglomeration of the particles.

Thermoplastic NR (TPNR) is a physical modification of NR effected by blending with crystalline polyolefins such as high density polyethylene or isotactic polypropylene. Depending on the composition of the blends, TPNR can be used in many applications which require the use of a hard but flexible material, such as in flexible bumpers, sight shields and rubbing strips.

Chemical Modification

There are four main ways of chemically modifying NR. These are by:

B. C. Anderson · Y. Imanishi (Eds.)
Progress in Pacific Polymer Science
© Springer-Verlag Berlin Heidelberg 1991

o rearrangement of the bonds without the introduction of new chemical groups

o attachment of pendant functional groups by olefinic addition or substitution reactions

o grafting of another polymer onto the NR molecule

o reducing the molecular weight by scission.

Isomerised and cyclized rubber are chemically modified NR produced by the rearrangements of the bonds. Of these only cyclized rubber is available commercially. It is used in shoe soles, hand mouldings, heavy-duty industrial rollers, adhesives, bonding agents, reinforcing resins, corrosion-resistant surface coatings and printing inks.

A vast number of derivaties can be obtained by the addition of pendant groups to the NR molecule. These include hydrogenated, halogenated, hydrohalogenated, alkyl halogenated and epoxidised natural rubbers. Except for epoxidised natural rubber (ENR) which is commercially produced in Malaysia, none of these modified NR have significant commercial value. ENR, like NR, can undergo stress-induced crystallization and, thus, has high strength properties. It is resistant to swelling by hydrocarbon solvents, and has a low air permeability and high damping characteristics. It is suitable for use in tyre tubes and inner liners of tubeless tyres, oil seals and applications requiring high damping.

The most commercially successful graft copolymerisation of NR is that with poly(methyl methacrylate)(PMMA). The resulting derivative is usually referred to as MG rubber. It is a hard rubber which blends well with NR in all proportions and is use as a reinforcing agent and in rigid mouldings.

Liquid NR is a recent development. It is obtained by scission of the rubber main chain to reduce its molecular weight from about 1 million to between 10,000 and 30,000. It is expected to have applications in co-curable plasticisers for rubbers, adhesives and sealants.

Conclusion

Although an excellent elastomer in its own right, the modification of NR either physically or chemically has enabled it to be used as the base polymer in applications where the unmodified rubber would not be suitable. Of the more recent developments in this area, it is expected that TPNR and ENR will be the most commercially important.

Positive Photoresist Chemistry

M. Murata, M. Koshiba and Y. Harita

Electronics Research Laboratory, Japan Synthetic Rubber Co., Ltd.
3-5-1 Higashi-Yurigaoka, Asao-ku, Kawasaki 215 Japan

Abstract: The present paper describes chemistry of naphthoquinone diazide (NQD) - novolak type positive photoresists focusing on the reactions of NQD. NQD is a highly reactive compound and has been known to render various kinds of reactions, among which photochemical, thermal and base catalyzed reactions are very important for the positive photoresist chemistry.   Upon light exposure it releases nitrogen and undergoes Wolff rearrangement to transform into a ketene, which then reacts with water to produce an acid.   This series of well-known reactions is regarded as the principle of pattern formation of the positive photoresist.   Recent photochemical works seem focused on revealing more detailed mechanisms and kinetics of the reactions.   Thermal and base catalyzed reactions give crosslinked structures to the resist which play important roles in several novel resist processes for improved resist performances such as deep-UV cure, REL, LENOS and DESIRE processes.   The present paper elucidates the mechanisms of these processes based on the results of our model experiments on thermal and base catalyzed reactions.

INTRODUCTION

NQD-novolak type positive photoresists have been widely used in the field of high resolution optical microlithography for years.   An example of the resist composition is shown in Figure 1.   Esters of polyhydroxybenzophenone and 1,2-naphthoquinone diazide-5-sulfonic acid are commonly used as a photoactive compound (PAC).

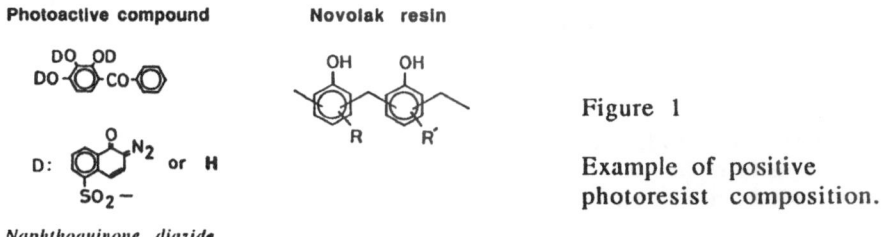

Photoactive compound   **Novolak resin**

D: ... or **H**

Naphthoquinone diazide

Figure 1

Example of positive
photoresist composition.

B. C. Anderson · Y. Imanishi (Eds.)
Progress in Pacific Polymer Science
© Springer-Verlag Berlin Heidelberg 1991

Figure 2 shows how the resist is used in the microchips manufacturing. The fine pattern is designed on the transparent mask, and the resist on the substrate is exposed to light through the mask and then it is developed in an aqueous alkaline developer. During the development the exposed part of the resist dissolves away in the developer so that only the unexposed part remains on the substrate. Thus a positive relief image is obtained. This resist image works as a mask in the following process like etching or doping of the substrate. After these processes the resist pattern is stripped away and another resist is coated on the processed substrate to undergo the next process.

Formation of the resist pattern is due to the solubility change of the exposed part of the resist which results from the photochemical reaction of NQD. Figure 3 shows a change of the dissolution rate of the resist. The dissolution rate of a novolak film itself in an alkaline developer is approximately 100Å/sec. Once NQD is added to the novolak resin, the rate decreases drastically by the order of one thousand, which means the unexposed part scarcely dissolves in the alkaline developer. This is called the "dissolution inhibition effect" of NQD (1). And then upon exposure NQD decomposes to produce indenecarboxylic acid which makes the exposed region even more soluble than the novolak itself.

As the design rule of microchips has been getting finer and finer, manufacturing of microchips requires the resist more than mere pattern formation. The most advanced silicon device requires resolution of less than half micron with a good rectangular wall angle and thermal stability. The improvement of the resist performance can be approached from the two aspects. One is from the material side and the other is from the process side (Figure 4). The material aspect involves such factors as PAC, resin, additives; and the process aspect involves exposure, bake, development. These two aspects are not independent but closely related each other. The present paper focuses on the process aspect.

Recently several novel photoresist processes have been proposed mainly by microchips manufacturers, some of which are listed in Figure 4 for example. A common feature of these novel processes is that they utilize non-photochemical reactions of NQD which take place either thermally or in the presence of base. The authors investigated these reactions using model compounds. The present paper elucidates the mechanisms of these processes based on the results of the model reactions (2).

REACTIONS OF NAPHTHOQUINONE DIAZIDE

Photochemical Reaction

Photochemical reactions of NQD have been extensively studied for nearly half a century (3), and it is generally accepted as shown in Figure 5a (4) that upon exposure NQD generates carbene by releasing nitrogen and undergoes Wolff rearrangement to transform into indeneketene although the carbene itself has not been detected yet. In the presence of water, the ketene reacts with it to

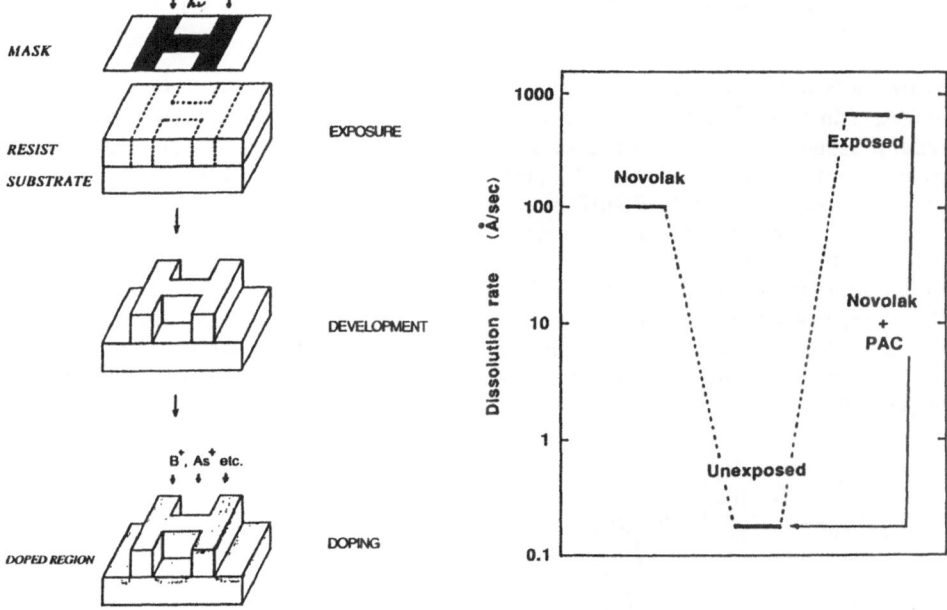

Figure 2   Resist process.   Figure 3   Change of dissolution rate.

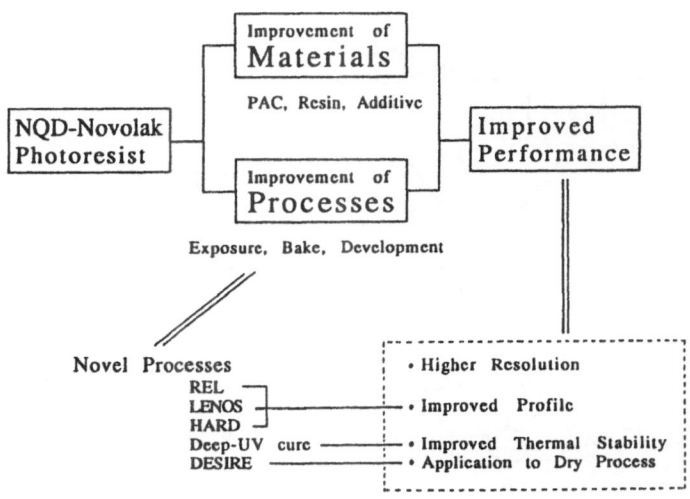

Figure 4   Improvement of resist performance.

produce indenecarboxylic acid. This is regarded as the principle of the alkaline solubility change of the positive photoresists.

More detailed information on the reaction has been obtained by several recent works, some of which even seem to make this scheme questionable. One of the recent issues on the photochemistry is whether oxirene exists or not as an intermediate. Tanigaki et al. performed a laser photolysis study using 1,2-naphthoquinone diazide-5-sulfonic acid sodium salt and assigned one of the intermediate species as an oxirene (see Figure 5b) (5). Tsuda et al. carefully calculated the transition state of the transformation from benzoquinone diazide to cyclopentadienyl ketene by using a MINDO/3 method (6). They claim from their results that benzoquinone diazide does not undergo the oxirene state because there is only one saddle point for the transformation. They concludes by anology that the oxirene state does not exist for the photodecomposistion of NQD, either.

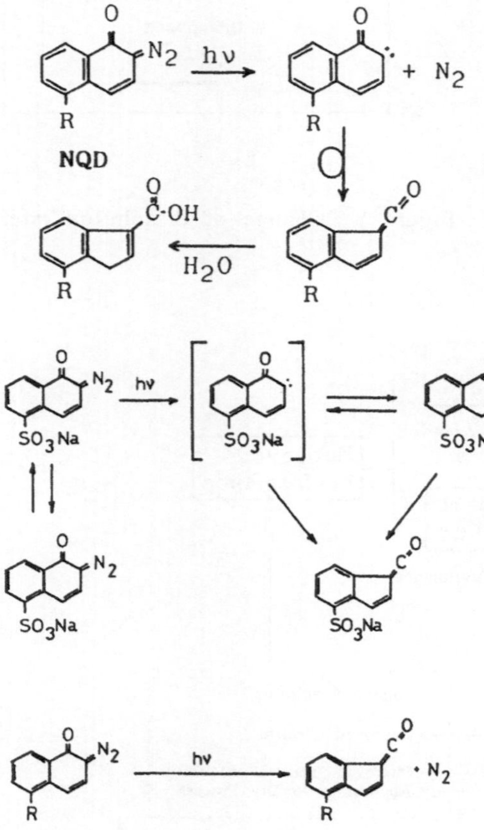

(a) generally accepted scheme.

(b) Scheme proposed by Tanigaki et al.

(c) scheme proposed by Tsuda et al.

Figure 5   Photochemical reaction schemes.

Further, their calculation even denies formation of the carbene; that is, the ketene is formed directly after extraction of nitrogen without undergoing the carbene state as shown in Figure 5c. They also claim that this would be supported by an experience that the positive photoresist is insensitive to atmospheric oxygen unlike other radical existing systems such as bisazide-cyclized rubber negative photoresist: the exposure of the positive photoresist is usually carried out in air without any problem.

Another photochemical issue is how the ketene reacts with water. Delaire et al. and Shibata et al. independently carried out very similar laser photolysis studies. As a result of Delaire et al. the decay of the ketene is second order with respect to water, and they have proposed a reaction scheme shown in Figure 6a (7). Shibata et al. conclude in a different way (8). They have observed that the proton concentration affects the decay rate of the ketene and claim that the ketene hydrate is formed first and then this intermediate transforms into indene carboxylic acid either through an interaction with water or the intramolecular rearrangement as shown in Figure 6b.

Although photochemistry of NQD is still a matter of discussion, it does not seem to be of great importance from the resist point of view because it would be difficult to improve the resist performance from the photochemical aspect. As far as the resist is concerned, the other reactions of NQD are more important.

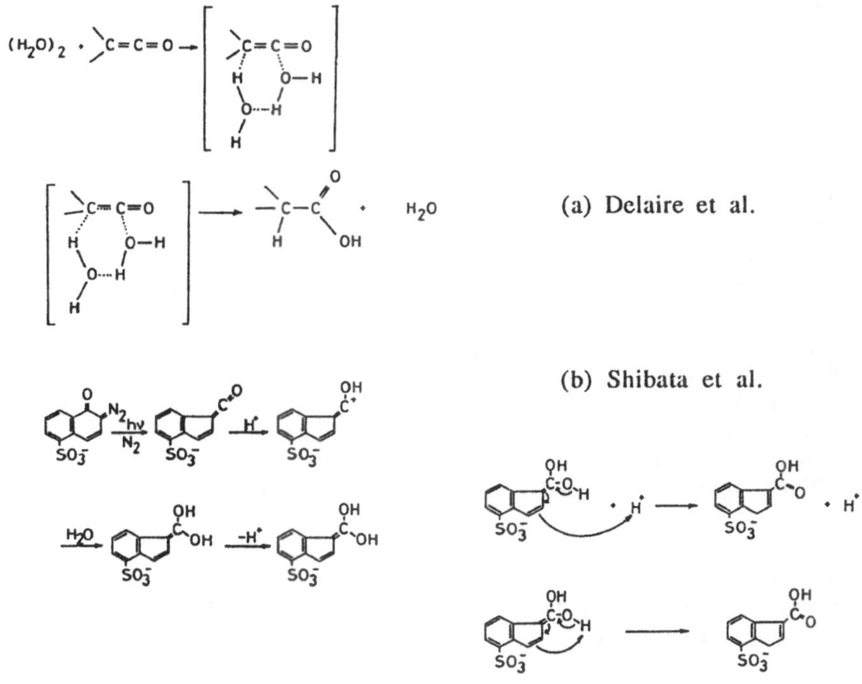

(a) Delaire et al.

(b) Shibata et al.

Figure 6   Reaction of ketene with water.

Thermal and Base Catalyzed Reactions

Although thermal and base catalyzed reactions of NQD are becoming important, there is limited information on these reactions which take place in the positive photoresist. The authors carried out model experiments using 1,2-naphthoquinone diazide-5-sulfonyloxybenzene (DAM) and p-cresol. DAM was selected as a model for the PAC and p-cresol was a model for the novolak resin. These two compounds were mixed together and reacted either by heating or in an alkaline developer which is 2.38wt% tetramethylammonium hydroxide (TMAH) aqueous solution.

DAM   p-cresol

Thermal reaction: As a result of liquid chromatographic analysis were observed seven different reaction products, out of which three major products were isolated. Structure determination of those products was performed, and the most probable structures of these three products are shown in Table 1. It is noticed here that all these products are coupling compounds of DAM originated group and p-cresol.

Figure 7 shows the relative production yield, caluculated from the liquid chromatograms, of each compound plotted against the reaction time. The production yields of TR-F4 and TR-F6 were found to increase as the reaction proceeds, and the same phenomenon is observed for the formation of TR-F7 when the reaction is carried out below 130°C. On the other hand, the azo compound, TR-F7, gradually decomposes at higher temperature than 150°C even though it is once formed at the initial stage of the reaction. In Figure 8 is plotted the final production yield (normalized by the final production yield of the reaction at 100°C) of each compound against the reaction temperature. The final production yields of TR-F6 and TR-F7 are temperature dependent while that of TR-F4 remains constant within the whole temperature range from 100 to 170°C.

Figure 9 shows the most probable reaction scheme. Upon heating DAM generates carbene (9), which might be transformed into a dipolar intermediate, which may react with p-cresol to form TR-F4. It is widely accepted a ketene reacts with p-cresol to form an ester compound, TR-F6. And an addition of another DAM to the product, TR-F6, forms an azo compound, TR-F7. Taking these products into account, crosslinking structure in the real resist layer upon heating is deduced as shown in Figure 10.

Table 1   The results of structure determination of the thermally
               induced reaction products.

	TR-F4	TR-F6	TR-F7
Chemical Structure	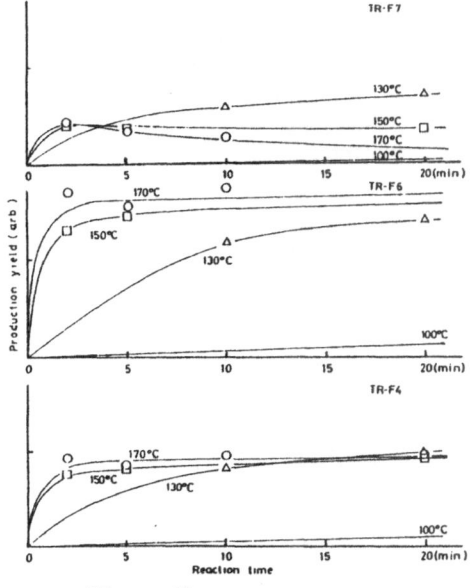		
FD-MS	m/z=406	m/z=406	m/z=732
ir	3600-3000cm^{-1}(-OH),2980-2840cm^{-1}(-CH$_3$), 1580cm^{-1} (aromatics), 1190cm^{-1} (-SO$_3$-)	3600-3000cm^{-1}(-OH),2920-2900cm^{-1}(-CH$_3$),1725cm^{-1} (-COO-),1590-1580cm-1(aromatics),1190cm^{-1}(-SO$_3$-)	3600-3000cm^{-1}(-OH), 2920-2900cm^{-1}(-CH$_3$), 1720cm^{-1} (-COO-), 1590-1580cm-1(aromatics),1190cm^{-1}(-SO$_3$-)
UV-vis	No absorption at λ≥400nm.	λ_{max} = 244nm (dioxane)	λ_{max} = 255, 510nm (dioxane)
^{1}H-NMR	2.29ppm (-CH$_3$) 9.28ppm, 9.97ppm (-OH)	2.37ppm (-CH$_3$) 3.93ppm (-CH$_2$-)	2.39ppm (-CH$_3$) 10.28ppm, 10.62ppm (-OH)
^{13}C-NMR	-	162.2ppm (-COO-), 40.3ppm (-CH$_2$-), 20.8ppm(-CH$_3$)	162.3ppm (-COO-), 20.3ppm (-CH$_3$)
EA:% ( )=theo.	-	C:65.9(68.0), H:4.4(4.5), S:7.7(7.9)	C:61.4(63.9), H:3.6(3.9), N:3.5(3.8), S:8.6(8.8)

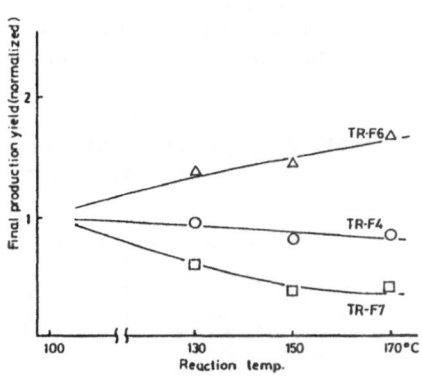

Figure 7

Production yields of the
three major reaction
products at several different
reaction temperatures.

Figure 8

Final production yields of the
three major reaction products
at several reaction temperatures.

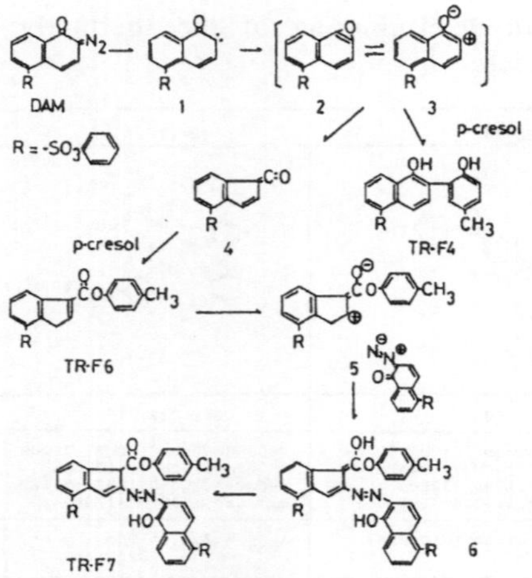

Figure 9 The probable reaction scheme of the thermal reaction.

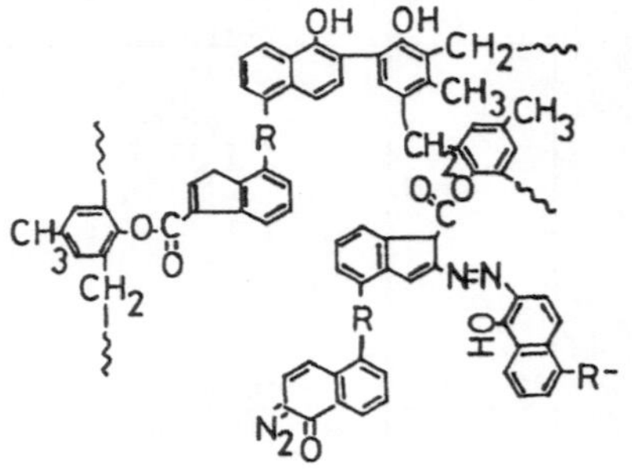

Figure 10 Possible crosslinking structure in the real positive photoresist layer.

Base catalyzed reaction: As for the base catalyzed reaction, two out of the four major products were isolated as shown in Table 2. Figure 11 shows the probable reaction scheme. Unlike photochemical or thermal reactions, base catalyzed reaction seems to take place upon the nucleophilic attack of hydroxy or phenoxy anions to the diazo group. It should be reminded that BC-F3 also implies a crosslinking structure. On the real resist surface NQD would crosslink with novolak resin as shown in Figure 12. This reaction is regarded as one of the mechanisms of the dissolution inhibition effect.

Table 2 The results of structure determination of the base catalyzed reaction products.

	BC-F1	BC-F3
Chemical Structure	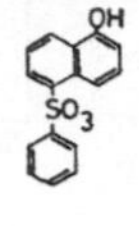	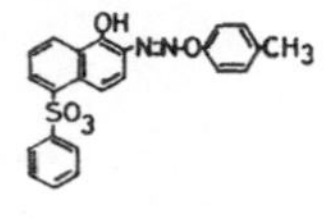
MS(EI)	$m/z=39(C_3H_3^+)$, $65((SO_2+H)^+)$, $115($ ... $^+)$, $143(C_{10}H_7OH^+)$ $300(parent)$	-
ir	$3440cm^{-1}(-OH)$, $1600,1580cm^{-1}$ (aromatics), $1190cm^{-1}(-SO_3-)$	$3000-2800cm^{-1}(-CH_3)$, 1600, $1580cm^{-1}$(aromatics, $1190cm^{-1}$ $(-SO_3-)$
UV-vis	-	$\lambda_{max} = 490nm(acetone)$
^{1}H-NMR	9.8ppm (-OH)	2.3ppm (-CH$_3$)

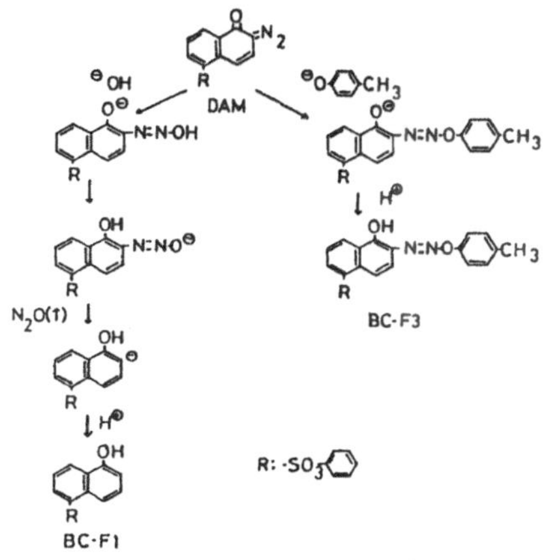

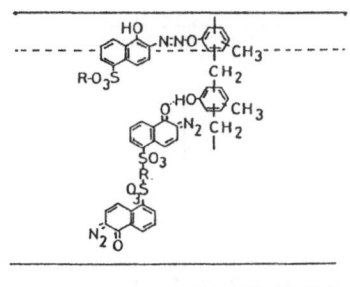

• Chemical crosslinks via azoxy bonds.

• Formation of the protective layer which is rich in naphthoquinone diazides.

Figure 11 The probable reaction scheme of the base catalyzed reaction.

Figure 12 Formation of the insoluble layer.

IMPROVEMENT OF RESIST PERFORMANCES

Improvement in pattern profile

Figure 13 shows the general idea to improve the pattern profile. In the conventional process the top of the unexposed part of the resist dissolves during development which results in a reduced wall angle. Therefore, if the top surface can be protected with an insoluble layer as shown in the figure, improvement of the pattern profile is expected. Treatment for this purpose can be applied either before or after the image exposure. One of the method to give the insoluble layer would be crosslinking.

Figure 14 shows the process flow of the Resolution Enhanced Lithography (REL) process (10). In this process, after the image exposure the resist is baked at around 100°C and exposed to deep-UV light to make the resist crosslink. This temperature is too low to decompose NQD rapidly, so the reaction needs to be induced by light exposure to take place. And at this temperature, there is less water in the resist and the novolak resin is in the rubbery state, which means the resin has higher reactivity than in the glassy state at room temperature. Therefore, upon deep-UV exposure, NQD may tend to react with the novolak resin to crosslink.

Latitude Enhancement NOvel Single layer lithography (LENOS) is another approach for profile improvement (see Figure 15) (11). In this process the insoluble layer is formed by the alkaline treatment. The mechanism is thought to be similar to that of the dissolution inhibition effect; that is, base catalyzed crosslinking and extraction of novolak resin to form the protective layer.

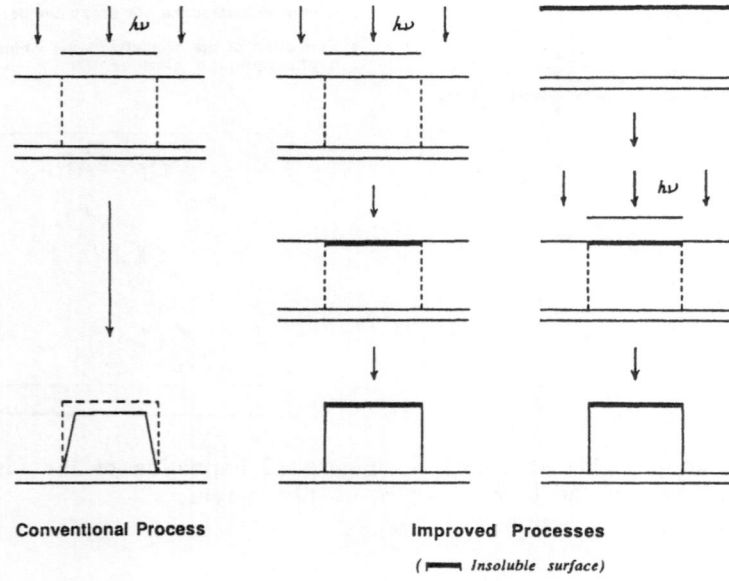

Conventional Process   Improved Processes

(▬ *Insoluble surface*)

Figure 13 General idea to improve the pattern profile.

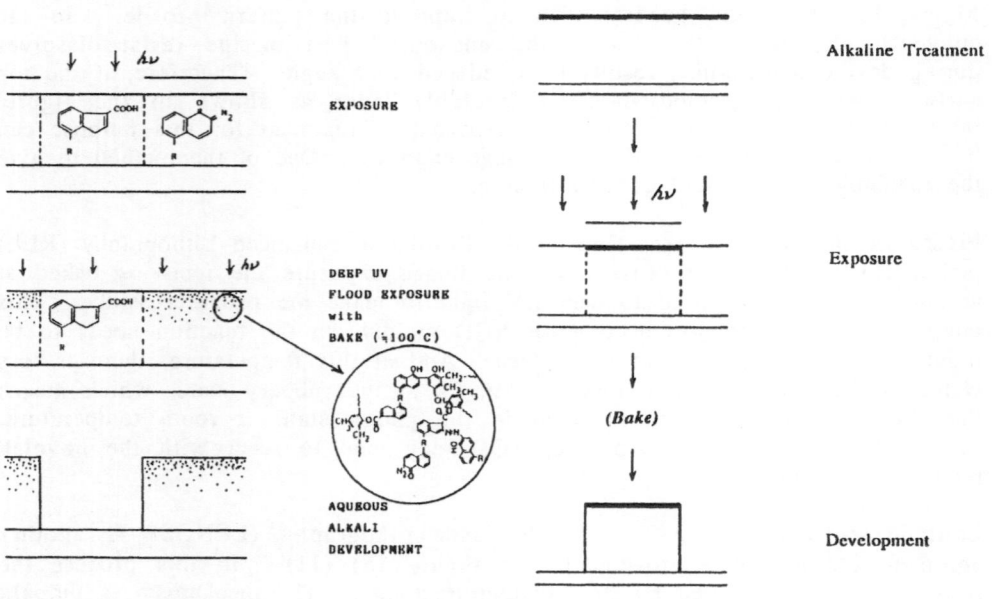

EXPOSURE

DEEP UV
FLOOD EXPOSURE
with
BAKE (≈100°C)

AQUEOUS
ALKALI
DEVELOPMENT

Alkaline Treatment

Exposure

(Bake)

Development

Figure 14   REL process.   Figure 15   LENOS process.

293

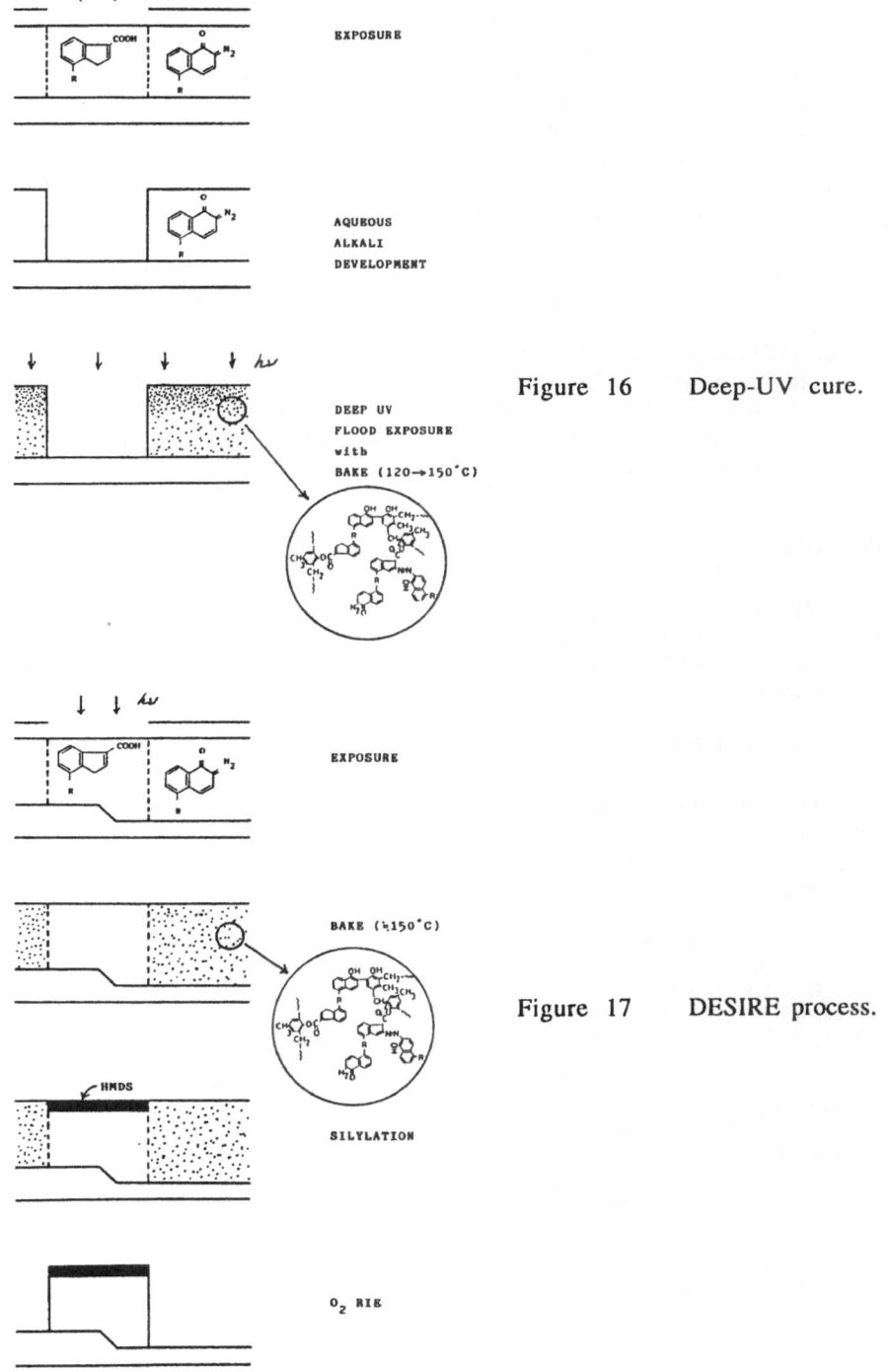

EXPOSURE

AQUEOUS
ALKALI
DEVELOPMENT

DEEP UV
FLOOD EXPOSURE
with
BAKE (120→150°C)

Figure 16   Deep-UV cure.

EXPOSURE

BAKE (≈150°C)

Figure 17   DESIRE process.

SILYLATION

O₂ RIE

Improvement in thermal stability

Figure 16 shows the process flow of deep-UV cure which is similar to REL process (12). The different point is that the baking and deep-UV exposure are applied at the same time after the development and the baking temperature is raised gradually to around 150°C. This is because the pattern will be damaged by resist flow if the high temperature is applied at the first stage.

Application of Positive Photoresist to Dry Process

Figure 17 shows the process flow of DESIRE (Diffusion Enhanced SIlylation REsist) process which is a single layer resist working like a bilayer resist (13). The point of this process is the selective diffusion of the silylating agent, hexamethyldisilazane at the silylation step. The crosslinking reaction is again the key to the process. When the resist is baked at around 150°C, NQD in the unexposed area crosslinks thermally. On the other hand crosslinking does not take place in the exposed area. This is why HMDS can diffuse into the resist only in the exposed area.

SUMMARY

Chemistry of naphthoquinone diazide was reviewed on the three different types of its reactions; i.e., photochemical, thermal and base catalyzed reactions. Mechanisms of the several novel resist processes were elucidated based on these reactions.

ACKNOWLEGEMENTS

The authors are grateful to Miss T. Imai, Miss M. Matsui, Miss N. Ando and Mr. T. Watanabe for their excellent analytical works.

REFERENCES

1 Murata M, Koshiba M, Harita Y (1989) Proceedings of SPIE 1086:48
2 Koshiba M, Murata M, Harita Y (1988) ibid 920:364
3 Sus O (1944) Ann 556:65
4 Pacansky J (1980) Polymer Eng Sci 20:1049
5 Tanigaki K, Ebbesen TW (1987) J Am Chem Soc 109:5883
6 Tsuda M, Oikawa S (1989) J Photopolym Sci Tech 2:325
7 Delaire JA, Faure J, Hassine-Renou F, Soreau M, Mayeux A (1987) Nouveau Journal de Chimie 11:15
8 Shibata T, Koseki K, Yamaoka T, Yoshizawa M, Uchiki H, Kobayashi T (1988) J Phys Chem 92:6269
9 Yates P, Robb EW (1957) J Am Chem Soc 79:5760
10 Okuda Y, Ohkuma T, Takashima Y, Miyai Y, Inoue M (1987) Proceedings of SPIE 771:61
11 Ogawa S, Uoya S, Kimura H, Nagata H (1988) Proceedings of 1st MicroProcess Conference :162
12 Hiraoka H, Pacansky J (1981) J Electrochem Soc 128:2645
13 Roland B, Vandendrissche J, Lombaerts R, Denturck B, Jakus C (1988) Proceedings of SPIE 920:120

Electronic Energy Transport in Vinyl Aromatic Polymers

K.P. Ghiggino, A.D. Scully, O. Vogl* and S.W. Bigger

Department of Physical Chemistry,
The University of Melbourne,
Parkville, 3052, Australia.

* Polytechnic University, 333 Jay St., N.Y., 1120, U.S.A.

Abstract: Ultrafast electronic energy transport phenomena occurring in macromolecules following absorption of light were investigated using picosecond laser-based time-resolved fluorescence instrumentation and computer-aided data analysis procedures. Analysis of the fluorescence data for copolymers of 2-naphthylmethacrylate (2NMA) and 2-(2'-hydroxy-4'-methacryloxyphenyl)-2H-benzotriazole (BDHM), indicate that the dominant energy transport mechanism is a one-step Forster-type process from both monomer and excimer sites to the BDHM trap sites in the polymer chain. The trapping of excitation energy by the highly photostable BDHM moieties results in enhanced photostability of the copolymers compared to poly(2-naphthyl methacrylate) (P2NMA).

INTRODUCTION

Fluorescence spectroscopy provides one of the most useful techniques for studying energy transport in synthetic polymers since it is capable of probing the relaxation pathways available to electronically-excited, aromatic moieties incorporated covalently on a polymer chain [1-3]. These electronic relaxation pathways include: (i) fluorescence from the initially-excited chromophore which is usually referred to as 'monomer' emission, (ii) radiationless relaxation of the initially-excited chromophore *via* internal conversion and/or intersystem crossing, (iii) energy migration of the electronic excitation to similar chromophores situated along the polymer chain which may then dissipate this energy through radiative and/or nonradiative relaxation, (iv) energy migration to sites which are capable of forming excimers (excited dimers) which may then undergo radiative and/or nonradiative relaxation, and (v) transfer of excitation energy from monomer and/or excimer donors to a suitable acceptor molecule having a lower energy than the donors.

B. C. Anderson · Y. Imanishi (Eds.)
Progress in Pacific Polymer Science
© Springer-Verlag Berlin Heidelberg 1991

An excimer may be formed by the association of an electronically-excited molecule with an unexcited molecule of the same species provided that they are in sufficiently close proximity and can adopt an orientation, within the residence time of the localised excitation, which favours the orbital overlap required for excimer formation. Alternatively, suitable preformed sites for excimer formation may be present in the polymer prior to the initial absorption of a photon. Excimer sites in a polymer can be envisaged to arise from several types of bichromophoric interaction, namely intermolecular excimer formation, which is due to the association of two chromophores on different chains, and/or intramolecular excimer formation which results from the interaction between two chromophores located on the same polymer chain. In the latter mechanism the two chromophores may be either nearest neighbours on the chain or distant chromophores brought into close proximity as a consequence of folding of the polymer backbone resulting in so called 'long-range' excimers [2,4]. The modes of electronic relaxation operative in synthetic polymers are shown schematically in Figure 1.

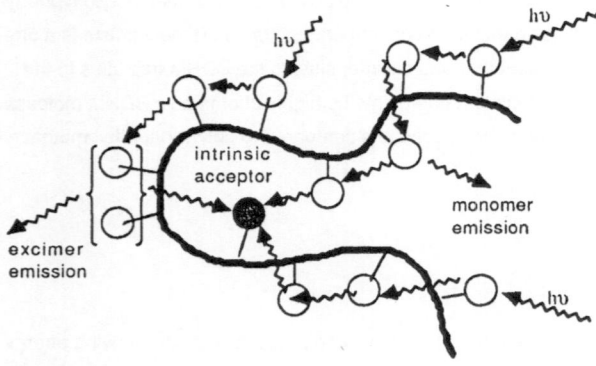

Figure 1. Pathways for electronic deactivation in synthetic polymers.

Poly(2-naphthyl methacrylate) (P2NMA) and copolymers of 2-naphthylmethacrylate (2NMA) with 2-(2'-hydroxy-4'-methacryloxyphenyl)-2H-benzotriazole (BDHM), provide a particularly interesting and useful example to illustrate energy transport processes. BDHM is a polymerizable photostabilizer which can act as an intrinsic acceptor for excitation energy in the polymers. The application of steady-state and time-resolved fluorescence techniques to investigate energy transfer in these polymers is described in this work.

THEORY

The rate constant for nonradiative energy transfer *via* a dipole-dipole mechanism, k_{et}, was derived by Forster [5,6] and is given in equation (1).

$$k_{ct} = \frac{9000 \ln(10) \, k^2 \phi_d J}{128\pi^5 n^4 \tau_d R^6 N} \qquad (1)$$

In this equation k is an orientation factor with an average value of $k^2 = 2/3$ for a random distribution of dipoles, ϕ_d is the quantum yield of fluorescence from the donor in the absence of acceptor, n is the refractive index of the solvent medium, τ_d is the fluorescence lifetime of the excited donor species in the absence of acceptors, R is the mean separation between the donor and acceptor species, N is the Avogadro constant and J is the spectral overlap integral which is defined in equation (2).

$$J = \int_0^\infty \frac{F_d(\bar{v}) \, \varepsilon_a(\bar{v})}{\bar{v}^4} \, d\bar{v} \qquad (2)$$

In this equation $F_d(\bar{v})$ is the fluorescence intensity of the donor emission and $\varepsilon_a(\bar{v})$ is the extinction coefficient of the acceptor at the wavenumber $\bar{v}$. The distance between the donor and acceptor species at which excitation transfer and spontaneous deactivation of the donor are of equal probability is known as the critical transfer distance, R_0. An expression for R_0 was also derived by Forster [5,6] and is given in equation (3).

$$R_0^6 = \frac{9000 \ln(10) \, k^2 \phi_d J}{128\pi^5 n^4 N} \qquad (3)$$

EXPERIMENTAL SECTION

Materials

All solvents used in this work were of spectroscopic grade and were free from fluorescent impurities. The monomer 2-(2'-hydroxy-4'-methacryloxy-phenyl)-2H-benzotriazole (BDHM) was prepared [7] and supplied by Professor O. Vogl (Polytechnic University, New York, U.S.A.) and was recrystallized from ethanol before use. The 2-naphthylmethacrylate (2NMA) monomer (Polysciences) was purified by three vacuum sublimations. Copolymers of 2NMA and BDHM were prepared by polymerization of the monomers in degassed benzene at 60°C for approximately 130 hours using 0.4% (w/w) azobisisobutyronitrile (AIBN) as the free-radical initiator. The polymers were precipitated from benzene into methanol at a minimum conversion of 40% and then purified by multiple reprecipitations from benzene into methanol. The number average molecular weights, expressed as poly(styrene) equivalents, of copolymers containing 1.1 and 6.2 mole% BDHM are 18,100 and 8,330 respectively. These were determined at the C.S.I.R.O. Division of Chemicals and Polymers by gel permeation chromatography (GPC) using a Waters GPC system equiped with microstyrogel columns.

Copolymer compositions were determined from the UV absorption spectrum of each sample dissolved in chloroform at room temperature. The extinction coefficients of the monomers, BDHM [7] and 2NMA, were used respectively as estimates for the extinction coefficients of the BDHM and 2NMA chromophores incorporated into the copolymers.

Fluorescence spectra were recorded from room-temperature, degassed solutions of the polymers on a Perkin-Elmer MPF-44A spectrofluorimeter using an excitation wavelength of 285 nm and an excitation bandwidth of 5 nm. Appropriate corrections were made for the competitive absorbance of light by BDHM chromophores in fluorescence yield calculations.

The fluorescence decay measurements were obtained from room-temperature, degassed solutions of the polymers in benzene by the technique of time-correlated, single photon counting using a laser-excited time-resolved fluorescence spectrometer described previously [8]. The fluorescence decay analyses required iterative reconvolution with the finite instrumental response function [8].

Degassed solutions of the polymers in benzene at room temperature were contained in quartz cells having a path length of 1 cm. The solutions were irradiated using 285-nm light with a bandwidth of 20 nm provided by a 150 W xenon lamp/monochromator combination of a Perkin-Elmer MPF-44A spectrofluorimeter. Changes in the absorption spectra of the solutions were recorded using an Hitachi 150-20 spectrophotometer/data processor.

RESULTS AND DISCUSSION

Steady-State Measurements

The fluorescence emission spectra of P2NMA and the 2NMA-BDHM copolymers in degassed benzene solution at room temperature, with a naphthyl group concentration of 1.7×10^{-4} M are shown in Figure 2. For P2NMA, a structured monomer emission with a maximum at 330 nm arising from single 2-naphthyl chromophores together with a broad, structureless emission with a maximum at 400 nm, attributable to intramolecular excimers, is observed. This is in distinct contrast to previous reports for dilute solutions of styrene/2-(2'-hydroxy-5'-vinylphenyl)-2H-benzotriazole (2H5V) copolymers in dichloromethane at room temperature with 2H5V concentrations of up to 7.5×10^{-6} M in which no quenching of the monomer or excimer fluorescence was detected [9,10].

The degree of spectral overlap between the absorption spectrum of BDHM and the P2NMA monomer and excimer fluorescence emission, indicates that the BDHM chromophore is capable of acting as an acceptor in a Forster-type energy transfer process from both the monomer and excimer species. The values of the Forster critical transfer distance, R_0, for energy transfer between the BDHM chromophore and the P2NMA monomer and excimer species were calculated using equation (3), where the spectral overlap integral, J, was obtained by using equation (2). The results of these calculations are presented in Table 1.

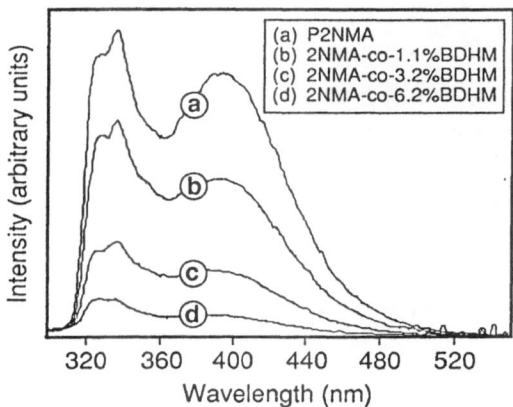

Figure 2. Fluorescence emission spectra of the polymers.

The critical transfer distance between the BDHM chromophore and P2NMA excimers is comparable with the estimated value of 17 Å which has been reported previously for Forster energy transfer between polystyrene (PS) excimers and the 2H5V comonomer [9,10]

Table 1. Forster energy-transfer parameters calculated using steady-state spectroscopic measurements from solutions of BDHM and P2NMA in benzene.

	ϕ_f	$J \times 10^{14}$ ($M^{-1}cm^3$)	R_0 (Å)	C_0 (M)
monomer	0.03	1.75	20	0.058
excimer	0.1	0.24	17	0.085

The theoretical quenching efficiency, f_{th}, can be calculated by using equation (4), if it is assumed that quenching of the fluorescence occurs via a one-step Forster transfer mechanism to a BDHM chromophore [5,6].

$$f_{th} = \pi^{1/2}.x.\exp(x^2)[1 - \text{erf}(x)], \text{ where } x = C/C_0 \text{ and } \text{erf}(x) = \int_0^x \exp(-t^2)\, dt \qquad (4)$$

In this equation, C is the quencher concentration and C_0 is the Forster critical concentration given by equation (5).

$$C_0 = 3000/(2\pi^{3/2}NR_0^3) \qquad (5)$$

The values of C_0 for energy transfer from P2NMA monomer and excimer species to the BDHM chromophore, obtained by applying equation (5), are presented in Table 1. However, the effective concentration of BDHM located within the polymer volume will be much higher than the bulk concentration of BDHM. An estimate of the effective concentration of BDHM in the polymer volume, C_{eff}, can be obtained by using the solution properties of P2NMA dissolved in benzene which have been reported previously [11] and classical Flory theory [12,13]. The value of C_{eff} for 2NMA-BDHM copolymers containing 1.1 and 6.2 mole% BDHM, was found to be 0.023 and 0.198 M respectively.

Figure 3 shows that a good correlation exists between the quenching of the monomer and excimer fluorescence which was determined from experiment and that predicted by the Forster energy transfer theory which was calculated using equation (4), where C_{eff} is used as the quencher concentration. This provides evidence that a single-step Forster energy transfer mechanism is able to explain the quenching of 2NMA fluorescence by BDHM chromophores in the 2NMA-BDHM copolymers.

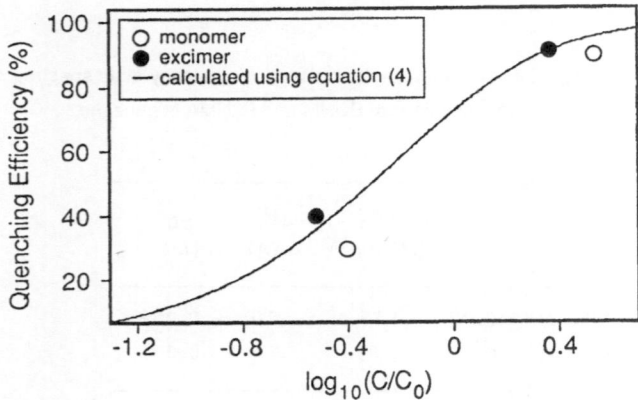

Figure 3. Experimental and calculated quenching efficiencies for 2NMA-BDHM copolymers.

Time-Resolved Fluorescence Measurements

The monomer and excimer fluorescence decay profiles obtained from P2NMA and 2NMA-co-6.2%
BDHM dissolved in benzene at room temperature are compared in Figures 4 (i) and (ii)
respectively. The steady-state quenching is accompanied by a significant decrease in both the
monomer and excimer fluorescence decay times as well as a reduction of the risetime of the
excimer fluorescence to one which is response-limited. The decay of the excimer fluorescence
from the copolymer is nonexponential but is fitted successfully by the Forster quenching
function [5,13] given in equation (6).

$$I_c(t) = I_0 \exp[(-t/\tau_0) - 2\gamma(t/\tau_0)^{1/2}] \qquad\qquad (6)$$

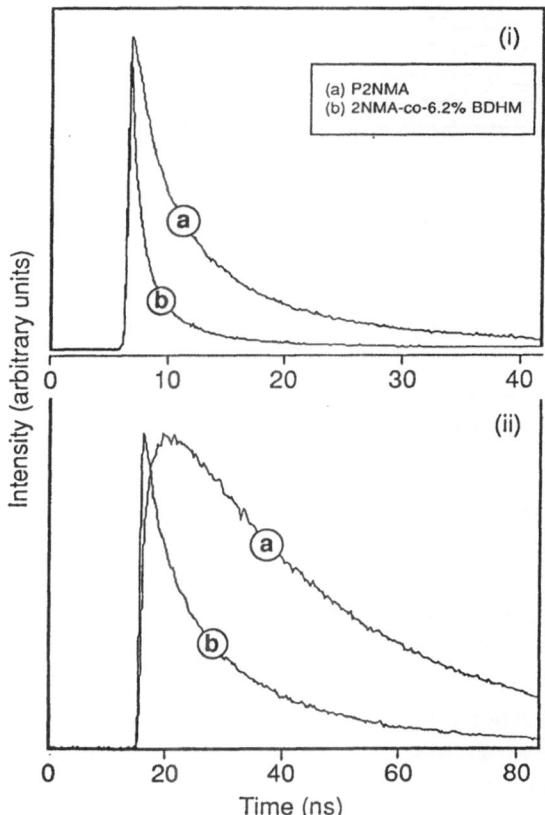

Figure 4. Fluorescence decay profiles for P2NMA and 2NMA-co-6.2% BDHM in benzene
monitored at (i) 330 nm (monomer region) and (ii) 400 nm (excimer region).

In equation (6), $\gamma = C/C_0$ and τ_0 is the fluorescence lifetime of the donor in the absence of quencher. The best-fit curve, calculated by using equation (6), and the excimer fluorescence decay of the copolymer determined by experiment are shown in Figure 5.

An estimate of C_0 for the nonradiative transfer of energy from the excimer to a benzotriazole acceptor can be obtained from the relation $C_0 = C_{eff}/\gamma$. After the substitution of the appropriate values of C_{eff} and γ into this equation, a value for C_0 of 0.109 M is obtained. Inserting this value for C_0 into equation (5) results in a Forster critical transfer distance of 16 Å which agrees well with the value estimated from steady-state measurements. This result provides further support for the proposal that the quenching mechanism for the excimer fluorescence from the 2NMA-BDHM copolymers involves one-step Forster energy transfer to a BDHM chromphore.

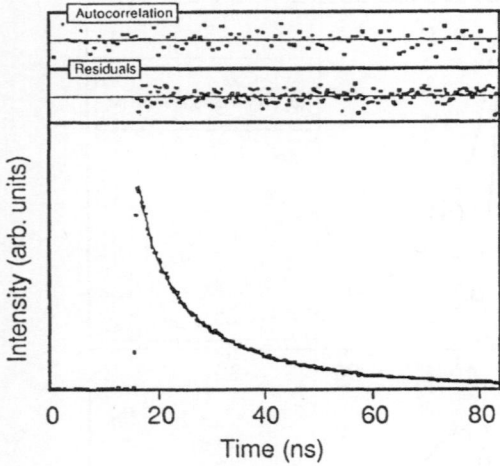

Figure 5. Curve (ii)b from Figure 4 fitted using equation (6) with best-fit parameters $\tau_0 = 62.7$ ns and $\gamma = 1.39$.

The absence of a detectable risetime for the time-resolved profile of excimer fluorescence from a solution of the 2NMA-BDHM copolymer suggests that the residence time of the excitation energy is too short to permit excimer formation by rotational reorientation of chromophores. This short residence time is attributed to efficient competition for the excitation energy by rapid one-step energy transfer to a benzotriazole trap and suggests that the excimer fluorescence originates from excimer sites which are formed prior to excitation.

The decay of the monomer fluorescence from the copolymer is also nonexponential but convergence was not obtained when analysed using equation (6) with all parameters adjustable. If migration of excitation energy occurs in P2NMA it could contribute to the poor fit obtained by using equation (6) to model the decay of monomer fluorescence from a solution of the 2NMA-co-6.2% BDHM copolymer. However, singlet energy migration among the naphthyl chromophores might be expected to increase the efficiency of energy transfer to BDHM to a value above that predicted by Forster theory. This was not observed in this work. The existence of excimer back-dissociation is also expected to contribute to the poor fit of equation (6) to the decay of monomer fluorescence. Additional heterogeneities in the fluorescence decay profiles may also arise as a result of the distribution of the degree of labelling of BDHM chromophores on the polymer chains.

Photodegradation of 2-Naphthylmethacrylate Polymers

Upon irradiation P2NMA readily undergoes degradation *via* a photo-Fries rearrangement [14,15] to form acenaphthol photoproducts with absorption bands at 320 and 360 nm. The appearance of similar product absorption bands is also observed upon irradiation of a degassed benzene solution of the 2NMA-co-6.2% BDHM copolymer, as well as a solution containing P2NMA with 6.7 mole% of the commercially used photostabilizer 2-(2'-hydroxy-5'-methylphenyl)-2H-benzotriazole (HMPB).

HMPB is structurally similar to BDHM and acts as a UV absorber and photostabilizer of polymeric materials due to the efficiency by which it can dissipate absorbed excitation energy [16,17]. At the concentration used, HMPB has only a minor screening effect at 285 nm and interactions with the polymer during the excited-state lifetime are unlikely. Figure 6 illustrates the rate of increase in the absorbance at 320 nm upon irradiation of these solutions. While the presence of HMPB in solution has only a minor effect on the product formation, the copolymer exhibits a dramatic increase in photostability compared to P2NMA.

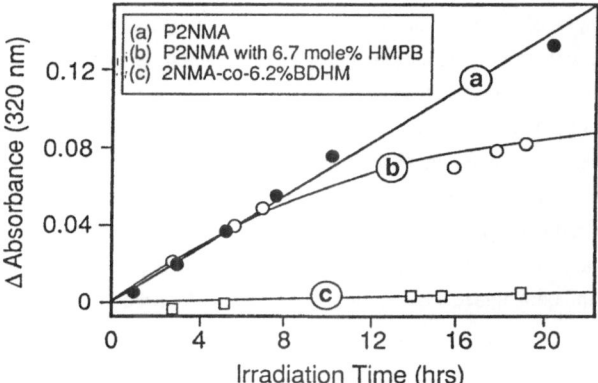

Figure 6. Plots of the change in absorbance at 320 nm upon irradiation at 285 nm of degassed solutions of the polymers.

304

A value for the quantum yield for acylnaphthol product formation, ϕ_p, from P2NMA in deoxygenated THF solution, upon irradiation using 280 nm light, has been estimated by Guillet and coworkers [14] to be 0.032. Using this value of ϕ_p as an estimate for the case of a degassed benzene solution of P2NMA irradiated using 285 nm light, the value of ϕ_p for the 2NMA-BDHM copolymer is calculated to be 0.001. This decrease may be attributed to the reduced lifetime of the excited singlet state of the naphthyl chromophores in the 2NMA-BDHM copolymer due to quenching by the BDHM traps.

The increase in the absorbance at 320 nm upon irradiation of a degassed benzene solution containing P2NMA and 6.7 mole% of HMPB displays an initial rate which is comparable with that observed for P2NMA indicating that the presence of HMPB in solution imparts little protection to P2NMA against photodegradation. Incorporation of the benzotriazole moiety covalently onto the 2NMA polymer backbone is required in order to increase the local concentration of the trap and thus inhibit photodegradation of the polymer.

CONCLUSIONS

The incorporation of BDHM into P2NMA results in significant quenching of both the monomer and excimer fluorescence intensity from the polymers in solution. The dominant mechanism for energy transport in room-temperature solutions of the 2NMA-BDHM copolymers in benzene is proposed to consist of rapid, one-step Forster-type energy transfer from both excimer and monomer species to a benzotriazole trap located on the same polymer chain.

The steady-state quenching of the fluorescence from the 2NMA-co-6.2% BDHM copolymer in solution is accompanied by a decrease in the fluorescence lifetime of both the monomer and excimer species. Analysis of the decay of excimer fluorescence from the 2NMA-co-6.2% BDHM copolymer, using the Forster theory, results in energy transfer parameters which are in good agreement with those calculated from steady-state measurements. Analysis of the decay of the monomer fluorescence from the copolymer is complicated. This is possibly a consequence of the existence of energy migration among the excited naphthyl chromophores and/or excimer back-dissociation. The 2NMA-co-6.2% BDHM copolymer displays a slower rate for the formation of acylnaphthol photoproducts, upon irradiation using 285 nm light, than P2NMA. This is attributed to the decrease in the lifetime of the excited naphthyl chromophores which is observed upon incorporation of the benzotriazole energy trap onto the 2NMA backbone.

ACKNOWLEDGEMENT

Financial support from the Australian Research Council is gratefully acknowledged.

REFERENCES

1. Phillips D (ed) (1985) Polymer Photophysics, Chapman and Hall, London
2. Guillet JE (1985) Photophysics and Photochemistry of Polymers, Cambridge University Press, Cambridge
3. Hoyle CE and Torkelson JM (1987) Photophysics of Polymers, ACS Symposium Series 358
4. Somersall AC and Guillet JE (1973) Macromolecules, 6:218
5. Forster Th (1949) Z Naturforsch, 4A:321
6. Forster Th (1959) Disc Faraday Soc, 27:7
7. Li S, Albertsson A-C, Gupta A, Bassett W and Vogl O (1984) Monatsch Chem, 115:853
8. Ghiggino KP, Bigger SW, Smith TA, Skilton PF and Tan KL(1987) In: Hoyle CE and Torkelson JM (eds) Photophysics of Polymers, ACS Symposium Series 358, Ch 28
9. Coulter DR, Gupta A, Yavrouian A, Scott GW, O'Connor D, Vogl O and Li S-C (1986) Macromolecules, 19:1227
10. Coulter DR, Gupta A, Miskowski VM and Scott GW (1987) In: Hoyle CE and Torkelson JM (eds) Photophysics of Polymers, ACS Symposium Series 358, Ch 22
11. Boudevska H, Brutchkov C and Astrug A (1979) Makromol Chem, 180:1113
12. Flory PJ (1969) Statistical Mechanics of Chain Molecules, Interscience, New York
13. Bennet RG (1964) J Chem Phys, 41:3037
14. Merle-Aubry L, Holden DA, Merle Y and Guillet JE (1980) Macromolecules, 13:1138
15. Holden DA, Shephard SE and Guillet JE, Macromolecules (1982) 15:1481
16. Ghiggino KP, Scully AD and Bigger SW (1988) In: Reichmans E and O'Donnell JH (eds) The Effects of Radiation on High-Technology Polymers, ACS Symposium Series 381, Ch 5
17. Ghiggino KP, Scully AD, Bigger SW, Yandell MD and Vogl O (1988) J Polym Sci, Polym Lett, 26:505

Technological Trend of Polymer Industry in the Age of New Chemistry

Masaaki Hirooka
Department of Economics, Kobe University
2-1, Rokkodai, Nada-ku, Kobe   657, Japan

TECHNOLOGICAL INNOVATIONS AND INDUSTRIAL RESTRUCTURING

While the .petrochemical industry has been in a matured state, a new chemical age has emerged to create the third generation of the chemical industry through the various technological innovations towards the 21st century.

According to Schumpeter's interpretation of Kondrachiev's long cycle, the economic prosperity is caused by a cluster of technological innovations inducing new industries. We are now in the course of the fourth cycles since the first industrial revolution. The present innovative age is distinguished from the previous ones by various characteristic features, especially in the sophisticated information-oriented society and the new chemical age.

MATERIALS TECHNOLOGY AND AGE OF NEW CHEMISTRY

The technological innovations are taking place in the fields of electronics-information technology, life science, new materials and new energy. The common feature of these high technology developments lies in analyzing and exploring at the molecular and atomic level. This induces the tendency that the chemical approach becomes more important and new materials play a critical role to the technological breakthrough. Thus chemistry will spread interdiciplinarily to new technological fields beyond the conventional territory to create a new era. The basic science is becoming an important part of the industrial research and chemistry is deeply involved in most innovative technologies of many industrial fields.

Outstanding advancements in performance and function of materials are very often achieved by the control of molecular conformation or higher ordered structure of known materials rather than by the inventions of novel compositions. Thus, the tailored design of materials is becoming important. The boundary between materials classifications such as ceramics, metals and polymers fades out and high performance and new functions are achieved beyond the conventional criteria. Especially, it has been revealed that control of material structure at nano-meter level leads to an extraordinary performance or a novel function.

POLYMER TECHNOLOGY IN THE FOURTH GENERATION

Now that the polymer industry has been in a matured state, newly commercialized polymers have hardly appeared and polymer technology seems to be saturated. From historical point of view, the first generation of polymer technology was the age of general purpose resins, the second was the development of engineering plastics, and it is now in the third generation of super engineering plastics. It could be said that we are approaching to the fourth generation in which polymers will be prepared at the molecular level on more practically at a nano-meter level. There will be a large gap between the third and fourth generations, and the development could be discontinuous. However, one of the practical evidence to be able to realize it was the gel-spun polyethylene having the extended chain structure in which an extraordinary mechanical strength was achieved. Liquid cuptalline polymers are another example.

B. C. Anderson · Y. Imanishi (Eds.)
Progress in Pacific Polymer Science
© Springer-Verlag Berlin Heidelberg 1991

308

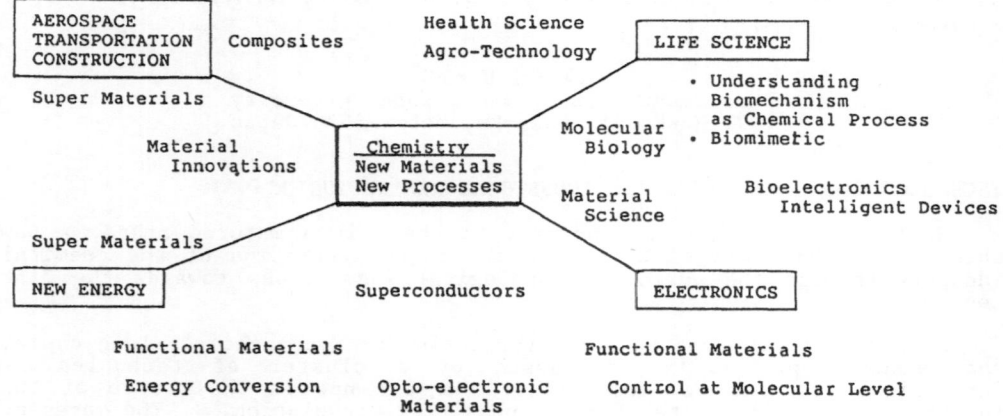

Figure 1.   Schematic View of the Age of New Chemistry

Conjugated polymers have been attracted in these days. Electrical conductivity of polyacetylene has reached at the level of 10^5 s/cm which almost corresponds to those of metals. Such high conductivity of the polymer is hardly explained in the conventional state of polymer conformation and could be realized by a more regular array of the polymers with the dopants. Epitaxially grown polyacetylene was successfully prepared and exhibited superior conductivity. Highly stretched poly(phenylene vinylene) showed superior mechanical strength to the conventional engineering plastics and was led to pure graphite film without any amorphous carbon by calcination at a higher temperature. These unexpected phenomena could be ascribed to the regular higher-ordered structure of polymers. Conjugated polymers are also expected to show excellent nonlinear optical properties by the regular arrangement of polymer conformation. Even superconductivity of conjugated polymers could be realized by a nano-scale control. These considerations encourage us to explore polymers of the fourth generation.

Furthermore, we should say that we are still in a premature stage of polymer technology. Let us look at the biomechanisms which are for more sophisticated than our level of technology. We have many to learn from such highly sophisticated bioorgans and are again encouraged to explore new polymers.

CHALLENGE OF POLYMER INDUSTRY

Finally, it should be stressed that polymer industry has been already involved in such basic research and is requested to realize many possibilities of new polymers under the collaboration with academia.

Petrochemicals for Malaysia

Ramlee Karim

Petroleum Research Institute

Lot 1026, PKNS Industrial Estate

Ulu Kelang 54200 Selangor, Malaysia

Abstract: A number of companies both foreign and local have expressed considerable interest to establish petrochemical operations in Malaysia. Proposals
for petrochemical complex to produce polyolefins and aromatics from crude oil
and naphtha has been announced; and joint-ventures to set up polypropylene and
MTBE plants have been formed. Interests in polyethylene have also been expressed
provided there is a stable supply of ethylene.

Malaysia does not have a large enough domestic demand for petrochemicals to
support a world scale economic size complex due to a small population and a
relatively undeveloped downstream plastic and fibre fabrication industries.

Petronas have embarked on the Peninsular Gas Utilization Project landing gas
from offshore, processing and distributing it to the South and to the West Coast.
Large quantities of ethane, propane and butane will be produced as a by-product
of processing the gas. The prospects are bright for a successful development of
the Petrochemical Industry in Malaysia.

1. INTRODUCTION

The topic of my lecture is "Petrochemicals for Malaysia." This, on first
examination, may be a contradiction in terms, and I will show why this is so
later. Although I make a special reference to Malaysia, this title could
equally apply to some of the neighbouring countries of Asean as well.

Once upon a time a British Resident brought some seeds of a plant from Kew
Botanical Gardens in London which originally came from Brazil to plant in
Malaysia. Very soon a network of Planters and estates was established throughout
the country - and in the process they made Malaysia the world's largest producer
of natural rubber for many years - and we still are. Then came petrochemicals;
and as you have heard from our previous presenter, we are trying our best to
make our natural product less biodegradable by cross-linking it with synthetic
materials in an effort to improve certain aspects of its natural qualities.

Then we thought we found silver and we started digging. In the process we made
ourselves the largest producers of tin for a long time. Then come petrochemicals
and as we know tin has lost its shine. We are still among the leading producers

B. C. Anderson · Y. Imanishi (Eds.)
Progress in Pacific Polymer Science
© Springer-Verlag Berlin Heidelberg 1991

of tin. Now we are also the largest producers of palm oil much to the discomfort
of some vegetable oil producers of the world. We are also dabbling with sugar
and spice; tea and rice - so against this backdrop of agricultural and mining
activities, petrochemicals not only seems out of character but also paradoxial.
But we kept on looking, we kept on digging and we now find ourselves a nett
exporter of crude oil. The quantities, however, are small; but fifteen years
ago we said oil would run out in fifteen years, and now we are saying that we
have another fifteen years. Hopefully in fifteen years time we will still be
able to say the same.

On the gas side, we export about 6 million tons of LNG annually to Japan from a
3-train LNG plant. We are putting up another three trains of similar capacity.
With the current anticipated utilization we expect the gas reserves to last
more than seventy years.

2. BACKGROUND

After years of lack-lusture growth, the global petrochemical industry has wit-
nessed a turn-around of fortune beginning from 1985. 1986 can be considered as
a boom year for the chemical industry. Petrochemical companies worldwide
experienced levels of profitability that prompted many experts to predict the
beginning of a new golden age for the petrochemical industry.

3. PROPOSALS

In Malaysia, a number of companies both foreign and local, have expressed considerable interests to establish petrochemical operations. The list of such
companies include Kellogg-Thyssen which recently announced that it will establish
a US$1 billion petrochemical complex producing polyolefines and aromatics from
crude oil and/or naphtha. Then there is Himont and C.T. Chao, a Taiwan-US joint
venture, proposing to set up a polypropylene plant in Pasir Gudang, Johor in
the south of peninsular Malaysia. Petronas, the National Oil Company, itself
has formed a joint-venture company with Neste Oy of Finland and Idemitsu Petrochemical Co. Ltd of Japan to establish an MTBE and Polypropylene Project in
Kuantan, Pahang on the East Coast of Peninsular Malaysia.

In the meantime, Petronas have also received a number of proposals from other
major petrochemical producers who are looking for a stable supply of ethylene
for the production of polyethylene and other derivatives. In this context BP
have very recently expressed their wish for a joint-venture partner to produce
ethylene. Companies are interested to locate their downstream polyethylene

plants in Malaysia if ethylene can be made available so as to position themselves to supply the Asia-Pacific regional market. These proposals call for
Petronas to set up an ethane cracker for the production of ethylene. It looks
most likely, therefore, that a joint-venture between BP and Petronas to
produce ethylene could emerge.

If all these proposals were to be implemented, then Malaysia is poised to become
one of the major petrochemical producers in the region. There is a possibility
that two ethylene crackers will be established. The one by Kellog-Thyssen will
be based on naphtha from crude oil, while the other by a Petronas joint-venture
will be based on ethane from natural gas. Also Petronas, Neste Oy and Idemitsu
Petrochemical will have a dehydrogenation facility in Kuantan, Pahang to produce
propylene and isobutylene. The facilities described above will have the combined potential capacity to produce approximately 550,000 mt of ethylene,
220,000 mt of propylene and 240,000 mt of isobutylene monomers. The Kellog-
Tyssen complex will also produce aromatics i.e. Benzene, Toluene and Xylenes.

Downstream, there will be two separate polyethylene units, one downstream of
Kellogg-Tyssen naphtha cracker and the other downstream of Petronas joint-
venture ethane cracker both of which will be capable of producing high density
polyethylene (HDPE) and linear low density polyethylene (LLDPE). As for polypropylene there may be three separate plants, namely a 150,000 MTPY downstream
of a naphtha cracker by Kellogg-Tyssen JV, a 80,000 MTPY downstream of a dehydrogenation complex by Himont-C.T. Chao JV; the latter based on imported propylene.

4. REGIONAL ACTIVITIES

Malaysia does not have a large enough domestic demand for petrochemicals to
support a world-scale economic size petrochemical complex. This is mainly due
to a small population and a relatively undeveloped downstream plastics and
fibres fabrication industries. The domestic demand for polyethylene is currently
estimated at 60,000 MTPY and that of polypropylene is around 35,000 MTPY.
Although the demand is expected to grow, the proposed combined production
capacities of the above projects is far in excess of the projected demand.
Kellogg-Tyssen is dedicating their entire production for the export market,
while the Petronas-Neste-Idemitsu JV will cater mainly for the domestic
market with the surplus for export.

Neighbouring countries - Indonesia and Thailand, which have relatively larger
domestic demand for petrochemical products, mainly due to a larger population
base, are also actively planning to develop their own petrochemical industries

to cater for the increasing domestic demand. Thailand is currently in the
midst of constructing an ethane-propane based petrochemical complex for the
production of ethylene and propylene derivatives, and is planning to build
a second petrochemical complex based on naphtha for the production of polyolefins and aromatics. Indonesia is planning to build three separate polypropylene plants and aromatics complex and may in the near future build a naphtha-
based petrochemical complex to supply the monomers required for the downstream petrochemical industries.

5. LOCATION OF PLANTS

Back in the early seventies, the global petrochemical industry was at its
lowest ebb. This was attributed to the global recession which had affected
many economies worldwide. During this period, there were no new petrochemical
facilities being built. In fact, the early 80's saw a shutdown of plants with a
total annual capacity of 8 million tonnes of surplus ethylene capacity in US,
Western Europe and Japan.

However, with an improvement in the global economic outlook, the demand for
petrochemicals started picking up in 1985/86. Consequently, this led to a
situation of tight supply, resulting in a sharp increase in the prices of
petrochemicals. Petrochemicals producers worldwide began to renew their
interest in petrochemical production. Fuelled by the surge in the profits,
they began to look for new investment opportunities.

5.1 Raw Materials

The present trend in the petrochemical business seems to be to locate the plants
close to its source of raw materials. This is especially so for the production
of bulk-commodity type petrochemicals such as polyolefins. Malaysia could
therefore take advantage of this trend and attract the potential investors in
view of its considerable reserves of natural gas and crude oil. Besides the
raw materials supply Malaysia is also located in a good geographical location
to supply the Asia-Pacific regional market.

5.2 Infrastructure

However, while availability of raw materials is one essential ingredient for the
successful implementation of a petrochemical project, it must be complemented
by good planning in terms of infrastructure development and other supporting
activities to make available this source of raw material at a competitive price.

It may be noted here that Petronas is taking a strong lead in ensuring that the conditions as stated above are fulfilled. Petronas have already incurred an expenditure in excess of US$10 billion in developing its oil and gas fields in the upstream sector of the industry. Although the production of crude oil is targetted mainly for the domestic fuel market, Petronas would consider favourably to allocate the local crude oil and refined products over and above Petronas requirements to those investors planning to establish downstream petrochemical industry in Malaysia.

In Peninsular Malaysia, Petronas have embarked on the Peninsular Gas Utilization (PGU) Project, landing gas from offshore and processing it at Kerteh, Trengganu in the East Coast of the peninsular, for distribution to the south including Singapore, and to the West Coast. Stage I (PGU I) of this project has already been completed at a cost of about US$200 million. Petronas is now implementing stage II (PGU II) of the project which is expected to cost about US$700 million. This involves the expansion of the existing gas processing facilities and the laying of 726 km transpeninsular gas pipelines. PGU II when completed by 1991 will have the capability of supplying in excess of 500 mmscf/d of natural gas. In addition, large quantities of ethane, propane and butane will be produced as a by-product during the processing of the gas. This means a steady supply of petrochemical feedstock will be made available from 1991.

To ensure this availability of feedstock for the development of petrochemical industry and to justify the substantial investment in the PGU project, the demand for gas in the country must be sufficiently large. In Malaysia, the power generating sector is potentially the biggest consumer of gas in the country. Plans are underway to convert the power stations now running on fuel oil to using gas.

6. CONCLUSION

The prospects are bright for the successful development of the Petrochemical Industry in Malaysia. Petronas is implementing one of the most comprehensive gas processing and distribution network in the region that would provide Malaysia with a comparative advantage with regard to the supply of raw materials. The success of the gas utilization programme is directly dependent on the country's energy policy and more specifically the choice of fuel for the electricity generation.

We are at a very exciting stage of petrochemical activities in the country. To misquote Churchill, we are at "The Beginning of the Beginning".

Recent Advances in the Determination of Crosslinking and Scission in Irradiated Polymers

David J.T. Hill, David A. Lewis[+], James H. O'Donnell[*],
Andrew K. Whittaker[++], Donald J. Winzor[x]
and Catherine L. Winzor[+++]

Polymer Materials and Radiation Group
Department of Chemistry, University of Queensland
Brisbane 4067, Australia

[x]   Department of Biochemistry, University of Queensland

Abstract: The sensitivity of polymer materials to radiation is
determined by the changes in properties, which reflect increases or
decreases in molecular weight.   The dose-dependence of M_z and M_w
for deriving yields of chain scission and crosslinking is described,
and use of the ultracentrifuge for simultaneous measurement of M_w
and M_z.   Evidence for the mechanism of crosslinking in polyolefins
is derived from NMR spectra.   Soluble fraction data for
poly(arylene sulfone) are analysed for end-linking.

INTRODUCTION

The changes in the properties of polymer materials resulting
from irradiation by gamma rays, electrons and other forms of high-
energy radiation are becoming of increasing technological
importance(1).   New and improved processes in various industries
involve the use of radiation as a key "reagent".   Development of

---

+ Present address: T.J. Watson Research Center, IBM, Yorktown
Heights, New York 10598, USA

++ Present address: BP Research Laboratories, Sunbury-on Thames,
Twickenham, Middlesex TW16 7LN, UK

+++   Present address; Dept of Chemical Engineering, University of
New Hampshire, Durham, New Hampshire 03824, USA.

B. C. Anderson · Y. Imanishi (Eds.)
Progress in Pacific Polymer Science
© Springer-Verlag Berlin Heidelberg 1991

advanced technologies utilizing radiation depends on an understanding of the chemical reactions resulting from irradiation and the relationship of these reactions to changes in the properties of the polymer materials(2).

The most important feature of gamma and electron radiation is that the energy of the radiation is sufficiently high to cause ionization of the substrate molecules through ejection of an electron from the valence shell to leave the parent radical cations, P_+^+. The electrons may be trapped at acceptor sites to form radical anions, P_-^-. Energy transfer from the radiation which is insufficient to produce ionization results in excitation of the polymer molecules, P^*. Recombination of electrons or radical anions with radical cations also produces excited molecules.

Dissociation of excited molecules, either homolytically to form neutral free radicals, or heterolytically to form anions and cations, and ion-molecule reactions provide the next step in the degradation mechanism(3). These reactions are shown in Fig. 1.

$$P \longrightarrow\!\!\!\!\wedge\!\!\wedge\!\!\wedge\!\!\!\!\longrightarrow P_+^+ + e_-^- \qquad\qquad e_-^- + P \rightarrow P_-^-$$
$$P \longrightarrow\!\!\!\!\wedge\!\!\wedge\!\!\wedge\!\!\!\!\longrightarrow P^* \qquad\qquad P_+^+ + e_-^- \rightarrow P^*$$
$$P^* \rightarrow R_1\!\cdot + R_2\!\cdot, \quad R_3\!\cdot + H\cdot, \quad A^+ + B^-$$

Fig. 1.   Reaction scheme for irradiation of polymers.

The reactive intermediate radicals and ions produced by the ionization and excitation steps shown in Fig. 1 undergo chemical reactions which cause changes in the molecular structure of the polymer. It is these changes which are mainly responsible for the modification of the properties of polymer materials. The chemical changes may be classified as shown in Table 1.

Table 1.   Classification of chemical changes in irradiated polymers.

1. scission of polymer backbone chains
2. crosslinking of polymer molecules
3. formation of small molecule products
4. structural changes in the polymer

The scission and crosslinking reactions have the greatest effect on the material properties of polymers and measurement of the radiation chemical yields of these two processes is very important for quantifying degradation of polymers by radiation(4). The main methods used to determine the yields of scission and crosslinking are listed in Table 2.

Table 2.   Methods for determination of scission and crosslinking in irradiated polymers.

1. soluble and gel fractions
2. molecular weight averages
3. molecular weight distributions
4. swelling ratio
5. modulus (of rubbers)
6. NMR spectroscopy

In this paper we present various improvements and extensions of some of these methods and examples of applications to selected polymers.

EXPERIMENTAL
Materials

Polysulfone, PSF, Udel P1700, was obtained from Union Carbide. The ethylene-propylene copolymers from Japan Synthetic Rubber Co. were provided by Dr S. Machi of the Japan Atomic Energy Research Institute.   Poly(methacrylic acid) was prepared by polymerization of methacrylic acid in aqueous HCl with persulfate initiator and purified by dialysis and freeze drying.
Irradiations

Gamma irradiations were performed in AECL Gammacell 200 and 220 units and in the pond facility of the Australian Nuclear Science and Technology Organization (ANSTO) at dose rates from 1 - 10 kGy/h. The samples were evacuated for 24 h at 25 $^{\circ}$C and for a further 12 h at above-ambient temperature to facilitate removal of solvent and water.
Characterization

Molecular weight distributions and soluble fractions were measured with a Waters gel permeation chromatograph. M_w and M_z of PMMA were obtained by sedimentation equilibrium using a Beckman Model E ultracentrifuge.

Solution NMR spectra were obtained with a JEOL GX400 spectrometer and solid state spectra with a Brucker CXP300 spectrometer using magic-angle spinning.
Measurement of Soluble Fractions

Soxhlet extraction is the classical method for dissolving soluble, uncrosslinked polymer out of a crosslinked network.   This method is restricted to relatively large blocks of polymer and is then limited by inefficient diffusion of large polymer molecules from the bulk of the sample into the surrounding solution.   It is necessary to carry out several cycles of extraction and drying and frequently the weight of insoluble gel continues to decrease, indicating disintegration of weak, lightly crosslinked gel.

An alternative procedure is to enclose the polymer sample in a stainless-steel mesh bag and immerse the bag in boiling solvent(5). This technique has the advantage that a number of samples can be extracted simultaneously.   It is useful for samples in the form of sheet, film or granules, but not for powders.   The polymer samples swell to many times their original volume at low crosslink densities, i.e. after doses not greatly in excess of the gel dose and the soft gel frequently extrudes through the mesh of the bags. Loss of gel through the extrudate breaking off and by adhesion to other samples and the container is then a problem.

We have shaken samples of irradiated polymer with solvent and then separated the soluble and gel fractions by centrifugation(6). Both phases can be recovered and determined gravimetrically after drying.   Advantages of this method are that (1) small samples (<50 mg) and (2) polymers in powder form can be used.   However, gravimetric methods are intrinsically slow.

The separation, drying and weighing operations can be avoided by using a gel permeation chromatograph to determine the soluble fraction(7).   A known volume of the liquid phase is removed after shaking the sample with solvent in a standard flask for 24 hours and allowing it to stand so that the swollen gel separates.   An aliquot is then filtered and injected into the chromatograph using a refractive index (RI) detector, which is insensitive to chromophores produced in the polymer by irradiation, to determine the concentration of polymer.   The RI detector is calibrated with unirradiated polymer.

A problem with this method, which is straight-forward, fast and accurate is that despite careful filtering, the GPC columns inevitably become cloggged with microgel.   This results in increasing pressure in the system and eventual cessation of operation and the need to rejuvenate or replace the columns.   A recent development which could overcome this problem is field flow fractionation in GPC(8).   There are no columns to become clogged and the plates can be separated and cleaned.

RESULTS AND DISCUSSION

Mechanism of Crosslinking

The crosslinks formed in irradiated polymers are usually attributed to combination of radicals on different molecules in close proximity to each other.   This proximity may result from C-H scission, followed by H abstraction from the adjacent molecule, or by migration of radical sites along the polymer backbone, e.g. by a hydrogen-hopping mechanism.   This conventional and widely accepted mechanism of crosslinking produces a H-type crosslink.   The reactive species may also be ions or radical-ions.

An alternative type of crosslink, known as a Y-link, has been periodically suggested, but only recently has evidence been obtained for formation of these crosslinks.   Randall et al.(9) identified Y-links in the solution ^{13}C NMR spectra of polyethylene irradiated below the gel dose.   They suggested that these crosslinks were formed by addition of a backbone radical to a terminal double bond on another molecule (mechanism II in Fig. 2).

Charlesby(10) suggested much earlier that Y-links could be formed in irradiated polymers by reaction of a chain-end radical, formed by scission of the main chain, with a backbone radical on another molecule. as shown in mechanism III.   We have proposed that crosslinks in irradiated polysulfone may be formed by this mechanism.

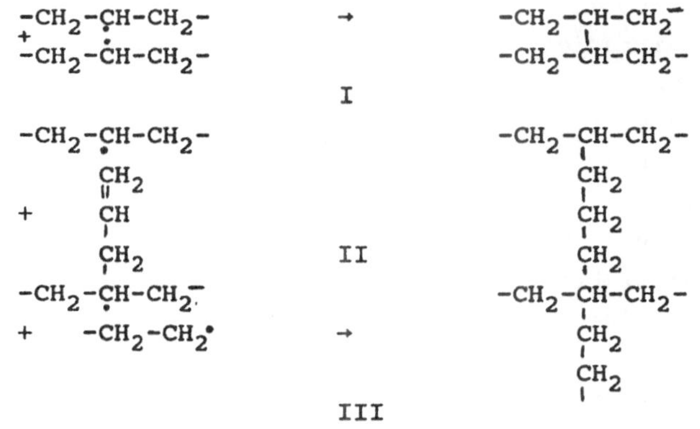

$$
\begin{array}{ccc}
-CH_2-\underset{+}{\overset{\bullet}{C}H}-CH_2- & & -CH_2-CH-CH_2^- \\
-CH_2-\overset{\bullet}{C}H-CH_2- & \rightarrow & -CH_2-CH-CH_2- \\
& I &
\end{array}
$$

$$
\begin{array}{ccc}
-CH_2-\overset{\bullet}{C}H-CH_2- & & -CH_2-CH-CH_2- \\
\quad CH_2 & & \quad CH_2 \\
+ \quad \parallel \; CH & & \quad CH_2 \\
\quad CH_2 & II & \quad CH_2 \\
-CH_2-\overset{\bullet}{C}H-CH_2^- & & -CH_2-CH-CH_2- \\
+ \quad -CH_2-CH_2^{\bullet} & \rightarrow & \quad CH_2 \\
& & \quad CH_2 \\
& III &
\end{array}
$$

Fig. 2   Mechanisms for crosslinking of polymers by radiation.
I: H-linking; II and III: Y-linking.

The yield, G(X), of H-type crosslinks can be obtained from a plot of $s + s^{1/2}$ versus 1/dose, where s is the soluble fraction. The Charlesby-Pinner(11) equation (1) predicts a linear relationship for an initial most-probable MWD and G(X) can be calculated from the

slope of the plot.   The yield of scission, G(S), is then obtained
from the intercept at 1/D = 0, i.e. at infinite dose.
Modifications of this equation for other initial MWD's, which can be
represented by Schulz-Zimm or Wesslau distributions, have been
derived by Inokuti(12) and by Saito et al.(13).

$$s + s^{1/2} = \frac{G(S)}{2G(X)} + \frac{4.82 \times 10^9}{G(X).M_n(0).D} \qquad (1)$$

where D is the dose in Gy, Mn(0) is the initial number-average
molecular weight, and G(S) and G(X) are the number of scissions
occurring and crosslinks formed per 100 eV (16 aJ) of energy
absorbed.

We have modified an equation derived by Saito(14) relating the
soluble fraction to radiation dose for Y-type crosslinking according
to mechanism III to obtain the relationship shown in eqn. 2 for an
initial most-probable MWD.   If $1 + 3s^{1/2}$ is plotted versus 1/D a
linear relationship should be obtained.   G(X) can be derived from
the slope and G(S) from the intercept of the plot.

$$1 + 3s^{1/2} = \frac{2G(S)}{G(X)} + \frac{4.82 \times 10^9}{M_n(0).G(X).D} \qquad (2)$$

We have tested eqn. 2 for polysulfone irradiated in vacuum at
temperatures from 30 °C to 150 °C and obtained linear relationships
for gamma and electron irradiation at moderate dose rates.   A
typical example is shown in Fig. 3.   However, irradiation with an
electron beam at high dose rates gave a curved plot, which was
attributed to an increase in temperature of the sample.
NMR Determination of Crosslinks and Scission

Measurements of soluble fractions, average molecular weights
and molecular weight distributions enable yields of crosslinking and
scission to be derived using appropriate equations.   However, these
relationships depend on assumptions which may not be valid, e.g.
that the crosslinks are distributed spatially at random in the
polymer.

Direct observation of crosslinks and of new chain ends
resulting from scission is possible from the NMR spectra of the
irradiated polymers.   The main difficulty with this method is that
NMR is relatively insensitive and only a small proportion of the
atoms of the polymer become crosslinked.   If crosslinking
predominates, i.e. G(X) > 4G(S), then the polymer will become

insoluble and solution NMR spectra are not possible. NMR spectra may then only be obtained by the solid-state, magic-angle spinning technique, when the resolution and sensitivity are greatly reduced - and increasingly so after high radiation doses.

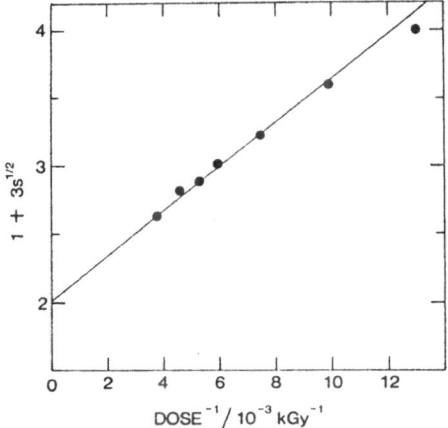

Fig. 3. Soluble fraction data for PSF irradiated in vacuum at 150 °C plotted according to eqn. 2 for Y-type crosslinking.

We have studied crosslinking and scission in ethylene-propylene copolymers containing 23-36% propylene(15). These copolymers gave narrow spectral lines in solution. Irradiation caused crosslinking and scission, with crosslinking predominating. The soluble fraction decreased with dose above the gel dose D_g and the NMR line widths rapidly increased in the solution spectra.

Fig. 4 shows a typical solution spectrum of EP copolymer after a moderate dose. The intensity of the spectrum decreases with increasing dose indicating that only part of the swollen, insoluble polymer is being observed. This is likely to contain chain ends, but not crosslinked regions of the polymer molecules. New peaks attributable to H-crosslinks (42 ppm) and methyl end groups (11-23 ppm) are evident. The intensities of the methyl peaks increased linearly with dose and a G value for scission could be derived. The intensity of the crosslinked carbons decreased at higher doses and was not suitable for estimation of G(X).

The estimate of G(S) can be compared with values obtained by measurements of soluble fractions, as shown in Table 3.

The unirradiated EP copolymers gave good solid-state NMR spectra with a line width of about 50 Hz. The spectrum could be simulated accurately with the appropriate numbers of different carbon atoms according to the composition of the copolymer, except that the intensity of the methyl peak at 20 ppm was less than expected, which is attributable to the greater mobility of the chain ends.

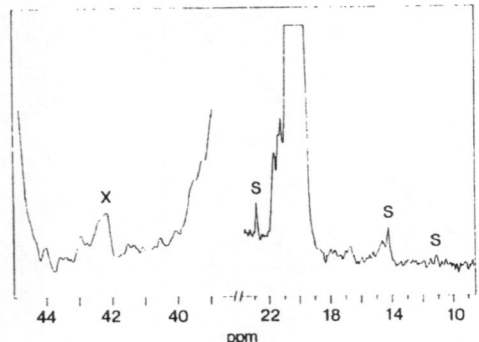

Fig. 4. ^{13}C NMR solution spectrum of ethylene-propylene copolymer after gamma irradiation in vacuum to 135 kGy.

The solid-state NMR spectrum showed increasing line width and decreasing resolution with increasing radiation dose. The spectrum could be simulated by a set of gaussian peaks with an increased line width (100 Hz after a dose of 5 MGy). However there was also a small peak at 22 ppm which we have assigned to methyl groups at the new chain ends formed by main-chain scission. Comparison of the experimental and simulated spectra also revealed a broad peak in the 40-46 ppm region, corresponding to H-type crosslinks, as observed in the solution spectra after low doses.

G values for scission and crosslinking were estimated from these peaks and are shown in Table 3. There is good agreement between the G values for scission and crosslinking derived from soluble fractions ($D>D_g \rightarrow >>D_g$, solution NMR (low doses $D<4D_g$) and solid-state NMR (high doses $D>>D_g$). This agreement is persuasive that the crosslinks and scissions occur spatially at random in these copolymers and that the yields are independent of radiation dose. The copolymers are almost completely amorphous. Different degradation behaviour could be expected in amorphous and crystalline regions.

Average Molecular Weights

G values for scission and crosslinking can be obtained from the dose dependence of two average molecular weights for the polymer. The averages, M_n and M_w, are normally used, and it is usually assumed that the polymer sample initially has a most-probable MWD. We have shown that the use of average molecular weights is only appropriate up to about 0.3 D_g due to the effect of the high-molecular weight tail of the distribution of the irradiated polymer. Soluble fractions cannot be used if G(X) < 4G(S) as the polymer will not reach a gel dose.

Table 3. G values for crosslinking and scission in EP copolymer containing 36% propylene determined by different methods.

Method	G(S)	G(X)
Solution ^{13}C NMR	0.37	–
Solid State $_{13}$C NMR	0.31	0.84
Sol fraction	0.38	0.82

The values of M_n and M_w are most frequently obtained from GPC measurements of the MWD. However, this method gives values, especially of M_w, which are increasingly susceptible to error with increasing G(X) and dose because the hydrodynamic volume of the polymer molecules will decrease with branching for the same molecular weight.

PSF irradiated in vacuum at 150 °C gave typical linear plots of $1/M_n$ and $1/M_w$ versus dose. This polymer has a most-probable initial MWD and crosslinking predominates over scission. The gel dose is given by extrapolation of $1/M_w$ to zero. There is a relatively small change in $1/M_n$ and a drift in $1/M_w$ above 0.3 D_g, which are disadvantages of this procedure.

If the initial MWD differs from the most-probable distribution then a modified eqn. should be used for M_w, although this is frequently neglected, which can lead to significant errors.

The variation in M_z with dose for an initial Schulz-Zimm MWD is given in eqn. 3. We have transformed this eqn. and utilized selective truncations of the mathematical series to obtain an equation amenable to linear analysis(16). A simplified form of the relationship for an initial most-probable MWD is given in eqn. 4.

$$M_z(D) = \frac{3M_n(0)[\phi_2(u\tau D,\sigma)/\phi_1(u\tau D,\phi)]}{[1 - (4\chi/u\tau^2 D\phi_1(D,\phi)]^2} \quad (3)$$

$$([M_z(O)/M_z(D)] - 1)/D = (\lambda - 8)u\chi + 16(u\chi)^2 D + \ldots \qquad (4)$$

M_w and M_z can be determined by sedimentation equilibrium in an ultracentrifuge. This procedure has the great advantage over GPC that the values of M_w are absolute and are not affected by crosslinking. Moreover, ultracentrifugation is particularly suitable for water-soluble polymers, especially polyelectrolytes, which are often not conveniently measured by GPC.

The use of M_w and M_z measurements on irradiated polyelectrolytes using sedimentation equilibrium(17) is shown in Fig. 5 for poly(methacrylic acid).

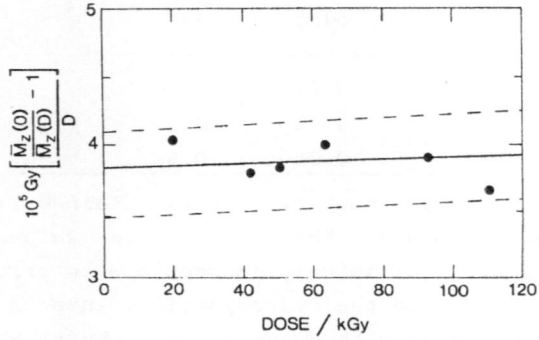

Fig. 5. M_z plot according to eqn. 4 for poly(methacrylic acid) irradiated in vacuum at 25 °C for determination of G(S) and G(X).

CONCLUSIONS
     The determination of radiation-induced yields of scission, G(S), and crosslinking, G(X), from measurements of average molecular weights is extended by utilizing M_z. It is useful to determine M_w and M_z simultaneously by sedimentation equilibrium in the ultracentrifuge.

     However, understanding of the chemical reactions occurring in the polymer to produce this scission and crosslinking requires structural characterization of the polymer. ^{13}C NMR spectroscopy of the irradiated polymer, in solution and the solid state, is a valuable technique.

Acknowledgements
     The authors are grateful to the Australian Research Council and the Australian Institute of Nuclear Science and Engineering for supporting their research, the Australian Nuclear Science and Technology Organization for providing irradiation facilities, Mr

D.F. Sangster for advice with the irradiations, Mr. J. Leeder for the ultracentrifuge measurements.

References

1 Reichmanis E, O'Donnell JH (eds) (1989) The Effects of Radiation on High-Technology Polymers, ACS Symposium Series 381, Washington
2 Thompson LF, Willson CG, Frechet JMJ (eds) (1984) Materials for Microlithography, ACS Symposium Series 266, Washington
3 O'Donnell JH, Sangster DF (1970) Principles of Radiation Chemistry, Edward Arnold, London
4 Hill DJT, O'Donnell JH, Pomery PJ, Sangster DF (1989) Radiation Chemical Yields: G Values, in Brandrup J, Immergut E.H. (eds) Polymer Handbook, 3rd Edn, John Wiley, New York
5 Dole M (1979) Polym Plast Technol Eng 13:41
6 O'Donnell JH, Rahman NP, Winzor DJ (1977) J Polym Sci 15:131
7 Hellman MY, Bowmer T, Taylor GN (1983) Macromolecules 16:34
8 Giddings JC (1988) Chem Eng News, Oct.10, 34
9 Randall JC, Zoepfl FJ, Silverman J (1983) Makromol Chem Rapid Commun 4:149
10 Charlesby A (1960) Atomic Radiation and Polymers, Pergamon, Oxford
11 Charlesby A, Pinner SH (1959) Proc Roy Soc A249:367
12 Inokuti M, (1963) J Chem Phys 46:3607
13 Saito O, Kang HY, Dole M (1967) J Chem Phys 46:3607
14 Saito O (1959) J. Phys. Soc. Japan, 14, 798
15 O'Donnell JH, Whittaker AK (1985) Br Polymer J 17:51
16 O'Donnell JH, Winzor CL, Winzor DJ (1990) Macromolecules 23:167
17 Hill DJT, O'Donnell JH, Winzor CL, Winzor DJ, Polymer, in press

New Polymer Coating for Reducing Ice and Snow Damages in Railways

Fujio Ohishi, Heihachi Murase*, Kiyoshi Nanishi*
Hideo Kogure*, and Manabu Hashiba

Railway Technical Research Institute
2-8-38 Hikaricho, Kokubunji, Tokyo, 185-JAPAN
*Technical Research Laboratory, Kansai Paint Co.,Ltd.
4-17-1 Higashiyawata, Hiratsuka, Kanagawa, 254-JAPAN

Abstract: The present study deals with the new polymer coating systems for prevention from railway accidents due to ice and snow accretion. The new polymer coatings developed for this purpose are a siliconemacromer-grafted fluorocarbon polymer(#10) and poly-perfluoroalkyl acrylate containing hydrophobic filler (#20). For the evaluation of surface properties of these materials the free surface energy and adhesive strength of ice and snow against them comparing with a new water-repellent clothes were measured. Furthermore, wind tunnel test was carried out and discussed the relationship between these results and surface properties. Applications of these systems to the railway use, i.e. to the pantograph, trolley wire and car body were tried.

Introduction: Recently there has been growing interest in and actual need for ice- and snow preventive materials. Though the actual application of such materials is limited to some restricted regions and periods in the field of navigation, aviation and ground traffics, the problem has been serious when accidents have taken place involving human life. In the railway, various means have been tried to prevent damages due to ice-, snow- and frost accretion. In the previous study[1], the authors reported the investigation upon the effects of a organo-polysiloxane modified with a lithium compound (#6) on the prevention from ice and snow adhesion, through monitoring tests by using pantograph. The statistical treatment of these results proved that the pantograph coated with the polymer coating #6 reduced accidents to one-

B. C. Anderson · Y. Imanishi (Eds.)
Progress in Pacific Polymer Science
© Springer-Verlag Berlin Heidelberg 1991

sixth in comparison with uncoated. Since then, this coating has been successfully applied to the object concerned.

In this study, two types of novel polymer coatings were developed, in order to clarify the different behaviour between ice and snow adhesion. Informations obtained will make feasible to apply them to the practical use in railway.

Tab.1 Test samples and their Free Surface Energies

Symbol	testing sample	surface energy (mN/m)
A	new coating ㉒-1	1 . 9 5
B	new coating ㉒-2	2 . 9 6
C	new coating ㉒-3	1 . 3 0
D	new coating ㉒	2 6 . 9
E	new coating ⑥	2 8 . 9
G	P T F E	2 4 . 3
O	b l a n k (Al)	3 9 . 6
H	MICROFT RECTAS	1 . 8 9
I	T 1 4 5 6	2 . 1 9

Experimental:

Materials: Materials used in this study are listed in table 1. A series of new polymer coating #20 are composed essentially of a copolymer having perfluoroalkyl chain and hydrophobic filler. Subordinate No.-1, -2 and -3 indicate 80, 100 and 200 phr of filler contents, respectively. The copolymer was synthesized by means of radical polymerization of comonomer, perfluoro(meth)acrylates and vinyltrimethoxysilane. Chemical structure of repeat unit is described as follows:

$$\left(\begin{array}{c} H \ \ Me \\ | \ \ | \\ C-C \\ | \ \ | \\ H \ \ C=O \\ \ \ \ | \\ \ \ OR_s \end{array} \right)_x \left(\begin{array}{c} H \ \ Me \\ | \ \ | \\ C-C \\ | \ \ | \\ H \ \ C=O \\ \ \ \ | \\ \ \ OR_{f1} \end{array} \right)_y \left(\begin{array}{c} H \ \ H \\ | \ \ | \\ C-C \\ | \ \ | \\ H \ \ C=O \\ \ \ \ | \\ \ \ OR_{f2} \end{array} \right)_z$$

where R_s is alkoxysilane, R_{f1} is perfluoroalkyl, x, y and z are number of repeated monomer units. As a hydrophobic filler, silica reacted with a hexamethyl-disilazane is used. Radii of silica powder treated are distributed 0.1 - 0.01 μm. Surface structure determined by XPS investigation is shown schematically in Fig.1.

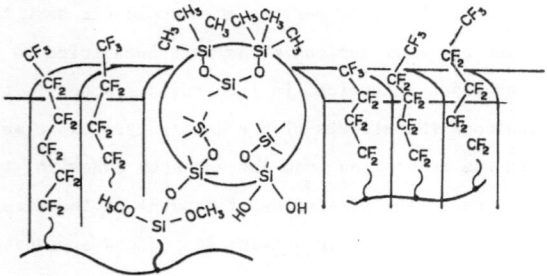

Fig.1 Schematic representation of coating #20

Another type of ice repellent polymer material(#10) was synthesized by grafting of organo-polysiloxane-macromer($\overline{M}w$ 5000-10000) to the copolymer consisted of
tetrafluoroethylene and hydroxylvinylether. Formulation of repeated unit of this
polymer is in following:

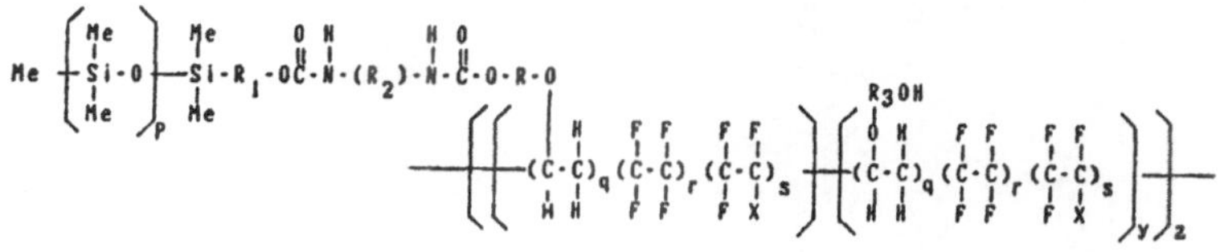

$$
Me \left(\begin{matrix} Me \\ Si-O \\ Me \end{matrix}\right)_p \begin{matrix} Me \\ Si-R_1-OC-N-(R_2)-N-C-O-R-O \\ Me \end{matrix}
$$

where, R_n is alkyl, X is halogen, p,q,r and s are number of repeated monomer units,
Y,Z are number of repeated segments. The product #6 is the coating based on organo-
polysiloxane modified with a lithium compound, which is precisely described in the
previous report[2]. PTFE is a commercial product(Nittoh Denko Co.) of polytetrafluoroethylene tape of 0.07 mm thick laminated on alminum plate was selected as the
typical hydrophobic material. As the control bare aluminum plate was used. Water
repellent clothes (Teijin Co.,MICROFT & T1456) made of ultra-thin fiber of polyester were set in reference.

Apparatus and test methods

1. Apparatus:

1-1. Goniometer: For the measurement of contact angle of sessile drop of water on
substrate coated we used a goniometer(Kyowa Kaimenkagaku Co.) consisting of a con-
actanglemeter, TV monitoring system.

1-2. Ice Adhesion Tester: For the measurement of adhesive strength of ice, Murase
deviced a new programming test apparatus with good accuracy and reproducibility.
The apparatus consists of a temperature regulated chamber, load cell, process con-
troller and recorder. The outside view of the apparatus is shown in Fig. 2.

1-3. Snow Adhesion Tester: For the measurement of adhesive force of snow, test
apparatus deviced by Ohishi was used. This apparatus,as shown in Fig.3, consists of
refregerator, blower, ultra sonic steam generator and load sensor.

1-4. Wind Tunnel Installation: For the evaluation of tendency toward snow accretion, the wind tunnel installation laid out by Fujii (R.T.R.I.) was used. This in-

330

Fig.2 Outside view of Ice Adhesion Tester

Fig.3 Outside view of Snow Adhesion Tester

Fig.4 Wind Tunnel Installation

stallation equipped snow stocker, mill wheel and blower, is built up in cold chamber.

2. Test Methods

2-1. Contact angle: Sessile droplet of $50\mu l$ distilled-deionozed water is set on the test panel, then contact angle thereof is measured at 20 ℃. In the same way contact angle of liquid paraffin is measured. From contactangle, thus obtained free surface energy of materials can be calculated from Young-Dupré's equation.

2-2. Adhesive strength of ice: Metal ring with inside section area 5 cm², height 1.5 cm is set on the specimen and 2 ml of distilled-deionized water is poured in the ring. After keeping for 1.5 hours at -10 ℃ the strength of ice is measured by shear.

2-3. Adhesive force of snow: Artificial snow which is made under - 5 ℃ atmosphere for three hours by super-cooled steam generated from ultra sonic boiler, is blowered toward test panel(90x180mm) set up inclined 45° against vertical line. The thickness of snow formed on the panel is measured and also adhesive strength of snow is evaluated as a removing force by means of spring balance in shear.

2-4. Wind tunnel test for the evaluation of snow accretion is carried out under condition at - 5 ℃, 5m/sec of wind speed, for 7min.of applied time by using stored ed natural snow. Two types of, i.e. dry- and wet(15% of water content) snow are prepared. This equippment makes it possible to deliver a definite amount of milled snow.

Results:

1. Contact angle and free surface energy

Fig.5 shows the contact angles measured. A series of new polymer coatings #20(A,B,C) has the highest value, over 150° compared with others. Among #20-1, -2, -3 it can be seen not so great differency. This shows that the contact angle depends not so strong on filler contents. Water repellent clothes have also considerably high value of contact angle, however get wettability by light fingertip friction.

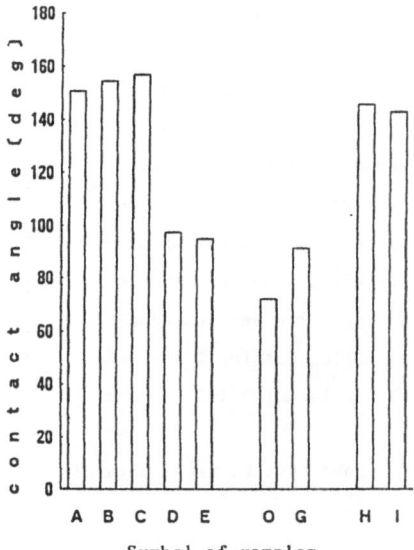

Fig.5 Contact angle of materials

Free surface energy calculated is tabulated in Tab.1. Extreme low values of free surface energy in the coatings #20 and water repellent clothes do not seem to indicate an intrinsic value, but contain some morpholosical effects in itself.

2. Adhesive strength of ice

Fig.6 shows adhesive strength of ice. The coating #20-1 has the minimum value so far as resistered. #20-2, #20-3 and #6 give also quite low ice adhesion strength. Graft polymer #10 shows a considerable high value. this is, however substantially lower value in comparison with that of PTFE. Aluminum and water repellent clothes.

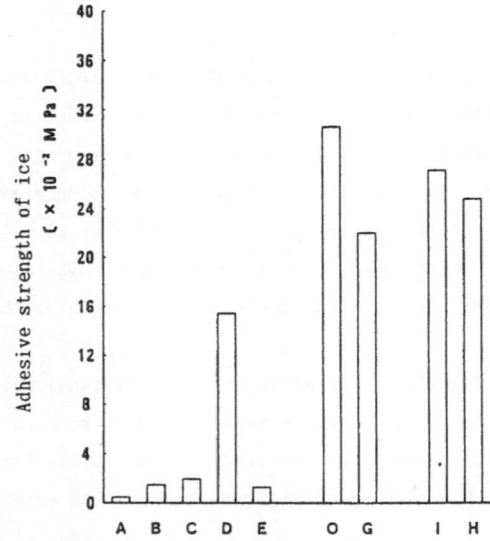

Fig.6   Adhesive strength of ice

3. Adhesive force of artificial snow

Fig.7 shows the force to remove artificial wet snow. As shown in this figure. there is not so distinct deviation among all samples than contact angle and ice adhesive force. The coating #20-3 is measured for lowest value.

4. Wind tunnel test by using natural snow

Results are summarized in Tab.2.

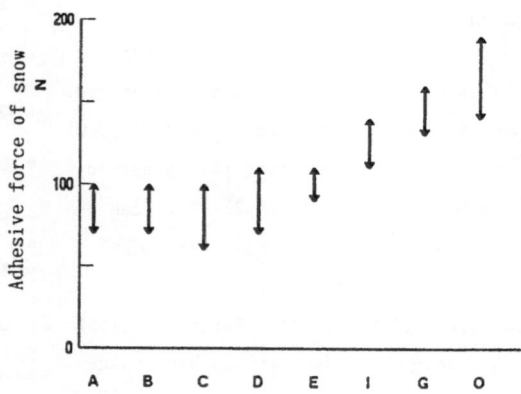

Fig.7   Adhesive force of snow

Tab.2 Amount(weight) of accreted snow (g/300x600mm)

Test samples	Amount of dry snow	Amount of wet snow
Coating #20-2	0	130
Coating #10	40	154
Coating # 6	24	143
PTFE	72	177
Aluminum	84	174
MICROFT RECTAS	53	260

Accretion amounts of wet snow are much higher than that of dry snow.
It is remarkable that the dry snow hardly adhere to the surface of the coating
#20-2, while wet snow keeps to accumulate on it, but that can be removed very
easily. It is also noticeable that the coating #6 can not prevent snow accretion,
even if the coating has very low ice adhesion strength. Repellency against snow
seems to depend on the intensity of hydrophobicity, it is however not valid, so
far as in the cace of water repellent cloth. Fig.8, 9 and 10 show snow accretion
of #20-2, Alminum and water repellent cloth, respectively.

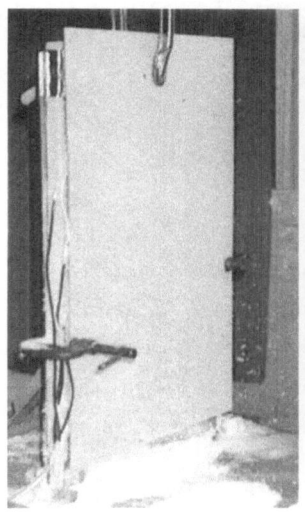 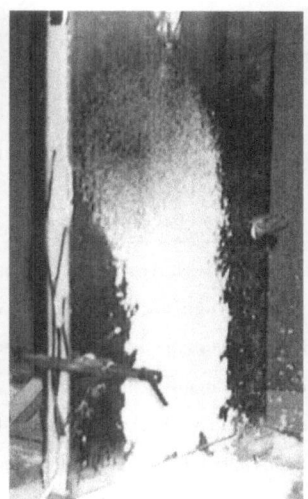 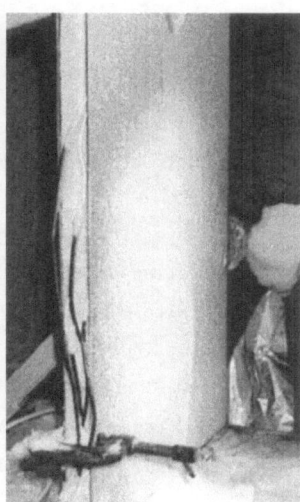

Fig.8 Coating #20-2   Fig.9 Aluminum   Fig.10 Water Repellent Cloth

334

Trial Applications and Discussions:

Based on the evaluation of ice and snow repellent performance of the new
polymer coatings, we have tried their applications to railways as shown bellow:

1. Prevention of pantograph from accident caused by snow adhesion

Arc generation by depression of pantograph under condition as like sherbet-snow
accretion causes the accident of pantograph. We applied the coating #6 to pantograph shoes for the purpose to remove sherbet easily, by down-up motion of pantograph. Field tests were carried out, using about one hundred of pantographs for
express trains in Kanazawa area, rear part of Japan (Kanazawa Car Operation Center)
As a conclusion, the coating #6 proved to be effective for prevention from accidents. We established new coating system using #6 to pantograph for practical uses.
Fig.11 shows the pantograph coated with #6 under monitoring test. Fig.12 shows TV
monitoring system for pantograph.

Fig.11 Pantograph under monitoring   Fig.12 TV Monitoring System

2. Prevention of trolley wire from accident caused by frost adhesion

Arc generation by frost adhesion to trolley wire causes accident of pantograph
and trolley wire. We applied the coatings #6 and #10 to trolley wire for the purpose to remove frost easily, when pantograph slided on the wire. Field monitoring
tests on several occasions were carried out, however any arc generation was not observed, under tested conditions. Artificial ice-adhesion test by dropping water on
trolley wire under icing condition proved to be effective for prevention from generation in the case of coatings #6 and #10. In this winter, we applied #10 to Fuku-
chiyama line for trial use, however unordinal mild weather interuppted us in evalu-

ation for this coating material.
Application of #10 to trolley
wire is shown in Fig.13.

3. Prevention of express train from
   accident due to snow accretion
   Tokaido Shinkansen and other
express trains have weak points
against snow. Snow blocks adhered
to car body drop down and hit bal-
lasts, resulting damages of window
glass and equipments of body.
We have tried the coating #6 to car

Fig.13 Application of #10 to trolley wire

body for prevention from snow. From this preliminary field tests using #6, it has
been recognized that the difficulties lie in the way of following:

1) Under parts of car body are complicated in shape. When the snow particle having
inertial force hit to the car body, it results in packing and bridging phenomena
of snow. From this reason, the effect of snow repellency by coating is reduced.

2) Car body, especially under part is suffered with oil, dust and abrated products
under running and weathering conditions. Surface performance of coating changes
unfavourably. We will try furthermore to apply the coatings #10 and #20 to car body
of Shinkansen in coming winter.

Application of these coatings
to the linear motor car MAGLEV is
proposed (Fig.14).

References:

1. Ohishi F. Murase H et al.(1987)
   Seppyo(J.Japanese Society of
   snow and ice) Vol.1,49 No.1, 9
2. Murase H. et al.(1988)
   XIX FATIPEC Congress, Aachen
   Congress Book IV, 203-222

Fig.14 View of testing on MAGLEV in Sapporo

"What Can Polymer Chemistry Do to Other Fields of Chemistry?" – Direct Examination of Solution Structure and Collision Theory by Visual Imagery

N. Ise, H. Kitano, H. Matsuoka and K. Ito
Department of Polymer Chemistry, Kyoto University,
Kyoto, Japan 606

In solid crystals, the arrangements of atoms, molecules or ions are fairly ordered, whereas those in gases are almost completely at random. In liquids including solutions, the situation is in-between but still obscure.

Small-angle X-ray scattering (SAXS) and neutron scattering (SANS) study show a distinct single broad peak for salt-free and low-salt solutions of ionic polymers such as synthetic polymers, proteins, polynucleotides and virus particles [For convenient reviews, see for example 1 and 2]. Based on our earlier work on thermodynamic activity of ionic polymers, the peak was taken as implying an ordered distribution of the macroions in solutions. The Bragg distance obtained from the peak $(2D_{exp})$ was much smaller than the average spacing derived from concentration $(2D_0)$. This implies that the ordered structure is localized and non-space-filling.

Such an inheterogeneity in solute distribution is highly unexpected. Thus we tried to make the situation doubly sure. For this purpose, we availed ourselves of polymer latex particles, which are large enough to see under an ultramicroscope by the technique of Hachisu et al.[3] Figure 1 is a micrograph (exactly, the computer-treated binary image), which shows the ordered structure and coexisting disordered regions (the two-state structure). The detailed Kossel line analysis proved that a fcc structure is stable at high concentrations whereas a bcc structure is favored at low concentrations [4]. From the picture the interparticle distance, $2D_{exp}$, was determined and confirmed to be smaller the average spacing, $2D_0$, for high charge density latices and at relatively low concentrations, in other words, the two-state structure was maintained, as was photographed by Hachisu for the first time [3].

By taking advantage of the video imagery and image-data-analyzer, we determined trajectories of latex particles in the two-state structure. From Figure 1 [5], we see a sharp contrast in the mobilities of latices in the ordered and disordered regions. In the latter, the motion was almost Brownian. On the other hand, the particles in the ordered structure displayed lattice vibration, though almost indiscernible in Figure 1. The amplitude of the motion amounts to about 25 % of the interparticle spacing at latex concentration of 1 %. Nonetheless the motion is restricted in an area around the lattice point. In other words, the average displacement of the ordered particles is practically zero. This is in a contrast with the free particles, whose average displacement is large and can be described by the Einstein theory on Brownian motion.

According to the very basis of scattering theory, the scattering profile is related to the density function by Fourier transformation. Fortunately we have density functions in the form of micrographs for latex systems. Thus, by using the above axiom and a computer, the scattering profile can be determined [6]. Although the figures were omitted, Fourier patterns were discrete spots for highly ordered arrangement whereas no spots nor halos could be observed for disordered cases. The two-state structure, however, gave one or two halos. According to the widely accepted interpretation, such a scattering profile implies the "liquid" structure. However, the present study undoubtedly shows that there exist small-sized ordered structures in coexistence with disordered structure, namely the

B. C. Anderson · Y. Imanishi (Eds.)
Progress in Pacific Polymer Science
© Springer-Verlag Berlin Heidelberg 1991

338

two-state structures. This point must be duly considered in the analysis of scattering curves of liquid. The single broad peak obtained by SAXS or SANS for ionic polymers thus corroborates the existence of the two-state structure.

The process of crystallization can also be discussed by using the latex systems [7]. The growth of the localized ordered structure was found to obey the Ostwald ripening mechanism [8], which testifies to the presence of an attractive interaction between particles.

A theoretical calculation of the scattering intensity from 3D-distorted paracrystals was carried out by Matsuoka et al.[9] Three parameters were demonstrated to be crucial;paracrystalline distortion, Debye-Waller effect and crystal size. The theory was compared with experimental data on ionic micelles, silica particles, and latex particles and fairly large degrees of distortion were obtained.

The latex particles further allow us to examine directly the Smoluchowski equation on binary collision, which is a basis of chemical kinetics [10]. By counting the number of dimeric pairs of two different kinds of latices under the microscope as a function of time, the rate constant of the collision was estimated. Table 1 shows the numerical data for a set of cationic and anionic latices. The experimental value for the forward rate constant (k_f) was of the same order of magnitude as the theoretical value shown in brackets, but always smaller than the latter, though the latices were oppositely charged. The reason for this difference is being considered.

Table 1. Rate constants of binary collision between cationic and anionic latices.

temp.	MATA-2 + SS-40	
^{o}C	$10^{-9} k_f$	k_b
15°	$1.9 M^{-1} s^{-1}$	$0 s^{-1}$
	(5.8)	(85)
20	2.4	0
	(6.5)	(96)
25	3.0	0
	(7.7)	(110)

( ):theoretical

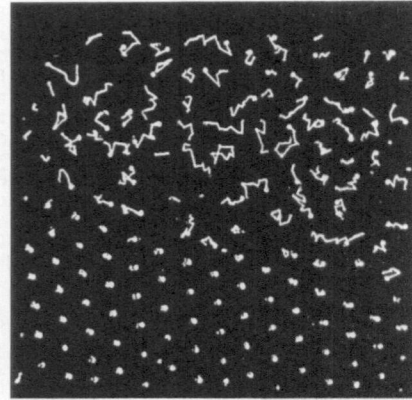

Figure 1. Trajectories in 11/15 s of latex particles (diam.:300 nm, concn.:2 %) in the ordered (below) and disordered regions. The interparticle spacing in the ordered structure is about 1000 nm. Taken from Ref. 5

[1]N. Ise;T. Okubo, Acc. Chem. Res. 13 313, (1980).[2]N. Ise, Angew. Chem. Int. Ed. 25 323, (1986). [3]S. Hachisu et al.J. Coll. Interface Sci. 44 330, (1973).[4]T. Yoshiyama;I. Sogami;N. Ise, Phys. Rev. Lett. 53 2153, (1984).[5]K. Ito; H. Nakamura; H. Yoshida; N. Ise, J. Amer. Chem Soc. 110 6955, (1988). [6]K. Ito; N. Ise, J. Chem. Phys. 86 6502, (1987).[7]K. Ito;H. Okumura;H. Yoshida;N. Ise, J. Amer. Chem. Soc. 111 2347, (1989). [8]Wil. Ostwald, Z.physik. Chem. 34 495, (1900).[9]H. Matsuoka;H. Tanaka;T.Hashimoto;N.Ise, Phys. Rev. B 36 1754, (1987). [10]H. Kitano; S. Iwai;N. Ise, J. Amer. Chem. Soc. 109 1867, (1987) and H. Kitano;N. Ise, "Polymer Association Structures" M. A. El-Nokaly ed., ACS Symposium series 384, Chapter 17, 1989.

Update on Technology of Thermotropic Liquid Crystalline Polymers

G. W. Calundann, L. F. Charbonneau, and J. P. Shepherd

Hoechst Celanese Corporation, Research Division,
Robert L. Mitchell Technical Center,
86 Morris Avenue, Summit, NJ 07901, U.S.A.

Thermotropic liquid crystalline polymers (LCPs) are exceptionally strong materials possessing thermal stability and resistance to flame and solvents. The rigid molecular structure of LCPs allow them to achieve a high degree of order in the melt which is maintained when the material is cooled. The high degree of molecular orientation achieved during the spinning of LCP fibers results in remarkably high tenacity and modulus. The properties of LCP fibers such as Vectran[R] (Hoechst Celanese Corporation) permit their use in a variety of applications including strength members for optical fiber cables, fishing line and nets, and high strength fiber reinforced composites.

Recent advances in the development of LCP technology have made it possible to produce melt spinnable polyester fibers with even higher modulus (>1000 g/d). Two of these materials, Ekonol (Sumitomo Chemical Company, Ltd.[1]) and COTBP (Hoechst Celanese Corporation[2]) are composed largely of the monomers p-hydroxybenzoic acid, terephthalic acid, and 4,4'-biphenol (BP). COTBP contains varying amounts of 6-hydroxy-2-naphthoic acid (HNA) and Ekonol contains small quantities of isophthalic acid. These monomers disrupt the crystalline order of the polymer chains and allow the materials to be processed at reasonable temperatures.

While the as-spun properties of Ekonol and COTBP are better than most conventional polyesters they may be improved by heating the fibers at an elevated temperature (below the melting point) for several hours.

During heat treatment of an LCP fiber the molecular weight, orientation and crystallinity of the polymer are increased leading to improved tensile strength. The modulus of LCP fibers may also be improved by heat treatment. For 4,4'-biphenol containing LCPs the

B. C. Anderson · Y. Imanishi (Eds.)
Progress in Pacific Polymer Science
© Springer-Verlag Berlin Heidelberg 1991

340

increase in modulus is dramatic (from 600 g/d to 1000 g/d for a typical COTBP), however with Vectran[R] the increase in modulus is very
small (530 g/d to 560 g/d is typical). The effects of heat treatment
on Vectran[R] and COTBP can be seen in the stress-strain curves shown
in Figures 1 and 2.

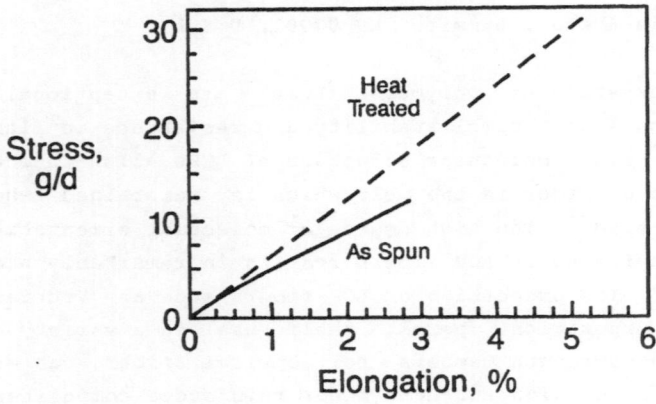

Figure 1. Stress-strain Curve of Vectran[R]

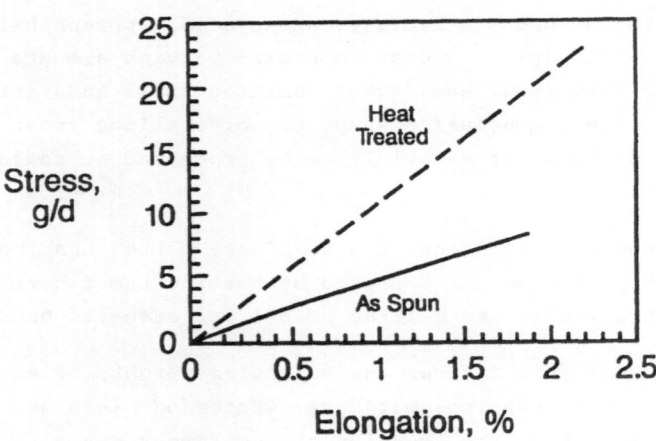

Figure 2. Stress-strain Curve of COTBP

Research at Hoechst Celanese has been directed toward optimizing the properties of heat treated COTBP fiber. This includes a study of the effects of altering the monomer composition of the polymer. For example, lowering the HNA content of the polymer tends to improve the heat treated tenacity and modulus but this benefit must be balanced against the higher spinning temperatures required. The effect of the fiber heat treatment conditions on fiber properties has also been investigated.

References:

1 Ueno K, Sugimoto H, Hayatsu K, U.S. Patent 4,503,005 Sumitomo Chemical Company, Ltd., (1985)

2 Calundann G, Charbonneau L, Benicewicz B, U. S. Patent 4,473,682, Celanese Corporation, (1984)

Light Switching Effects of Polymer/(Liquid Crystal) Composite Systems

T. Kajiyama, H. Kikuchi, J.C. Hwang, A. Miyamoto,
S. Moritomi and Y. Morimura

Department of Applied Chemistry, Kyushu University,
6-10-1 Hakozaki, Higashi-ku, Fukuoka 812, Japan

Abstract: Aggregation states and electro-optical effects based on light scattering have been investigated for polymer/(liquid crystal) composite films. A continuous liquid crystalline domain is embedded in a three-dimensional spongy polymer matrix. The composite film composed of poly(methyl methacrylate)(PMMA) and nematic liquid crystal (LC) with positive dielectric anisotropy exhibited reversible (light scattering)-(light transmission) switching upon off- and on-a.c. electric fields, respectively. A light scattering state was dependent on optical heterogeneities such as a spatial distortion of nematic directors and/or mismatching in refractive indices of the components. Reversible and bistable electro-optical effects were also recognized for a smectic phase of a binary composite system composed of liquid crystalline polymer (LCP) and nematic LC. In the case of a smectic mesophase, turbid and transparent states remained unchanged as it was, even though after turning off an electric field (memory effect).

Introduction: Recently, functional characteristics of low molecular weight LCs have been studied in many fields because of their unique orientation behaviors and hydrodynamic properties.

The authors have applied polymer/LC composite films as novel permselective membranes [1-8]. The polymer/LC composite film exhibited a distinct and rapid light switching between turbid and transparent states upon off- and on-a.c. electric fields, respectively [9-13]. Also, we proposed a novel type of reversible and bistable light switching by the use of the LCP/LC composite system[14,15].

B. C. Anderson · Y. Imanishi (Eds.)
Progress in Pacific Polymer Science
© Springer-Verlag Berlin Heidelberg 1991

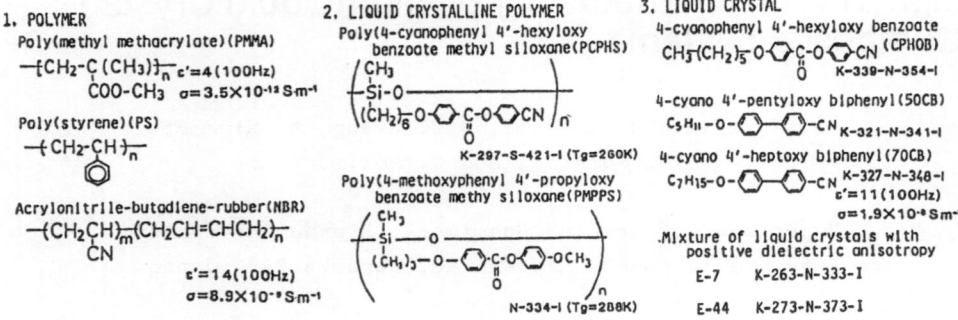

Figure 1. Chemical structures of polymers(1), liquid
crystalline polymers(2) and low molecular weight liquid
crystals(3).

Experimental

For electro-optical studies, several kinds of polymer/LC and
LCP/LC composite films were prepared from solutions on a glass
plate. A weight ratio of polymer/LC was 40/60 and the thickness of
the film was about 40 μm. The chemical structures of polymers,
side chain type LCPs and LCs are given in Figure 1. Nematic LCs
used were CPHOB, 50CB, 70CB and commercial available E-7 and E-44,
which exhibited positive dielectric anisotropy. The phase
transition behaviors and the aggregation state of the polymer/LC
and LCP/LC composites were investigated on the basis of
differential scanning calorimetry (DSC), polarizing optical
microscopy (POM), scanning electron microscopy (SEM) and X-ray
diffraction study.

In order to investigate the electro-optical effect, the
PMMA/LC and LCP/LC composites were sandwiched by the two ITO-coated
glass plates which were transparent electrodes. The distance
between the two electrodes was maintained by the PET spacer of 4-10
μm thick, if necessary. An incident light source was He-Ne laser.
The electro-optical effect of the composite film based on light
scattering was investigated under the various conditions of the
magnitude and the frequency of an a.c. electric field.

Results and Discussion

1. Possible Origins of Light Scattering in the Polymer/LC Composite
   System

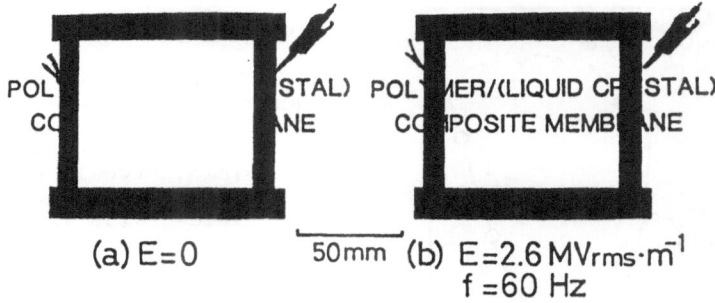

POL... ...STAL)   POL... ...ER/(LIQUID CR...STAL)
C...   ...NE   C...POSITE MEMB...ANE

(a) E=0   50mm   (b) E=2.6 MVrms·m^{-1}
f =60 Hz

Figure 2. Reversible turbid and transparent changes of
the PMMA/E-44(40/60) composite film upon (a) off- and
(b) on- a.c. electric field.

From the SEM observation for the PMMA/E-44 (40/60)
composite film before and after extracting E-44 with methanol at
300 K, it is apparent that the matrix polymer forms a three-
dimensional spongy network and the LC of E-44 is embedded as a
continuous phase in the polymer matrix. Then, the PMMA/E-44
(40/60) composite is a phase-separated system. A degree of light
scattering is dependent on an aggregation state such as an optical
interfacial state, the dimension of LC domain and so on.
Therefore, a combination of polymer and LC materials is an
important factor because it varies an aggregation state of the
composite film.

Before we will start to discuss the origin of light scattering
of the polymer/LC (40/60) composite system, we will show a typical
example of reversible turbid and transparent changes of the PMMA/E-
44 (40/60) composite film corresponding to (a) off- and (b) on-
a.c. electric fields (Figure 2). The electro-optical properties of
the composite system will be discussed in detail in Chapters 2-4.

Figure 3 shows the E-44 weight fraction dependence of
transmittance of the PMMA/E-44 (40/60) composite film in the
absence of an a.c. electric field (curve 1) and also, in the
presence of 1 kHz a.c. electric field of 250 V_{p-p} (curve 2). The
maximum optical heterogeneity to scatter visible light and a highly
transparent state appear in an E-44 weight fraction range of 0.5-
0.7. Then, Figure 3 indicates that a sufficient and strong
interaction between polymeric walls and LC molecules is necessary
to obtain a strong contrast between light scattering-transparent
states, because an E-44 weight fraction of 0.5-0.7 means the
maximum formation of interfacial area between two components.

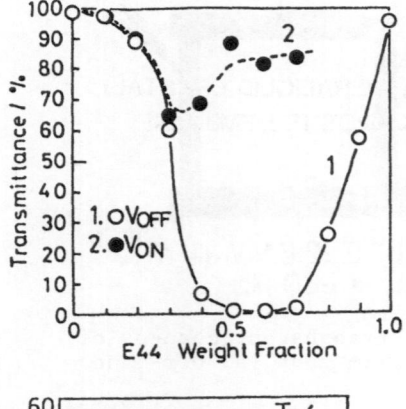

Figure 3. LC weight fraction dependence of transmittance of the PMMA/E-44(40/60) composite film in the absense of a.c. electric field(curve 1) and in the presence of 1 kHz a.c. electric field of 250 V_{p-p}.

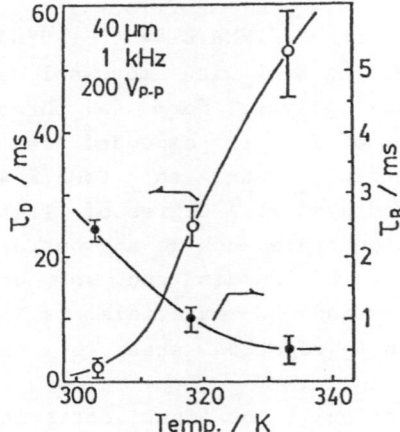

Figure 4. Temperature dependence of the rise time, τ_R and the decay time, τ_D for the PMMA/E-44 (40/60) composite film. An a.c. electric field of 1 kHz and 200 V_{p-p} was applied.

Figure 4 shows the temperature dependence of the decay time, τ_D and the rise time, τ_R under the conditions of off- and on- a.c. electric field of 1 kHz, 200 V_{p-p}, respectively. The conventional rise time, τ_R was defined as a time period to change from 10 % to 90 % transmission and τ_D is the time period for the reverse transmission phenomena after turning off an electric field. The magnitudes of τ_R and τ_D at room temperature range were in a few ms. The magnitude of τ_D for the composite film was fairly faster than that of a twisted nematic (TN) cell. Since the composite film has much polymer-LC interface, it is reasonable to consider that the random orientation of nematic directors may be induced and enhanced compulsorily by a strong polymer-LC interaction in the case of off- a.c. electric field. The magnitude of τ_D increased

with an increase of measuring temperature. This temperature
dependence of τ_D was contrary to the case of a TN cell. Since τ_D
corresponds to how fast LC molecules return in a interacting region
of polymeric walls, the decay characteristics might be concerned
with the temperature dependence of the surface interaction between
the components as well as that of viscosity of LC. Since the
strength in intermolecular interaction becomes weaker with an
increase of measuring temperature in similar manner to solubility
of low molecular weight molecules on a polymeric surface, τ_D
increases with an increase in measuring temperature. Also, since
τ_R corresponds to how fast LC molecules escape from an influence
from polymeric walls, the magnitude of τ_R decreases with an
increase of measuring temperature due to decreases of both the
viscosity of LC and an interfacial interaction between the two
components. The opposite temperature dependences of τ_D and τ_R
indicate that a strong interfacial interaction between polymeric
walls and LC molecules is indispensable to obtain a light
scattering state for the polymer/LC composite system without any
application of a.c. electric field.

Figure 5 shows the schematic representation of the turbid and
transparent states for the composite film. With respect to light
scattering phenomena of the polymer/LC composite film in the
absence of an a.c. electric field, three origins may be proposed
as shown in Figure 5 (a); 1) a spatial distortion of nematic
directors induced compulsorily by non-parallel matrix walls, 2) a
discontinuous change of nematic directors among neighboring LC
domains separated by thin polymeric walls and 3) an optical
boundary owing to the difference of refractive indices between LC
and matrix polymer.

The composite film in an LC state is remarkably turbid due
to its strong light scattering in the case of off- a.c. electric
field (Figure 5 (a)). On the other hand, the composite film in an
isotropic state becomes highly transparent in spite of off- a.c.
electric field (Figure 5 (c)). These results indicate that an
anisotropic nature of LC molecules is indispensable to exhibit
strong light scattering. That is, a nematic director plays an
important role of light scattering. A turbid state of the
composite film (Figure 5 (a)) becomes to a transparent one (Figure
5 (b)) with a response time of ms by the application of an a.c.
electric field. Nematic directors well orient along the direction

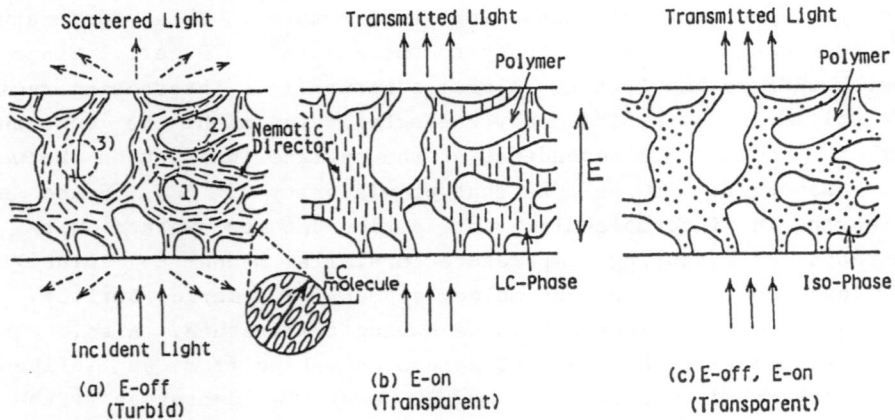

Figure 5. Possible origins of light scattering in the composite film: (1) a spatial distortion of nematic directors, (2) a discontinuous change of nematic directors among different LC domains and (3) the mismatch in refractive indices of the components. Also, the turbid and transparent changes are shown under the conditions of off- or on-a.c. electric fields, and mesomorphic- or isotropic- phases.

of an applied electric field due to positive dielectric anisotropy of LC molecules. Therefore, in the case of off- a.c. electric field, it is reasonable to consider that the polymer-LC interfacial interaction may induce a random orientation or vigorous spatial fluctuation of nematic directors with various curvatures. The dimension of the spatial fluctuation of nematic directors must be fairly comparable to the wavelength of visible light because of its strong light scattering. It is reasonable to conclude that the spatial distortion of nematic directors is one kind of optical heterogeneity.

Based on the morphological observations of the PMMA/E-44 (40/60) and polystyrene PS/E-44 (40/60) composite films, the PS/E-44 composite film has more structural boundaries compared with the PMMA/E-44 one. However, as shown in Figure 6, a light scattering intensity ($\theta=0°$) of the PS/E-44 composite is weaker (higher magnitude in transmittance) than that of the PMMA/E-44 one. This comparison can be explained by the difference of refractive indices (mismatch in refractive indices) between the components, that can be as one of possible origins of light scattering. In the case of off-a.c. electric field, the difference between the refractive index of polymer, n_p and the average refractive index of LC, n_{ave} is important to decide a light scattering intensity. On the other hand, in the case of on- a.c. electric field, the

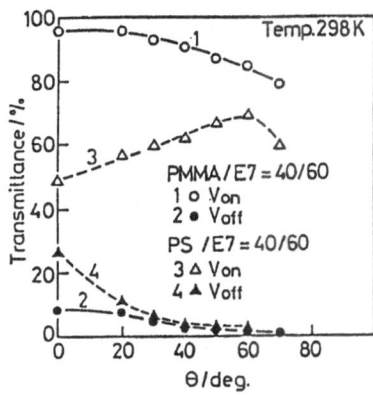

Figure 6. θ dependence of transmittance for the PMMA/E-7 (40/60) and the PS/E-7(40/60) composites. θ is the angle between the normal direction to the film surface and the direction of an incident light.

difference between n_p and the ordinary refractive index (perpendicular to the LC molecular axis) of n_o contributes to a degree of transparency. The magnitude of (n_p (PMMA) - n_{ave} (E-7)) is greater than that of (n_p (PS) - n_{ave} (E-7)) and also, the magnitude of (n_p (PMMA) - n_o (E-7)) is smaller than that of (n_p (PS) - n_o (E-7)). These calculated results on the mismatch in refractive indices are completely comparable to the experimental results of transmittance at θ =0°. Therefore, Figure 6 indicates that the mismatch in refractive indices between the components plays an important role as an origin of optical heterogeneity as well as various spatial distortions of nematic directors as discussed above.

The term of dynamic scattering was introduced to describe the phenomenon of a light scattering state which is generated by hydrodynamic shear forces leading to turbulent flow upon an application of an a.c. electric field to nematic material with negative dielectric anisotropy. The dynamic scattering phenomenon has potential usefulness in a variety of devices, such as shutters, matrix displays, reflective TV, and so on. A strong interfacial interaction between polymeric walls and LC molecules generates a light scattering state without any application of a.c. electric field, as shown in Figure 5 (a). The optical heterogeneity in the PMMA/E-44 composite film may be mainly induced by a compulsory distortion of nematic directors as well as mismatch in refractive indices, as mentioned above. Therefore, a light scattering state of our polymer/LC composite system under off- a.c. electric field can be termed "a static dynamic scattering state".

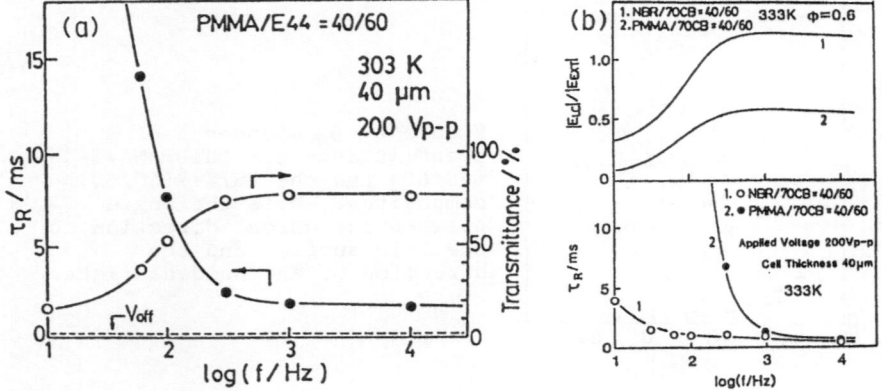

Figure 7. (a) Frequency dependence of transmittance and τ_R for the PMMA/E-44(40/60) composite under the conditions of 200 V_{p-p} and 303 K, and (b) frequcney dependence of E_{LC} / E_{EXT} and τ_R for the NBR/70CB(40/60) and the PMMA/70CB (40/60) composites under the conditions of 200 V_{p-p} and 333 K.

2. Electro-Optical Properties of the Polymer/LC Composite System

Figure 7 (a) shows the frequency dependence of τ_R and the transmittance for the PMMA/E-44 (40/60) composite film under the condition of a.c. electric field of 200 V_{p-p} at 303 K. The transmittance decreased and the magnitude of τ_R strikingly increased with a decrease in frequency, in particular, below around 300 Hz.

In order to eliminate the problems of both a sharp increase of τ_R and a decrease in transmittance in a low frequency range as shown in Figure 7 (a), we adopted a series combination of polymer and LC phases as a dielectric model to calculate the ratio of the amplitude of effectively applied electric field to an LC phase, $|E_{LC}|$ to that of external electric field, $|E_{EXT}|$. When an a.c. electric field is applied to such a dielectrically series combination of the two composites, an interfacial polarization is generated below a certain frequency. The distribution of E_{EXT} to a polymer phase and an LC one, E_P and E_{LC} varies during generation and development of an interfacial polarization. Therefore, the ratio of $|E_{LC}|$ / $|E_{EXT}|$ can be expressed as follows:

$$\frac{|E_{LC}|}{|E_{EXT}|} = \frac{\sigma_P}{\phi\sigma_P + (1-\phi)\sigma_{LC}} \quad \text{(in a low frequency range)}$$

$$= \frac{\varepsilon_P'}{\phi\varepsilon_P' + (1-\phi)\varepsilon_{LC}'} \quad \text{(in a high frequency range)}$$

Where σ , ε' and φ are conductivity, dielectric constant and the volume fraction of an LC phase, respectively. The above equations indicate that the magnitude of E_{LC} is strongly influenced by the magnitudes of σ and ε' of the matrix polymer in a range of low and high frequency ranges, respectively. Then, it is apparent that E_{EXT} can be applied effectively to the LC phase in a low frequency range if the matrix polymer with large σ is used. Then, acrylonitrile-butadiene rubber (NBR) was used as a matrix polymer instead of PMMA, because the magnitudes of σ and ε' of NBR is about 2500 and 3 times as much as those of PMMA, respectively. Figure 7 (b) shows the frequency dependences of $\left|E_{LC}\right| / \left|E_{EXT}\right|$ and τ_R for the NBR/70CB (40/60) and PMMA/70CB (40/60) composite films. The NBR/70CB composite film shows higher rise-response speed in a wide frequency range, owing to the considerably high magnitude of E_{LC}/E_{EXT}. Figure 7 (b) indicates that the electro-optical response speed can be remarkably improved in a wide frequency range by using the matrix polymer which has the greater magnitudes of both σ and ε'.

3. Bistable Light Switching of the PCPHS/CPHOB Composite System

Thermotropic liquid crystalline polymer (LCP) with mesogenic side chain group has both characteristics of polymer and liquid crystal. Recently, electro-optical properties of thermotropic LCP have been studied extensively. Unfortunately, the magnitude of response time of LCP must be much longer than that of corresponding LC molecule due to its high viscosity in a mesomorphic state. In this chapter, we prepared the composite system composed of the side chain type LCP and LC of which chemical structure is similar to that of the mesogenic group. PCPHS was used as LCP and CPHOB as LC as shown in Figure 1.

The phase diagram of the PCPHS/CPHOB composite was obtained on the basis of DSC, POM and X-ray studies. The glass transition temperature, Tg of PCPHS decreased with an increase in CPHOB weight faction. This decrease in Tg might be caused by plasticizing effect of CPHOB. Since only one endothermic peak attributed to the mesophase-isotropic transition was observed, the composite forms a homogeneously mixed mesomorphic phase (molecularly dispersed state). Therefore, CPHOB is miscible over a whole concentration range of PCPHS in both isotropic and mesomorphic states. The sharp low angle

X-ray diffraction with a d-spacing of about 3.06 nm corresponding to a smectic layer spacing, was observed in a weight fraction range of CPHOB below 40 wt%. Then, a smectic phase certainly exists in a mesophase region with the PCPHS fraction above 60 wt%.

The PCPHS/CPHOB (60/40) composite system in a smectic state transmitted 86 % of an incident light of He-Ne laser in the case of the as-cast film of 10 µm thick. The transmittance strikingly decreased to 5 % after applying 1 Hz a.c. electric field of 100 V_{p-p} due to a remarkable increase of light scattering. Furthermore, an application of 1 kHz a.c. field of 100 V_{p-p} made transmittance increased to 94 % within a few seconds. These two states of light scattering and non-scattering (transparent) could change reversibly by imposing an electric field in a low (1 Hz) and a high (1 kHz) frequencies, respectively. Even if an electric field was removed, the turbid or transparent state was stably remained (memory effect). In the case of the PCPHS/CPHOB (40/60) composite system being in a nematic state, the stable turbid state could not be maintained after an electric field was removed. These results indicate that there is stable memory effect in the case of the PCPHS/CPHOB composite film in a smectic phase. Figure 8 illustrates schematically the molecular aggregation states for the turbid and transparent cases. Application of a low frequency electric field induces an ionic current throughout the composites. An induced turbulent flow of the PCPHS main chain caused by an ionic current collapsed a fairly large smectic layer into many small fragments, resulting in a decrease of transmittance up to 5 %. Since a high frequency field does not induce an ionic current, large scale homeotropic alignment is easily formed by dielectric force, increasing transmittance up to 94 %. Such a bistable and reversible light switching driven by the two different frequencies can be realized by connecting the turbulent effect of the main chain of the LCP with the high speed dielectric response of the low molecular weight LC [14,15].

4. Induced Smectic Structure and Bistable Light Switching of the PMPPS/5OCB Composite System

It has been reported that a binary mixture of nematic LCs with both strong and weak polar ends gives an induced smectic phase. In this study, it has been investigated that this rule can be applicable

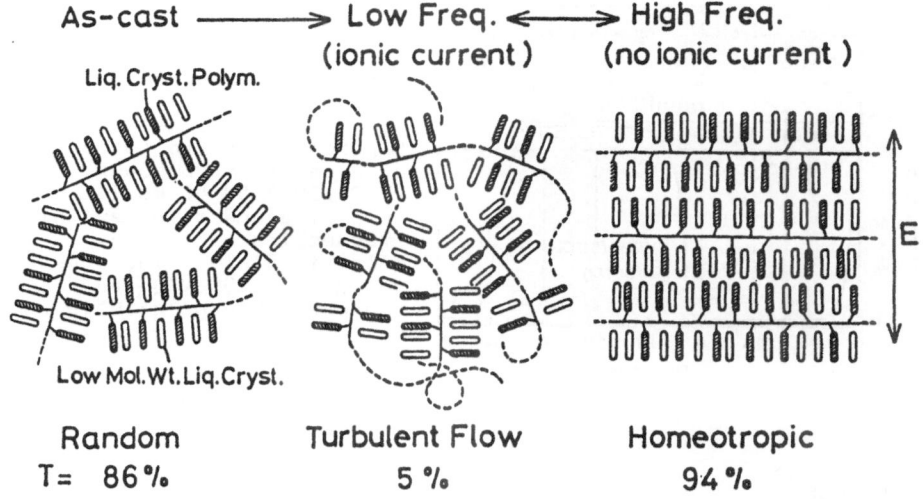

As-cast ⟶ Low Freq. ⟷ High Freq.
(ionic current) (no ionic current)

Liq. Cryst. Polym.

Low Mol. Wt. Liq. Cryst.

Random   Turbulent Flow   Homeotropic
T = 86%   5%   94%

Figure 8. Schematic illustration of the turbid and transparent
cases for the LCP/LC composite film under the conditions of
different frequencies.

to a binary mixture of a side chain type LCP with weak polar
end along the side chain and an LC with strong polar group, which
individually exhibit a nematic phase. LCP was a polysiloxane type
(PMPPS) and LC was 50CB, of which chemical structures were shown in
Figure 1. The characteristic X-ray diffraction pattern of the
PMPPS/50CB (50/50 mol%) composite exhibited sharp Debye rings which
confirmed an induced smectic phase. The sharp X-ray diffraction
rings resulting from a smectic layer were observed for the
PMPPS/50CB composite systems of which the fraction ranges 80/20-
20/80 (mol%).

Figure 9 shows the electro-optical effects based on light
scattering for PMPPS, 50CB and the PMPPS/50CB (50/50 mol%)
composite. The PMPPS/50CB (50/50 mol%) composite exhibited a
bistable and reversible light switching driven by two different
frequencies a similar fashion to the PCPHS/CPHOB composite system.
Both transparent and turbid (light scattering) states could be
stored stably, even though an electric field was turned off. Both
rise and decay response times were in a range of several seconds.
The above results indicate that the rule of an induced smectic
phase to a binary LC mixture is also held to the PMPPS/50CB mixture
system and an induced smectic phase of the PMPPS/LC composite
system makes us expect that its electro-optical effect may be
promised as a novel type of memory display (bistable light
switching).

354

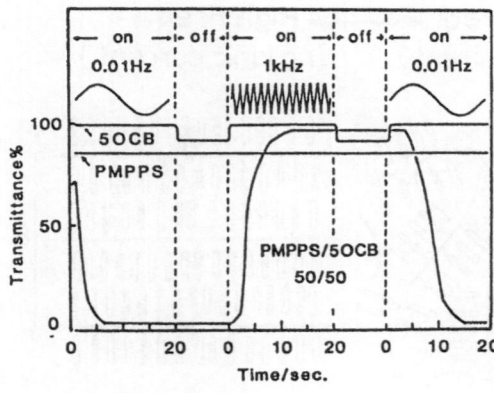

Figure 9. Electro-optical effects of PMPPS, 50CB, PMPPS/50CB(50/50 mol%) composite under low and high frequencies of an a.c. electric field. The applied voltage is 200 V_{p-p} and the cell thickness is 16 μm.

REFERENCES

1. Kajiyama T, Nagata Y, Maemura E, Takayanagi M (1979) Chem Lett 1979:679
2. Kajiyama T, Nagata Y, Washizu S, Takayanagi M (1982) J Membrane Sci 11:39
3. Kajiyama T, Washizu T. Takayanagi M (1984) J Appl Polym Sci 29:3955
4. Kajiyama T, Washizu T, Kumano T, Terada I, Takayanagi M, Shinkai, S (1985) J Appl Polym Sci Appl Polym Symp 41:327
5. Kajiyama T, Washizu T, Ohmori Y (1985) J Membrane Sci 24:73
6. Kajiyama T, Kikuchi H, Shinkai S (1988) J Membrane Sci 36:243
7. Kajiyama T, Kikuchi H, Katayose M, Shinkai S (1988) New Polym Mater 1:99
8. Kajiyama T (1988) J Macromol Sci-Chem A25:583
9. Kajiyama T, Miyamoto A, Kikuchi H, Morimura Y (1989) Chem Lett 1989:813
10. Craighead H G, Cheng J, Hackwood S (1982) Appl Phys Lett 40:22
11. Fergason J L (1985) SID Int Symp Dig Tech 16:68)
12. Drazic P S (1986) J Appl Phys 60:2142
13. Doane J W, Vaz N A, Wu B-G, Zumer S (1986) Appl Phys Lett 48:27
14. Kajiyama T, Kikuchi H, Miyamoto A, Morimura Y (1989) Frontiers of Macromol Sci IUPAC 505
15. Kajiyama T, Kikuchi H, Miyamoto A, Moritomi S, Hwang J C (1989) Chem Lett 1989:817

Priorities for Research and Economic Development in the Polymer Industry in Australia

Peter M Robinson
Invetech Operations Pty Ltd
Sydney, 2000, Australia

The polymer industry in Australia is a significant economic sector which over the past decade has been growing at a rate significantly exceeding general economic growth. With a few exceptions, the industry has been orientated towards domestic markets. Expansion has been the result of sustained economic expansion, creating a large demand for infrastructure and consumer goods, and of continuing product substitution for other materials.

The polymer manufacturing industry is dominated by subsidiaries of trans-national companies, producing a full range of basic polymers from domestic feedstock. The majority of speciality polymers are imported. The turnover of the polymer manufacturing sector was approximately $1.7B in 1987.

The polymer processing/fabricating industry in Australia is diverse and broadly based, producing products ranging from domestic consumer goods to aircraft and aircraft components such as wing flaps for Boeing. The industry had a turnover of approximately $4.0B in 1987.

The polymer manufacturing industry, as a result of ownership patterns, operates largely using proven international technology for the production of bulk resins. However, there is significant research and development in primary process optimisation, process control to maximise the efficiency of downstream fabrication operations and in tailoring polymer properties to meet the harsh environmental conditions imposed by the Australian climate.

The majority of research and development on polymers in Australia is focused on processing and fabricating techniques and on the optimisation of polymer performance in the final products. The topics involved range from the development of orientation in PEEK polymers by deformation and annealing, reactive processing, the novel processing and fabrication of polymer blends to orientation of PVC to produce light, high fracture toughness pipes. The developments are spread between many organisations and are often in response to distinctive Australian market requirements.

For example, in the rapidly expanding trunk optical communication system in Australia, the requirements to directly plough and bury optical cables in transcontinental lines of 2,500 and 2,000 km over a wide variety of terrain has placed a heavy commitment on polymer material developments for strength members, insulation and outer jacketing. Similarly, the 2,500km oceanic optical cable link between Australia and New Zealand, which decends to depths of 2.5 kms, has created a similar demand for specialist materials and fabrication technologies.

The need to avoid environmental degregation in the hot Australian climate has driven a significant proportion of the research and development effort over the past decade. The periodic devastation by bush fires in the urban and rural areas of the southern and eastern states has initiated much work on the development of fire retardent polymer alloys, especially for electrical power cables, These materials are now widely used in applications such as public buildings and urban transit systems. Similarly, the need to avoid UV degregation has lead to the development of a firm research and technological base in environmentally stable coatings and in the appropriate formulation of bulk polymers.

Research and development on polymers in Australia is carried out in Universities and other tertiary education institutes which exist in each State, within the development laboratories of companies and within the divisions of the Commonwealth Scientific and Industrial Research Organisation (CSIRO). For many decades, Australia has been unique amongst Western countries in that the bulk of research and development has been carried out in the public sector, mainly by CSIRO which is an organisation of some 7000 staff and annual R & D budget of over $400M.

Basic research and development in Australia has traditionally been strong but fragmented, with generally a lack of resources to bring a broad range of projects to the commercialisation stage. As evidence of this state of affairs, Australia over the past decade has contributed 1% of the world's scientific literature but has made an infinitesimal contribution to the equivalent patent literature. The situation is the same in the polymer area.

B. C. Anderson · Y. Imanishi (Eds.)
Progress in Pacific Polymer Science
© Springer-Verlag Berlin Heidelberg 1991

In response to the need to stimulate industrial research and development and move the focus of activity to the commercialisation end of the spectrum, the Australian Federal Government in 1986 re-organised support for industrially orientated research, under the Australian Industrial Research and Development Board. (IRD Board).

The Board administers expenditure of approximately $250M per annum which is divided amongst a number of schemes.

- 150% tax incentive scheme for companies carrying out industrial Research & Development.

- A complementary Discretionary Grant scheme for entreprenerial start-up companies, and companies undergoing restructuring, which cannot take advantage of the 150% tax scheme.

- A Generic Technology Grants scheme to support focused research and development efforts in key technologies for Australia's future industrial growth.

At this stage, four generic technology areas have been identified.

- Materials Technology
- Biotechnology
- Information Technology
- Communication Technology

In materials technology, a number of specific target areas have been selected for preferential support including engineering polymers, advanced structural and composite materials, biomaterials and devices and surface engineering. The aim is to develop:

- More market driven research and development in Australian industry.

- Greater interaction between public and private sector research and development through the formation of industry/government/university consortia in identified areas of key technologies by selective grant support.

- Better balance between market orientated and curiousity driven research and development in CSIRO and Universities.

- Encouragement of entrepreneurial companies in technology intensive, high value added markets.

The objective is to focus development work on polymers in a few target areas, utilise the available expertise and resources in different institutions in consortia projects of above critical mass and provide a framework for successful commercialisation on the basis of collobration between R & D and industrial partners. This approach is providing additional stimulus to an R & D based, innovative polymer processing and fabrication industry which has been involved in projects such as:

- the production of the world's first polymer-based banknote; the bicentenary $10 bill.
- the production of novel ion exchange resins for effective water purification.
- the production of some of the first integral polyethylene fuel tanks for mass produced automobiles.
- the sole supply of wing and tail assemblies to North American aircraft manufacturers.
- the development of ultra high modulus, high density polyethylene rods and plates for stabilisation of walls and roofs of mines and for concrete reinforcing in construction.
- the development of computer software for polymer injection moulding designs and tooling which is used world-wide by automotive companies and major appliance manufacturers.

Control of Addition and Ring-Opening Polymerizations with Metalloporphyrin Catalyst

Shohei Inoue, Takuzo Aida, Masakatsu Kuroki, Hiroshi Sugimoto,
Yoshihiko Watanabe

Department of Synthetic Chemistry, Faculty of Engineering,
University of Tokyo, Hongo, Bunkyo-ku, Tokyo 113, Japan

Abstract : Aluminum and zinc porphyrins are excellent initiators of wide
applicability for the living polymerizations of heterocyclic and vinyl monomers
such as epoxides, episulfides, lactones, methacrylates, acrylates, and
methacrylonitrile. Most characteristic of the metalloporphyrin systems is the
remarkable accelerating effect of visible light, as observed in the polymerizations
of acrylic monomers with aluminum porphyrin and of epoxide with zinc
porphyrin. Substituents on the porphyrin ligand exhibit profound influence on
the reactivity of initiating and growing species, to provide the possibility of
molecular design of highly active systems.

INTRODUCTION

Control of molecular weight in polymerization reaction is a central subject of
synthetic polymer chemistry in fundamental as well as practical aspects. We
have developed metalloporphyrins such as **1** and **2** as excellent initiator for living
polymerizations of a variety of cyclic and vinyl monomers such as epoxides[1],
episulfides, lactones[2], methacrylates[3], acrylates, and methacrylonitrile, and also
for living alternating copolymerization of epoxide and carbon dioxide[4] or cyclic
acid anhydride[5].

Some monomers give the polymers with narrow molecular weight distribution
even in the presence of a protic compound as a chain transfer agent. This is
different from living polymerization, and a new concept of 'immortal
polymerization' has been presented[6]. Living and immortal polymerizations
initiated with metalloporphyrins can provide various types of block copolymers[7]
and end reactive polymers[8].

B. C. Anderson · Y. Imanishi (Eds.)
Progress in Pacific Polymer Science
© Springer-Verlag Berlin Heidelberg 1991

The activity of metalloporphyrins as initiator is dependent on (1) nucleophilicity of metal-axial ligand bond, (2) Lewis acidity of metal, (3) electronic and steric effect of porphyrin group, and (4) photo-excitation of porphyrin group. In this article will be described some of the recent advances in our studies on polymerizations with metalloporphyrins.

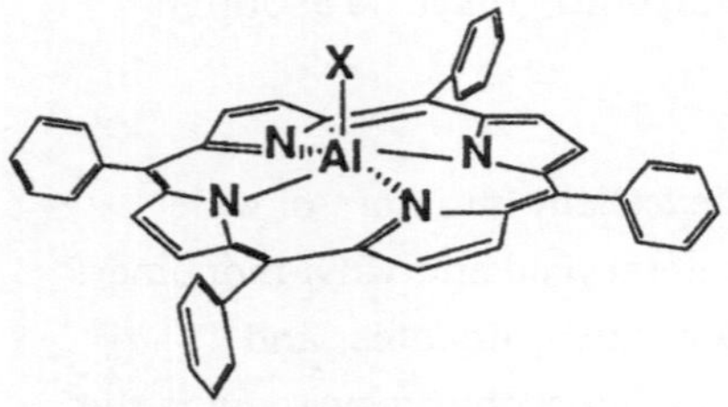

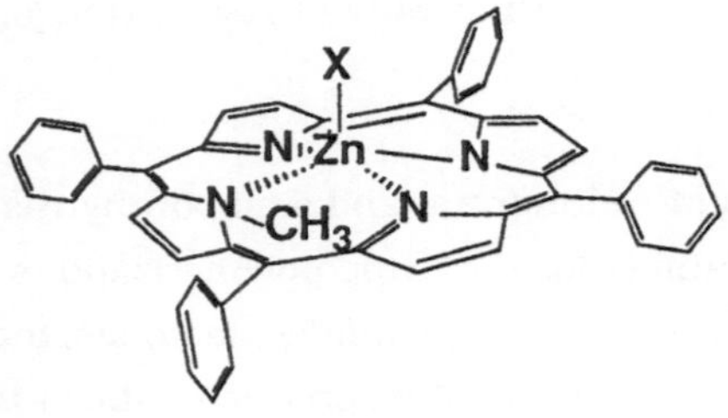

1 (TPP)AlX
   TPP: Tetraphenylporphinato

2 (NMTPP)ZnX
   NMTPP: N-methyl-
          tetraphenylporphinato

LIVING POLYMERIZATION OF β-LACTONE INITIATED WITH ALUMINUM PORPHYRIN. MOLECULAR DESIGN OF HIGHLY ACTIVE INITIATING SYSTEM

In the living polymerizations of heterocyclic and vinyl monomers initiated with aluminum porphyrins, the chain growth takes places at the central aluminum atom - axial ligand bond of the initiator. Thus, the rate of polymerization is considered to be affected by the structure of the porphyrin ligand. In the present study, the polymerization of β-propiolactone (β-PL)[2a] was investigated by using

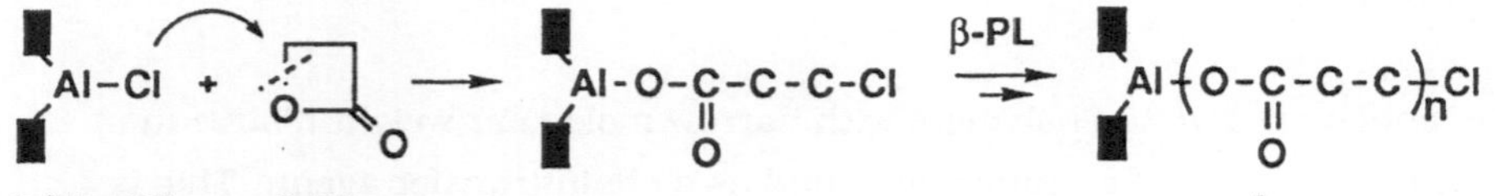

3 (X = Cl)   β-PL   4

as initiators the aluminum porphyrins bearing substituted phenyl groups on the porphyrin ring (3).

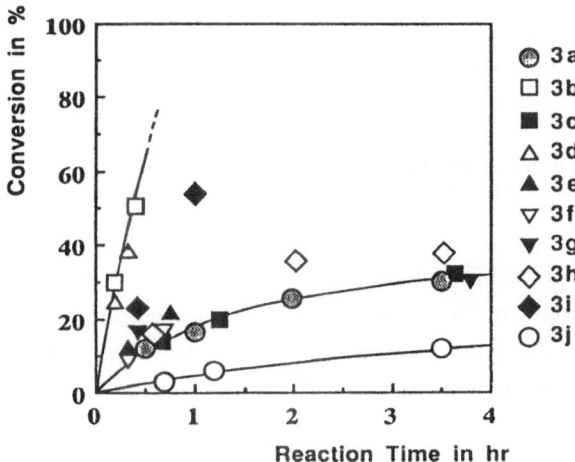

Ar=

b	c	h
a		
d	e	i
f	g	j

3

The polymerization of β-PL was carried out using aluminum porphyrins (**3a** - **j**) as initiators in CH_2Cl_2 at 30 °C with the mole ratio $[β-PL]_0/[3(X=Cl)]_0$ of 50. As clearly shown by the time - conversion curves of the polymerizations (**Figure 1**), some of the aluminum porphyrins brought about the polymerization of β-PL very rapidly.

Figure 1. Polymerization of β-Propiolactone Initiated with Aluminum Porphyrins (**3a** - **j**).
Time - Conv. Relationships.

For example, the polymerization with aluminum porphyrin carrying 2, 4, 6 (*ortho* and *para*) - trimethoxyphenyl groups (**3b**) was observed to proceed up to 51 % conversion in only 24 min, while the polymerization with **3a** under the same conditions proceeded to 13 % conversion. Among the series of aluminum tetrakis(MeO-substituted phenyl)porphyrins (**3b** - **g**) , **3d** also brought about the polymerization very rapidly, while the rates of polymerizations with **3c** and **3e** - **g** were comparable to that with **3a** irrespective of the number of MeO substituents on the phenyl rings. It should be also noted that all the polymers formed with **3b** - **g** as initiators were of very narrow molecular weight distribution (Mw/Mn = 1.06 ~ 1.28) with the number-average molecular weights (Mn) close to the values estimated from the mole ratio of the monomer reacted and initiators (**Figure 2**).

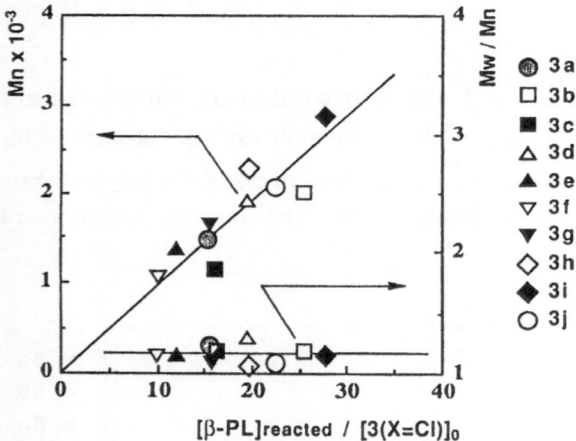

Figure 2. Relationships between **Mn** or **Mw/Mn** and the Mole Ratio of the Monomer Reacted and Initiators (**3a** - j).

This fact indicates the living nature of the polymerizations of β-PL initiated by **3b** - **g** with the 100 % initiation efficiency. Thus, the observed difference in the polymerization rates in **Figure 1** can be directly attributed to the difference in the activities of the growing species (**4**).

Taking into account the notably high rates of polymerizations initiated with **3b** and **3d** bearing *ortho*-disubstituted phenyl rings, the polymerization of β-PL was further investigated with **3i**, **3j**, and **3h** as initiators. The polymerizations with all these initiators also proceeded with living character (see **Figure 2**). On the other hand, with respect to the rate of polymerization, the polymerization initiated with **3i**, which carries 2, 6-dichlorophenyl groups, was found to take place as rapidly as those with **3b** and **3d**, while that with **3j** bearing 2, 4, 6-trimethylphenyl groups was much slower than with **3a**. The polymerization with **3h**, which carries perfluorinated phenyl groups, proceeded with a rate comparable to that with **3a** (**Figure 1**) . Polymerization of epoxides with a series of the above aluminum porphyrin initiators (**3a** - **j**) gave similar results.

Thus, The substituent effect of the peripheral phenyl groups on the activity of the growing species (**4**) is very remarkable particularly when the peripheral phenyl groups are *ortho*-disubstituted. When the *ortho*-substituents are MeO or Cl groups, the growing species exhibits high activity.

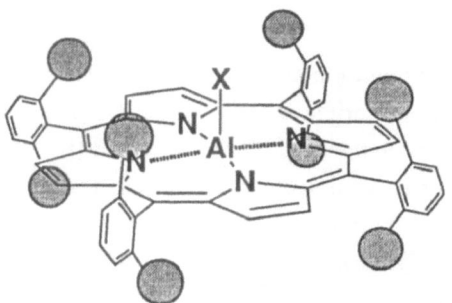

(*ortho*-Disubstituted TPP)AlX

LIVING POLYMERIZATION OF ACRYLIC MONOMERS BY ALUMINUM PORPHYRIN

Living Polymerization of Methacrylic Esters

For the living polymerization of methacrylic esters, aluminum porphyrins with X = alkyl and alkylthio (arylthio) (**1**, X = R', SR') are effective. The polymerization of

$$
x \quad \underset{\substack{\text{C} \\ \| \\ \text{C}=\text{C} \\ | \\ \text{CO}_2\text{R}}}{} \quad \xrightarrow[\text{r.t., 80 °C}]{\substack{\text{1: (X=R'), } h\nu \\ \text{(X=SR'), dark}}} \quad -\left(\underset{\substack{\text{C} \\ | \\ \text{C}-\text{C} \\ | \\ \text{CO}_2\text{R}}}{}\right)_x
$$

5

methacrylic esters with these initiators proceeds at room temperature to give the polymers of narrow molecular weight distribution (Mw/Mn ~ 1.1) with the number-average degree of polymerization close to the monomer-to-initiator mole ratio. The sequential polymerizations of different methacrylic esters with aluminum porphyrin initiators afford tailored block copolymers. The living polymerization of methacrylic esters with alkylaluminum porphyrins (1, X = R') as initiators provides the first successful example of visible light - mediated living polymerization[3].

Living Polymerization of Acrylic Esters

Successful examples of the living polymerization of acrylic esters are very limited compared with those of methacrylic esters, due to the presence of an active C-H group adjacent to the carbonyl group. We have found that the polymerization of acrylic esters initiated with aluminum porphyrin (1, X = SR') proceeds with living character to afford the polymers of narrow molecular weight distribution (Mw/Mn = 1.1 ~ 1.4). The living polymers of methacrylic esters prepared with 1 also bring about the polymerization of acrylic esters very rapidly at room temperature to afford the methacrylate - acrylate block copolymers of uniform block lengths.

Living Polymerization of Methacrylonitrile

The polymerization of methacrylonitrile from the living polymer of methyl methacrylate takes place in the presence of a Lewis base such as pyridine at room temperature to give the corresponding block copolymer of uniform, controlled block lengths. In this case, the acceleration effect of visible light irradiation is remarkable.

Addition - Ring-Opening Living Polymerizations

Addition polymerization of methacrylic esters (**5**), followed by ring-opening polymerization of heterocyclic monomers such as epoxides and lactones, is successful by aluminum porphyrin initiators to give novel polyvinyl - polyether and - polyester block copolymers of tailored block lengths[7].

Mechanistic Aspects - Structure of the Growing Species

The polymerization of methacrylic esters with aluminum porphyrins (**1**; X = R', SR') is initiated by the conjugate addition of the Al-X bond of **1** to the monomers, generating Al-enolates (**6**) as the growing species. For example, the ^1H NMR spectrum of the polymerization system of tert-butyl ester of perdeuteriated methacrylic acid initiated with **1** (X = CH$_3$) in benzene-d^6 clearly shows the ap-

pearance of a characteristic singlet signal at δ -0.3 ppm (9H) due to the *tert*-butyl group of the enolate moiety and the disappearance of the signal at δ -5.8 ppm (3H) due to the methyl group of the initiator.

ZINC (N-SUBSTITUTED)PORPHYRIN AS NOVEL INITIATOR FOR LIVING POLYMERIZATIONS OF EPOXIDE AND EPISULFIDE

Recently, we have found that a zinc complex of N-substituted porphyrin such as zinc N-methyltetraphenylporphyrin ((NMTPP)ZnX, **2**) can initiate the living and immortal polymerizations of episulfide *via* a (NMTPP)Zn-thiolate (**2a**) as the growing species. The present section describes the novel visible light - mediated living polymerization of epoxides by using (NMTPP)ZnX (**2**) as initiator.

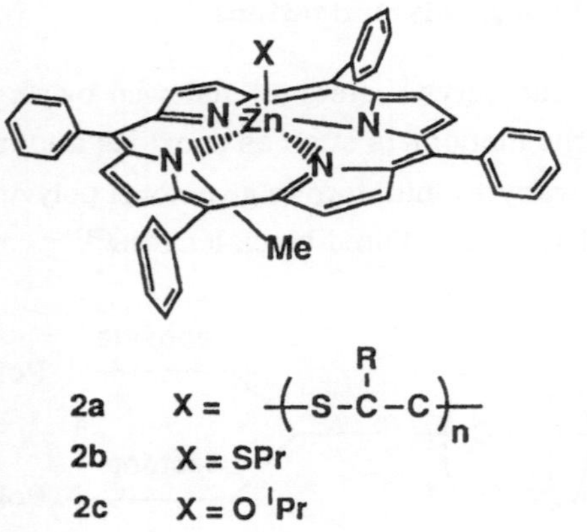

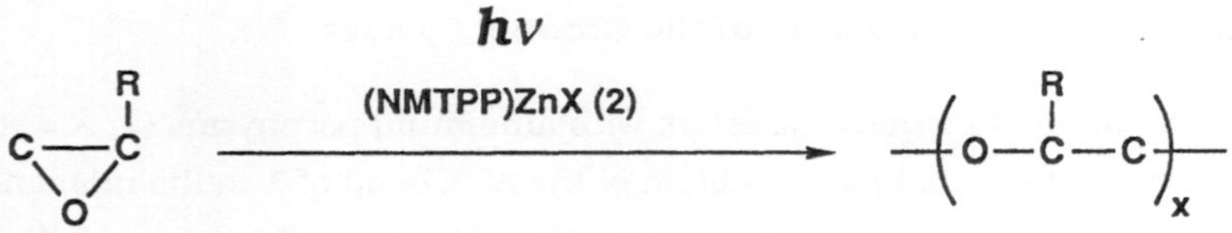

$$2a \quad X = \quad \left(S - \overset{\overset{\textstyle R}{|}}{C} - C \right)_n$$

2b   X = SPr

2c   X = O ^{i}Pr

Living Polymerization of Epoxides

An (NMTPP)Zn complex carrying a Zn-SR bond (**2**, X = SR) was found to initiate the living polymerization of epoxides under the irradiation by visible light (Xenon lamp (λ > 420 nm)). The representative example is shown by the polymerization of propylene oxide (PO) with (NMTPP)ZnSPr (**2b**) ($[PO]_0/[2b]_0$ = 40) in benzene-d^6 at 26 ^oC, (**Figure 3**), which was initiated upon visible light irradiation and proceeded rapidly up to 100 % conversion. The polymer prepared with **2b** as initiator was of very narrow molecular weight distribution with the ratio of the weight- and number-average molecular weights (Mw/Mn) being 1.05. The number-average molecular weight (Mn) of the polymer was increased as the polymerization proceeded, with the degree of polymerization (Dp) close to the mole ratio $[PO]_{reacted}/[2b]_0$. When the monomer (PO) was added again to the system after the completion of the first-stage polymerization, the second-stage polymerization ensued, and the molecular weight of the produced polymer was increased, retaining the narrow molecular weight distribution. Thus, the visible light - mediated polymerization of propylene oxide initiated with **2b** is of a character of "living" polymerization. The polymerization of ethylene oxide with **2b**

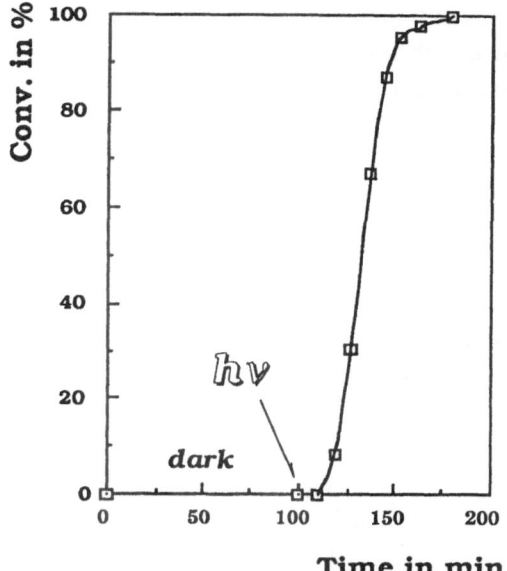

Figure 3. Polymerization of Propylene Oxide (PO) with **2b** as Initiator in Benzene-d^6 at 26 °C ([PO]$_0$/[2b]$_0$ = 40).

as initiator similarly proceeded under the irradiation by visible light to give the polymer with narrow molecular weight distribution (Mw/Mn < 1.1).

Polymerization Mechanism

The NMR studies have demonstrated that the polymerization proceeds *via* an (NMTPP)Zn-alkoxide (**7**) as the growing species.

The photoinduced polymerization of propylene oxide from the living polymer of propylene sulfide initiated with **2b** (**2a**, R = Me) was successful to give poly(thioether) - poly(ether) block copolymer with narrow molecular weight distribution.

Block Copolymer

With respect to the effect of visible light irradiation, (NMTPP)ZnSPr (**2b**) did not initiate the polymerization of epoxides in the dark under the conditions described above. Thus, the irradiation of visible light is essential for the initiation step. On the other hand, the polymerization, once photoinitiated to generate **7** as the growing species, did not subside upon turning the light off. In connection with this observation, the attempted polymerization of propylene oxide using an (NMTPP)Zn-alkoxide (**2c**) as initiator did occur even in the dark. It should be further noted that the polymerization with (NMTPP)Zn-alkoxide (**7** or **2c**) was also accelerated by visible light irradiation.

SYNTHESIS OF BLOCK COPOLYMERS

By the virtue of the wide applicability of metalloporphyrin as initiator for the living and immortal polymerizations of various monomers, a wide variety of block copolymers with controlled chain lengths can be synthesized (**Table 1**).

Tabel 1. Examples of Block Copolymers Obtained with Metalloporphyrin as Initiator

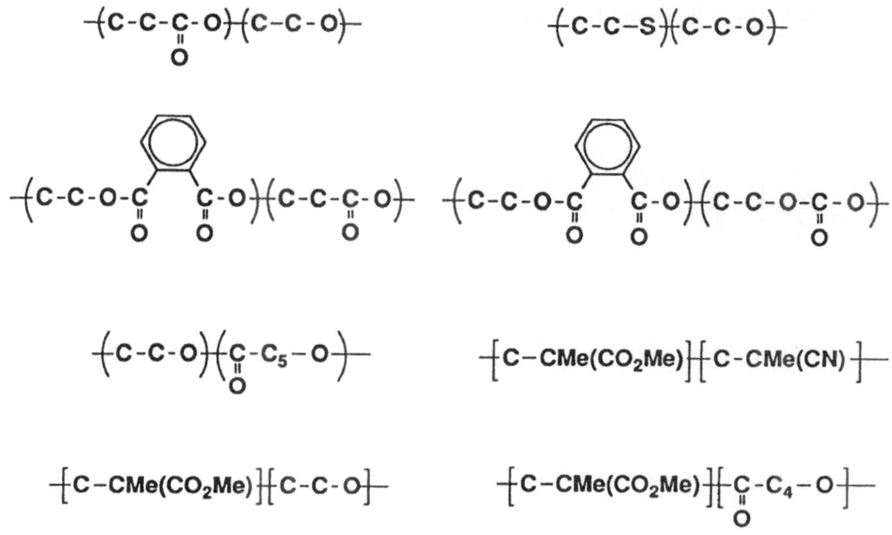

+(C-C-C-O)(C-C-O)+
      ‖
      O

+(C-C-S)(C-C-O)+

+(C-C-O-C C-O)(C-C-C-O)+
      ‖   ‖   ‖
      O O   O

+(C-C-O-C C-O)(C-C-O-C-O)+
      ‖   ‖   ‖
      O O   O

+(C-C-O)(C-C$_5$-O)+
         ‖
         O

+[C-CMe(CO$_2$Me)][C-CMe(CN)]+

+[C-CMe(CO$_2$Me)][C-C-O]+

+[C-CMe(CO$_2$Me)][C-C$_4$-O]+
                           ‖
                           O

The table includes the above-mentioned example of particular interest : The synthesis of block copolymers from methyl methacrylate and propylene oxide or δ–valerolactone by the direct, successive addition and ring-opening polymerizations.

References

[1] Aida T, Inoue S (1981) Macromolecules 14 : 1162, 1166

[2] (a) Yasuda T, Aida T, Inoue S (1984) Macromolecules 16 : 1792; 17 : 2217 (b) Shimasaki K, Aida T, Inoue S (1987) Macromolecules 20 : 2076 (c) Endo M, Aida T, Inoue S (1987) Macromolecules 20 : 2982 (d) Trofimoff L, Aida T, Inoue S (1987) Chem Lett 991

[3] Kuroki M, Aida T, Inoue S(1987) J Am Chem Soc 109 : 4737

[4] Aida T, Ishikawa M, Inoue S (1986) Macromolecules 19 : 8

[5]   Aida T, Inoue S (1985) J Am Chem Soc **107** : 1358

[6]   Aida T, Maekawa Y, Asano S, Inoue S (1988) Macromolecules **21** : 1195

[7]   Kuroki M, Nashimoto S, Aida T, Inoue S (1988) Macromolecules **21** : 3114

[8]   Yasuda T, Aida T, Inoue S (1984) J Macromol Sci Chem **A21** : 1035

Synthesis and Properties of a New Fluoropolymer Obtained by Cyclopolymerization

M.Nakamura T.Kawasaki M.Unoki K.Oharu N.Sugiyama I.Kaneko G.Kojima

ASAHI GLASS Co. Ltd., Research Center
1150, Hazawa-cho, Yokohama-shi, KANAGAWA, 221, JAPAN

INTRODUCTION

Unconjugated 1,5- and 1,6-dienes are known to undergo intramolecular-intermolecular polymerization or cyclopolymerization to form polymers with cyclic mainchain structure[1], but a very few are reported about cyclopolymerization of fluoro- or perfluorodienes. Perfluorodienes with a general formula $CF_2=CF-(CF_2)_xCF=CF_2$ (x=1 to 4) and $CF_2=CFCF_2CFClCF_2CF=CF_2$ are reported to be cyclopolymerized under extreme high pressure (e.g.13.6katm) and gamma radiation, to yield amorphous soluble polymers[2]. Perfluorinated divinyl-ether with the formula of $CF_2=CFOCF_2CF_2OCF=CF_2$ is also reported to be cyclopolymerized, but to form a gel containing polymer under the monomer concentration exceeding 12%[3] (Scheme 1).

The authors have found a new perfluorinated vinylether monomer (AVE: $CF_2=CFOCF_2CF=CF_2$) which

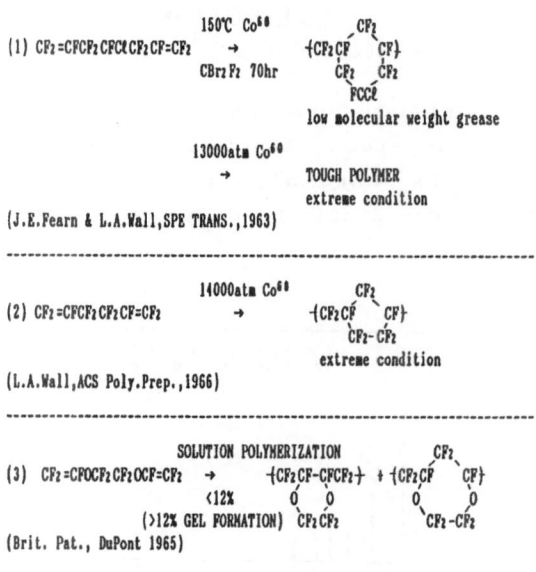

Scheme 1 Previous Work

B. C. Anderson · Y. Imanishi (Eds.)
Progress in Pacific Polymer Science
© Springer-Verlag Berlin Heidelberg 1991

yields a completely soluble polymer under mild conditions by radical cyclopolymerization.

This paper reports synthesis and properties of this new fluoropolymer.

SYNTHESIS OF PERFLUOROALLYLVINYLETHER (AVE)

AVE can be obtained through the following two path ways, i.e. (1) through the direct pyrolysis of sodium-perfluoro-5-oxa-6-heptenoate or (2) through the pyrolysis of the chlorine adduct[4,5]. The latter gave the higher yield since the the Claisen rearrangement to perfluoropentenoilfluoride is avoided.

(1) $CF_2=CFO(CF_2)_3COOCH_3 \xrightarrow{NaOH} CF_2=CFO(CF_2)_3COONa \xrightarrow{\Delta} CF_2=CFOCF_2CF=CF_2$

yield=20%

(2) $CF_2=CFO(CF_2)_3COOCH_3 \xrightarrow{Cl_2} CF_2-CFO(CF_2)_3COOCH_3 \xrightarrow{NaOH} CF_2-CFO(CF_2)_3COONa$

$\overset{|}{Cl}\ \overset{|}{Cl}$ $\overset{|}{Cl}\ \overset{|}{Cl}$

$\xrightarrow{\Delta} CF_2-CFOCF_2CF=CF_2 \xrightarrow{Zn} CF_2=CFOCF_2CF=CF_2$

$\overset{|}{Cl}\ \overset{|}{Cl}$ yield=63%

Fig.1 shows the ^{19}F-NMR spectrum of AVE where signals of fluorine atoms attached to two double can be seen. IR spectrum (Fig. 2) shows also two different vibration peeks due to the vinylether double bond and the allyl double bond.

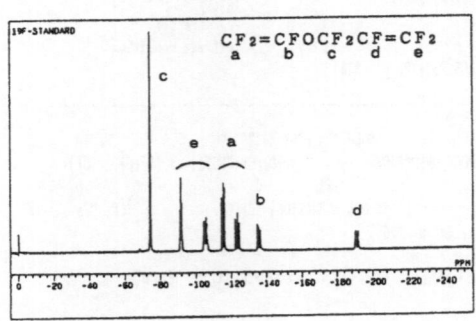

Fig. 1 19F-NMR Spectrum of AVE

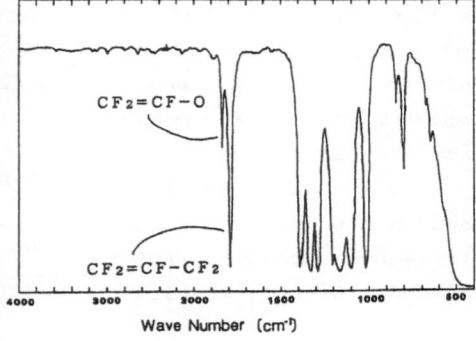

Fig. 2 IR Spectrum of AVE

POLYMERIZATION OF PERFLUOROALLYLVINYLETHER

The radical polymerization of AVE was attempted under different conditions, i.e. in bulk, solution or suspension. Table 1 shows the conditions and results of these polymerizations[6].

Table 1 Polymerization of AVE

polymerization	monomer(g)	initiator(mg)		temp.(°C)	time(hr)	yield(g)	int.visc.$[\eta]$
bulk	30	IPP*	300	25	16	4.5	0.50
solution**	10	IPP	10	40	14	6.1	0.37
suspension***	25	$(C_3F_7COO)_2$	25	30	5	16.0	0.53

*diisopropylperoxydicarbonate **50wt%trichlorotrifluoroethane solution
***in 54g water

The polymerization of AVE can result in two different polymers. One has a cyclic structure in the main chain caused by cyclopolymerization, the other is a linear polymer with a double bond containing pendant which leads to gel formation.

In order to analyze the polymer structure, solubility of the polymer, infrared, ^{13}C- and ^{19}F-NMR spectra are investigated. The absorption peak due to the double bond of either allyl or vinyl moiety in the monomer disappeared in the IR Spectrum (Fig.3).

BULK, SUSPENSION or SOLUTION
25~40°C INITIATOR: IPP, $(C_3F_7COO)_2$

$$R\cdot CF_2=CF \overset{CF_2}{\underset{O-CF_2}{\diagup\diagdown}} CF \rightarrow R-CF_2CF \overset{CF_2}{\underset{O-CF_2}{\diagup\diagdown}} CF\cdot$$

cyclopolymerization: soluble

$$R-CF_2-CF\cdot \rightarrow R-CF_2-CF-CF_2-CF\cdot$$

pendant: gel formation

The polymer was completely soluble in perfluorobenzene. ^{13}C- and ^{19}F-NMR spectra of the polymer in perfluorobenzene agree with a polymer structure having cyclic recurring units obtained by cyclopolymerization (Fig. 4,5). So the scheme below is assumed.

$$CF_2=CFOCF_2CF=CF_2 \rightarrow (CF_2-CF \overset{CF_2}{\underset{O-CF_2}{\diagup\diagdown}} CF)_n$$

AVE   PAVE

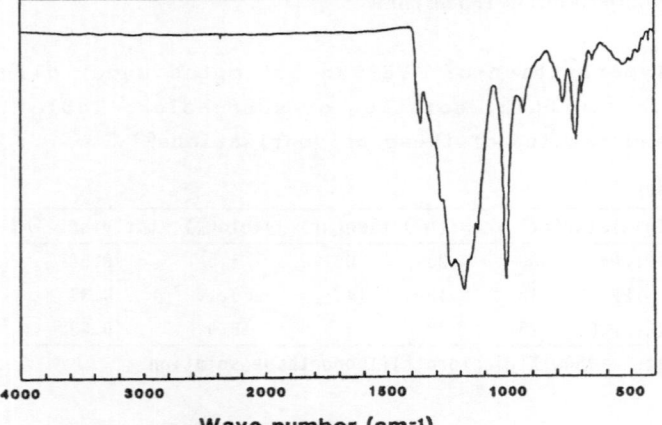

Fig.3
IR Spectrum of
PAVE

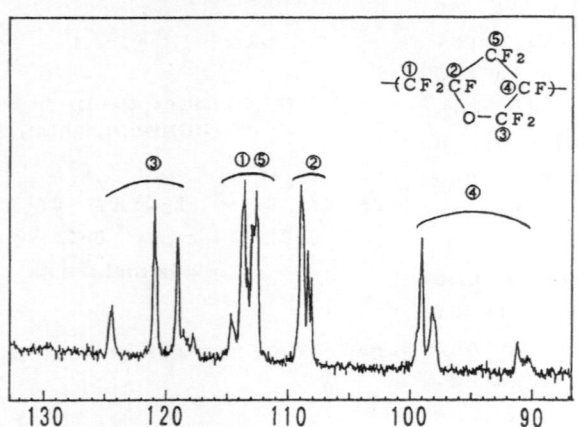

Fig.4
13C-NMR Spectrum
of PAVE

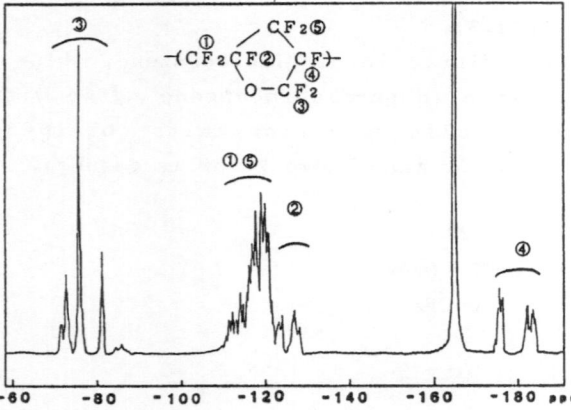

Fig.5
19F-NMR Spectrum
of PAVE

PROPERTIES OF POLY(PERFLUOROALLYLVINYLETHER) (PAVE)

Morphology

      We studied the morphology of PAVE by X-ray diffraction and DSC
thermogram. X-ray diffraction of PAVE is compared with that of PTFE
and PFA (Fig.6). PAVE has no sharp intensity peaks due to crystals
of repeating tetrafluoroethylene units. DSC thermogram of PAVE shows
a glass transition at 78℃ , but no melting point till 300℃ ; at a
temperature where PFA shows a melting point (Fig.7). Therefor PAVE
is considered to be a completely amorphous polymer.

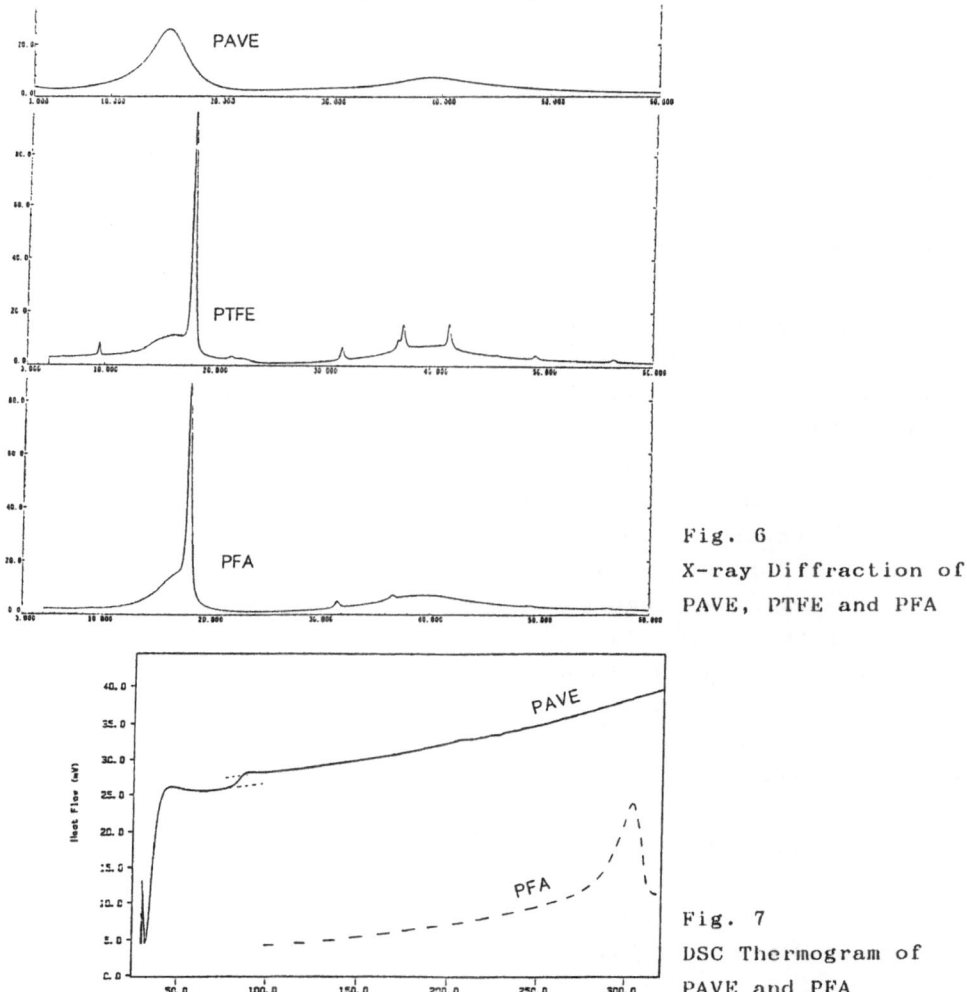

Fig. 6
X-ray Diffraction of
PAVE, PTFE and PFA

Fig. 7
DSC Thermogram of
PAVE and PFA

Optical Properties

Fig. 8 shows light transmission vs wavelength of a 200μ thick PAVE film. The film is clear and not hazy because of the entirely amorphous morphology of the polymer.

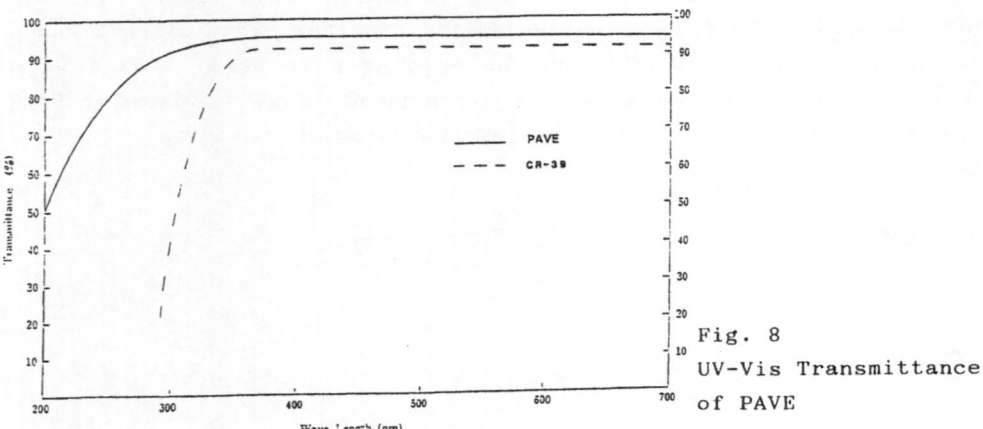

Fig. 8
UV-Vis Transmittance
of PAVE

The refractive index (Table 2), at 1.34, is very low. The low refractive index is probably the result of the lower density, because the fluorine content is lower than PTFE. It's density, at 1.84, is lower than the estimated density of completely amorphous PTFE at 2.00[7]. It is assumed that the free volume of PAVE is increased by the cyclic main-chain structure. The Abbe number of PAVE is 90 and suggest minimal color aberration.

Table 2 Optical Properties of PAVE

	PAVE	PTFE	PE
Transmittance' (%)	95	Opaque	Opaque
Refractive Index	1. 34	1. 35	1. 51
Fluorine Content(wt%)	67	76	0
Density (g/cm³)	1. 84	2.1~2.2	0. 91

*:Visible Light Region

Electrical Properties

Dielectrical constant of PAVE is low and decrease by increasing temperature (Fig.9); the bend at about 75℃ indicate the glass

transition. Volume resistivity is higher than 10^{17}. These electrical
properties are similar to that of other perfluoropolymers.

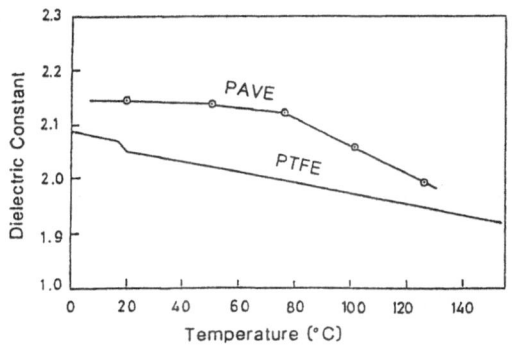

Fig. 9
Dielectric Constant
of PAVE (at 100KHz)

Chemical Properties

PAVE shows very low water absorptivity at <0.01%, and high chemical resistance. No degradation, no weight change are observed in contact with strong acid and bases (Table 3). It is not attacked by solvents based on hydrocarbons or chlorocarbons. But particularly, PAVE is soluble in specific perfluorinated solvents such as perfluorobenzene, perfluorohexane and perfluoroalkylamines (Table 4). It is stable in air to temperatures above 400°C (Fig. 10).

Table 3 Chemical Resistance

Chemicals	Volume Change (%)	Appearance
Acids		
12N HCl	0	No Change
98% H₂SO₄	0	No Change
Alkalis		
10% NaOH	0	No Change
44% NaOH	0	No Change

Immersion Condition : 60°C x 1 Week

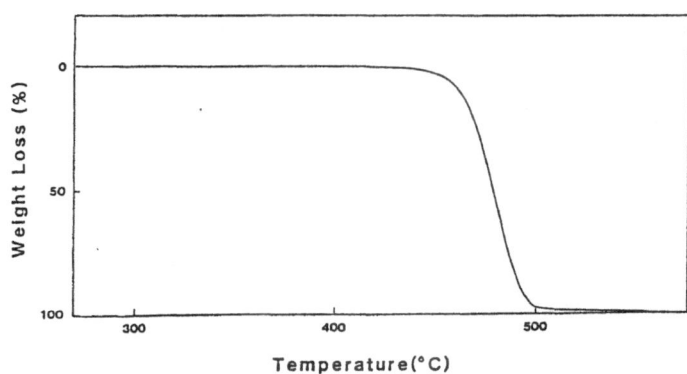

Fig. 10
Thermal
Gravimetry
of PAVE
(Air 10°C /min)

Table 4 Solvent Affinities of PAVE

Solvent		Volume Change	Appearance
Perfluoro Benzene	(F)	Dissolved	Clear Solution
Perfluoro Hexane	C₆F₁₄	Dissolved	Clear Solution
Perfluoro 2-Butyl THF	(F)O〈C₄F₉	Dissolved	Clear Solution
Perfluoro Tributyl Amine	N(C₄F₉)₃	Dissolved	Clear Solution
Hexane	C₆H₁₄	0%	No Change
Benzene	C₆H₆	0%	No Change
Acetone	CH₃-C-CH₃ O	0%	No Change
MEK	CH₃-C-C₂H₅ O	0%	No Change
Trichloro Ethylene	CCl₂=CHCl	0%	No Change

Conditions : 1~5% at Room Temperature ,7days

Mechanical Properties

Tensile properties of PAVE are: tensile strength 310kg/cm², elongation at break 170%, tensile modulus 10,000kg/cm², yield strength 280kg/cm² (Fig. 11). Tensile modulus and yield strength of PAVE are higher than those of PFA.

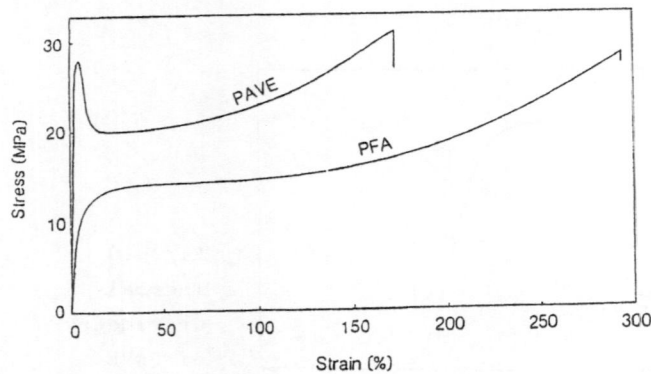

Fig. 11
Tensile Properties
of PAVE

Dynamic mechanical properties are shown in Fig. 12. Dynamic modulus E' is higher than 10^{10} dyne/cm^2 till 50°C , then drops sharply above 70°C . This temperature corresponds to the glass transition temperature of 78°C determined by DSC measurement.

The viscosities of PAVE at higher temperature is similar to that of polycarbonate (Fig. 13).

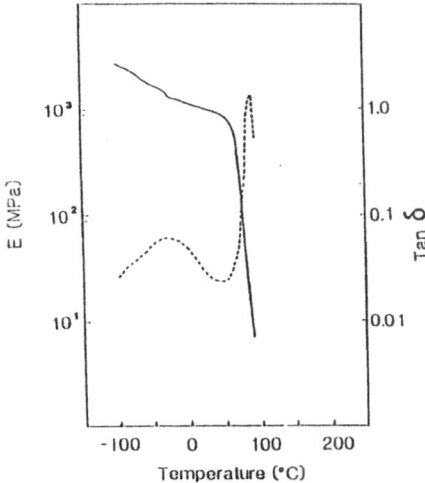

Fig. 12   Dynamic Mechanical Properties of PAVE

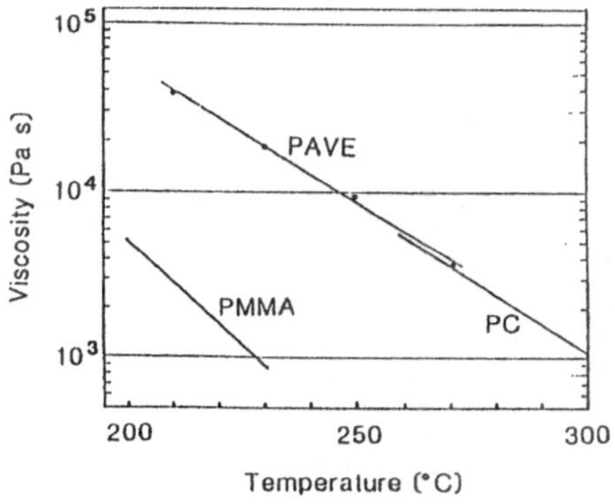

Fig. 13 Viscosities of PAVE

SUMMERY

We synthesized a new perfluoropolymer PAVE by cyclopolymerization of a new monomer AVE. The properties of the new polymer are unique due to the entirely amorphous morphology. PAVE is highly transparent to visible and ultraviolet light, has a low refractive index and dielectric constant, is highly resistant to chemicals but soluble in some specified solvent such as perfluorobenzene, is stable at high temperature ($\sim$ 400°C) and has a higher modulus of elasticity compared with other perfluoropolymers.

References:
1)Butler GB, Encyclopedia Polymer Sci. Techn., 4
2)Wall LA, Fluoropolymers,Wiley-Science,4,High Pressure
   Polymerization,127

3)USP 3418302 Du Pont
4)Japan Patent 60-45619 Asahi Glass
5)Japan Open 1-113328 Asahi Glass
6)Japan Open 63-238111 Asahi Glass
7)Satokawa T et al, Fussojushi,Nikkan Kogyo Shinbunsha, p.29

Some Hypothetical Allotropes of Carbon

Roald Hoffmann
Department of Chemistry, Cornell University, Ithaca, NY 14853

We are so used to diamond and graphite that it is difficult for us to think that there might exist different allotropes of carbon. But there are many reasons to ponder this problem, among them the following:

(1) For other non-metallic elements which form reasonably strong, electron pair bonds, there often are several allotropes. Look at sulfur. (2) There is a vast industry of carbons, formed under an incredible variety of conditions. What structures are hidden among these? (3) Following Russian and Japanese work, even Americans have discovered "amorphous" carbon, or diamond-like films, deposited epitaxially from discharges in hydrocarbon vapors. (4) Note the great interest in C_{60} and other gas-phase carbon clusters. (5) People are routinely (using diamond anvils!) studying materials under pressures around 1 megabar. There is evidence for molten diamond. Could there be carbon phases which are denser than diamond, not close-packed, but retaining some localized bonds?

It turns out that there are many space-filling networks of carbon that could exist as potential allotropes. All studied so far seem to be (theoretically) less stable than diamond. But then thermodynamic instability has never frightened a synthetic organic chemist.

I'll begin with a historical overview of carbon allotropes, summarizing the known forms. An account will be given of the controversial "karbin" structure, discussed at length in the Russian literature. It is supposed to consist of polyacetylene (C_n rather than $(CH)_n$ here!) needles. The verdict is still out on this form.

A rich source of hypothetical four-coordinate carbon allotropes is to be found in the known structures of inorganic four-coordinate compounds, especially silica and the silicates and zeolites. These generate perfectly reasonable, but rather open, not dense, carbon networks.

Are there any space-filling networks of sp^2 carbon atoms? You have to supply some further conditions, of angle strain (three bonds at 120°, or not), of conjugation (should the p orbitals perpendicular to the trigonal plane be conjugated, as in graphite, or will you allow arrays such as 1 shown below), and of dimensionality.

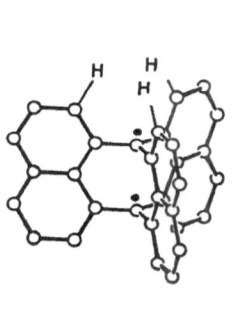

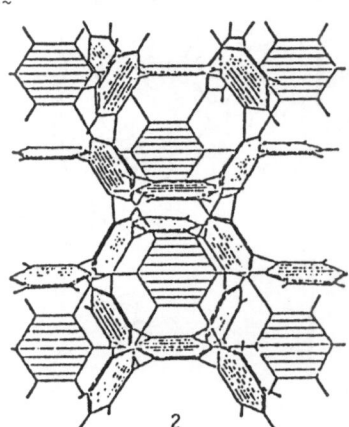

1

2

B. C. Anderson · Y. Imanishi (Eds.)
Progress in Pacific Polymer Science
© Springer-Verlag Berlin Heidelberg 1991

380

For instance, you can pave your bathroom floor with octagons and squares, just as well as hexagons (or Penrose tiles, if you want five-fold symmetry). But that introduces angle strain. In two dimensions the only way to get a strain-free, trigonal array is through the familiar honeycomb net of graphite. Does this mean that there are no alternatives to graphite? Not at all; what it means is that all such arrays, if they exist, must be inherently three-dimensional.

There's a multitude of them. One in the literature since 1942, is a beautiful open polyphenylene structure suggested for amorphous graphite by H. L. Riley, 2.

The net 3 is another one we came up with. Note in it one-dimensional polyene (polyacetylene, if you like) chains running in two dimensions. And no conjugation in the third dimension. The smallest ring has ten carbons in it. The structure was suggested by Peter Bird, calculations carried out by Tim Hughbanks and Miklos Kertesz.

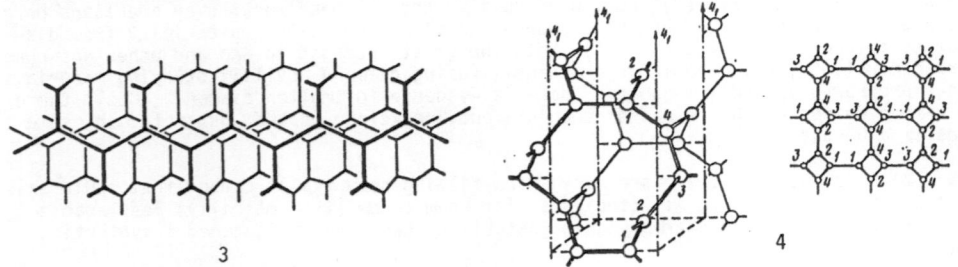

3

4

The structure (which incidentally is that of the silicon framework in $ThSi_2$) is likely to have a density between diamond and graphite. It is calculated, unreliably, to be 17 kcal/mole of C unstable relative to diamond. And predicted to be metallic. These two predictions are connected to each other.

These are not the only structures filling space strainlessly with sp^2 carbons. For instance one can replace each polyene chain by a polyacene. Or if one relaxes a little about complete conjugation one can generate the remarkable chiral structure of Stankevich and Bochvar, shown in 4.

Structures mixing sp^2 and sp^3 carbons are interesting because they allow the possibility of constructing one-dimensional ethylene stacks. Two of these (5,6, work of Ken Merz) are shown below; if the ethylenes are forced sufficiently close to each other, one again has the possibility of conductivity.

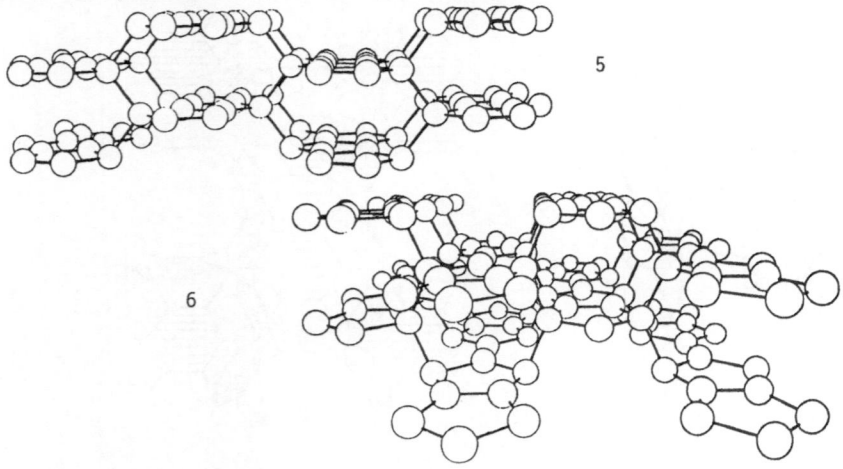

5

6

New Challenge of Polymer Industry Towards the 21st Century

Dr. Yoshikazu Ito

Chairman of the Board, Toray Industries,Inc.
Former President, Chemical Society of Japan

Abstract

Polymer chemistry will undoubtedly be the key-science for the 21st century, and the main action-point will be basic research based on human needs.

The importance of nurturing and supporting young researchers that are both curious and ertrepid. This is because the most important thing for the basic research is young, creative researchers with new ideas.

Introduction

The polymer industry as it is today, can be said to have started with the invention of nylon by Carothers in 1931. It has achieved spectacular growth over the last 50 years with the support of the development of the petrochemical industry after World War II.

Both the Japanese and the world production of plastics have now reached near the level of steel's production in absolute volume. This means the polymer industry has grown into a key-industry equivalent in importance to the steel industry. The world production of plastics is now at about 100 million tons annually, and the production of plastics in Japan is about 15 million tons a year. The rapid growth of the polymer industry resulted from the innovative progress of polymer sciences and technologies, as well as cost-reduction of commodity materials by mass-production. Plastics have helped improve the overall living standard of all of us.

The polymer industry should be able to respond to the future needs of society, because polymer chemistry deals with molecules and can create new materials of high performance and functions which fulfill human needs.

The unique characteristics of polymer materials when compared with metals and ceramics are;

B. C. Anderson · Y. Imanishi (Eds.)
Progress in Pacific Polymer Science
© Springer-Verlag Berlin Heidelberg 1991

(1) light weight
(2) ease of processing
(3) ease of designing at the level of molecules or
     molecular aggregates.

Structural Change of Industry

Through the changes caused by the two oil-crises and the recent concern for the earth environment, human needs are now moving in a more intellectual and humanistic direction. In the future, man will pursue personal affluence, realize truly global communication and enjoy a healthy, long life in a comfortable environment. New material such as high-performance polymers, high-functional polymers and advanced composite materials are closely related to the high-technology industries, global environmental protection, energy savings and health care, all of which are expected to undergo technical innovations.

More humanistic polymers should be developed in the future for the betterment of human life, such as modified proteins, biomimetic polymers or intelligent polymers. It is no exaggeration to say that without high-performance and high-functional polymers, few innovations and developments will be achieved in any industries which will have significant impact on any aspects of human life.

The polymer industry should change in the future; transforming from quantity to quality and from "materials" to "materials with intelligence", responding to the change in human requirements.

The New Chemistry

The driving force of this change is the New Chemical Technology. Moreover, the chemical technology and the influence of the chemical technology will spread worldwide. What we call the New Chemistry is the science which will satisfy new human demands in the 21st century. The New Chemistry is born of the fusion of chemistry, which deals with molecules and molecular aggregates, with other leading-edge sciences and technologies such as biotechnology and electronics. As is shown in Figure 1, protein engineering, bio-chips, advanced materials and intelligent materials are the examples of New Chemistry.

I believe new discoveries and findings are born in the area where these different disciplines overlaps, just like fish school, where warm currents cross with cold currents.

Figure 1   # Domain of New Chemistry

	1 μm	0.1 μm	10 nm
Computer	256K	4M 6·1M	Non-linear Optical Devices
			Bio-chip Molecular Lattice
Memories	Magnetic Record	Optical Memory	Photo Memory (PHB)
Polymers	Ultrafine Fiber	Polymer Alloy	Super-Structure
Ceramics		Fine Ceramics	Superfine Particles
Metals		Multiphase	New Metal Alloys / Amorphous
Biotech.	Bacteria	Virus Cell	Protein

▨▨ Domain of New Chemistry

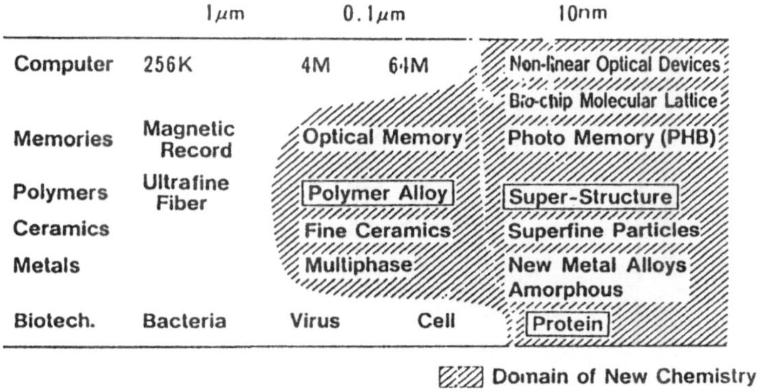

One goal of the New Chemistry is the creation of polymers with various functions. This will come about by studying the three-dimensional polymeric structure or the assembling of molecular aggregation on the molecular level. Polymer chemistry has the ability to create new materials that can meet the demands for the expanding human activities and, at the same time, conserve the world's valuable energy and other resources.

Advanced Composite Materials and Aircraft Use

Chemistry deals with molecules and allows man to organize these to build new materials. Chemistry gives us the ability to create new materials to meet with the demands of changing human activities and at the same time, to conserve the world's energy and other resources. Light weight advanced composites reinforced by carbon or aramid fiber are some of the leading high-performance materials. This highly oriented structure of the polymer is the key-technology to obtaining the properties of high-strength and high-modulus. New liquid-crystalline polyester fibers, or high-strength polyethylene fibers can also be developed, based on the this molecular design.

Japan is the world's top carbon fiber producer and consistent efforts are being made to improve the properties, cost competitiveness and processing technology. Figure 2 shows the progress of carbon fiber technology, with Toray's "Torayca" as an example. The level of tensile strength was only 3 GPa when Toray first commercialized it about 20 years ago. Last year we succeeded in the production of the fiber with tensile-strength of 7 GPa. This fiber is called a "Million fiber", as it has tensile-strength of 1 million p.s.i. To meet the high-modulus requirements of the aerospace industry, we developed this year "Type M 60J", the modulus of which is up to 600 GPa. Furthermore, we will continue efforts to improve these properties in these directions shown in Figure 2.

Figure 2

Direction of Carbon Fiber Improvement

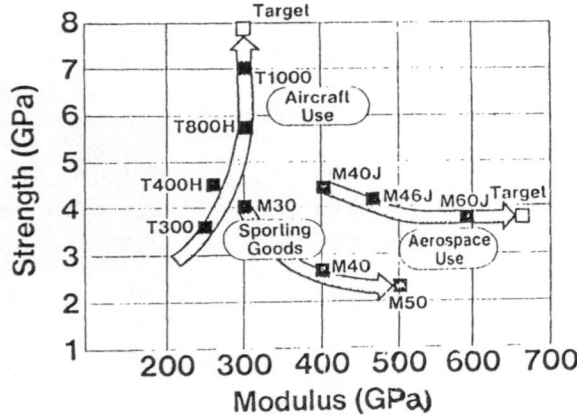

At present, the total market worldwide for carbon fiber is 6,000 tons annually and we estimate 60,000 tons a year in ten years. This growth will be driven by the development of improved products which will meet the various market needs.

In the aircraft industry advanced composites are penetrating into large markets. The first application of advanced composites was in the F14 Fighter in 1971. At that time, boron-fiber re-inforced composite was used. Today, the materialsused to make the "Harrier AV-8B" consists of 25 wt% advanced composites re-inforced by carbon or aramid fibers. For commercial aircraft, advanced composites were applied for the first time in 1977, for the secondary structural parts of the DC-10. Afterward, the proportion of advanced composites used has risen from 3 wt% on the Boeing 767 in 1982 to 20 wt% of the Air-Bus A320 in 1987. Moreover, the application of advanced composites has extended into the primary structural parts of aircraft. In the future, we expect, the next generation aircraft will consist of more than 30 wt% of advanced composites.

The reason for the use of advanced composites applied in aerospace and aircraft is not only because of fuel-saving through weight reduction, but also because of the increased reliability and the construction-cost reduction by one-piece fabrication. In the case of Aluminum-alloy, 17,000 elements are fabricated into 600 parts for the manufacturing of a single rudder. On the other hand, in the case of advanced composites, only 4,800 elements are needed for fabrication into 335 parts. The reduction of elements or parts by one-piece fabrication results in increased reliability and overall production-cost savings.   Our aim is to produce carbon fiber at a cost below 10$/kg. We call this fiber "ten dollar fiber". When the ten dollar fiber is developed, a new market for advanced composites will open in the automotive sector.

Liquid Crystalline Polyester

In order to obtain the highly oriented perfect crystal of an extended molecular chain, we make stride;
(1) to eliminate foreign particles and irregular structures which cause defects within the crystal,
(2) to obtain perfectly extended chain crystals to enable the closest-packing of molecular chain possible,
(3) to complete the orientation of crystals by extenting or stretching.

New liquid-crystalline polyester fibers and high-strength polyethylene fibers are also being developed, based on this molecular design. Rigid whole aromatic polyester or polyester, including a segment of a whole aromatic polyester is melted or dissolved in a nematic liquid crystalline phase, and then extruded through a die into a fiber under high shearing force. The fiber is then heat treated under tension or stretching, to perfect the crystallization and orientation of crystals of an extended molecular chain. In this way, a high-strength, high-modulus fiber is obtained. Figure 3 shows the fibril structure of a whole aromatic polyester fiber. The molecular structure is shown in Figure 3. It consists of naphthalic acid, two(2)-methyl-para-quinone and para-benzoic acid. The high orientation of the crystals, especially those of the core, is deduced from the fibrillar structure observed.

Figure 3

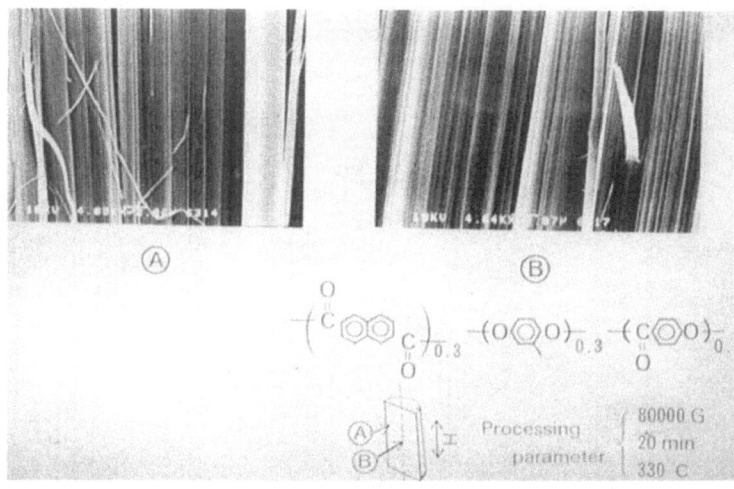

This result was obtained by the research group of the "Research Association for Basic Polymer Technology" of the "Research and Development Project on Basic Technologies for Future Industries", which was established by the Japanese government in 1981, to promote col-laboration between industry, universities and national research institutes. As the chairman of this Research Association, it is my pleasure that fundamental and innovative results are being obtained by the collaboration of the young researchers from various research institutions.

Polymer Alloy and Automobile Use

The largest potential user and target for the polymer industry is undoubtedly the automotive industry. The first application of polymer materials for automobiles was for the interior parts, such as instrument panels or console boxes, at the end of 1950s. Since the end of 1960s, engineering plastics have come to be used for functional parts, such as mechanical gears or electric parts under the hood. Light weight fiber reinforced plastics such as sheet molding compounds with improved strength and heat stability are applied for use in secondary structures such as oil-pans, radiator-tanks,or cylinder-headcovers. Sometimes there are also used in primary structure, such as outer panels, as energy saving caused by the two oil crises at the end of 1970s) is an important new requirement.

Later, polymer materials have penetrated into larger sectors of the automotive industry. Today, some advanced cars such as General Moters' "Fierro", Honda's "CR-X" and Nissan's "Be-one", have parts of the body panel which are constructed entirely from plastics, using polymer-alloy or glass-mat reinforced thermo-plastic sheet technologies. Plastics are applied today partly as exterior parts of automobiles ; for example, front-bumpers, air-intakes, roofs, spoilers and rear-bumpers. We aim at replacing the steel of the outer vertical panels, with thermo-plastics in the near future. The driving forces behind the replacement with plastic are leight weight, freedom of design to meet the individual tastes of customers, and safety assurance.

Figure 4

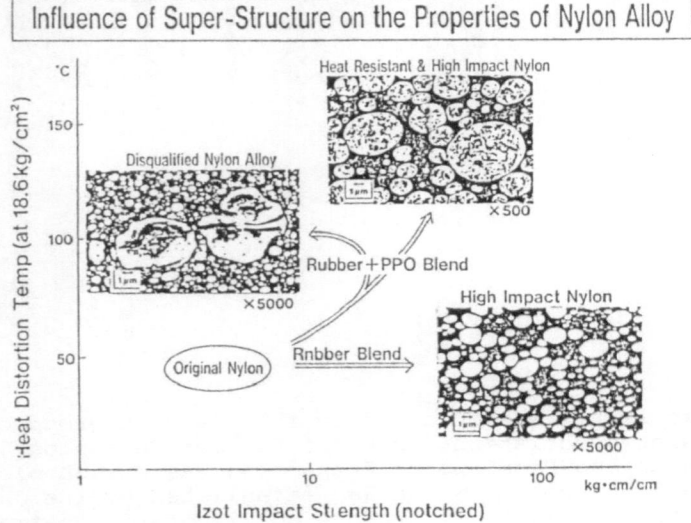

Influence of Super-Structure on the Properties of Nylon Alloy

Polymer-alloy technology is important, and has the potential bringing polymer materials in automobiles applications, together with fiber reinforcement technolgy. The technolgy is to blend or mix several kinds of polymers while controlling micro-structure. This provides polymer blends or mixtures with the desired properties.

Figure 4 shows an example of a nylon-alloy. In order to improve the impact-strength of nylon, we used to blend rubber with nylon, but the blend of rubber resulted in the reduction of tensile-modulus and heat-stability. We then, improve tensile-modulus and heat-stability by meansof blending the nylon-alloy with polyphenylene oxide, which possesses a higher glass transition temperature. When rubber and polyphenylene oxide are blended separately or independently in nylon, we could not obtain the desired properties. But when the micro-structure of the blend is controlled, as polyphenylene oxide is first blended in nylon and the rubber is finely dispersed in the polyphenylene oxide phase, a heat-stable, high impact-strength nylon is obtained.

Protein Enginnering

Another goal we must pursue is to understand and duplicate synthetically the working of living organisms. Protein engineering is, I think, a typical example. Protein engineering is considered to be "the third generation of biotechnology", followed by fermentation technology and gene engineering. The research target for protein engineering is to obtain new functions by artificially changing the arrangement of amino acids or the higher structure of proteins. Protein engineering has a large and attractive potential for the future. This is apparent by considering the fact that a wide variety of different organisms with complicated activity and structures consist of proteins that are made of a combination of only 20 amino acids.

For the basic research on protein engineering, the Protein Engineering Research Institute (PERI) was established by the cooperation of 14 companies, including non-Japanese firms, in 1986. I was asked to take the position of the president of the Institute, and it is my great pleasure to visit the Institute often and to hear of the recent progress of research from the young researchers. The total amount of research fund of the Institute is 17 billion yen over a ten-year plan, This consists of an investment ratio of 70 % government, from the "Japan Key Technology Center" and 30 % private, from industry.

Research subjects of the Protein Engineering Research Institute are;
(1) to elucidate the relationship between the structure and function of proteins,
(2) to develop computer-graphics and soft-ware for molecular dynamics,
(3) to develop new proteins with improved functions.

About 60 researchers are working in the Protein Engineering Research Institute, and 2/3 are from industries and 1/3 are from academia. The Protein Engineering Research Institute is also open to non-Japanese. A few researchers from foreign countries are already working in the Institute. Basic innovative research, such as protein engineering can not be achieved without interdisciplinary collaboration of researchers. I firmly believe that this type of basic research should be strengthened to attain overall progress of the New Chemistry.

Some modified human-lysozymes are designed and obtained by recombinant DNA technology. Some of them are crystalline proteins. The tertiary structures and their active sites have been investigated by means of X-ray analysis and molecular dynamics by computer-graphics as is shown in Figurer 5.

Figure 5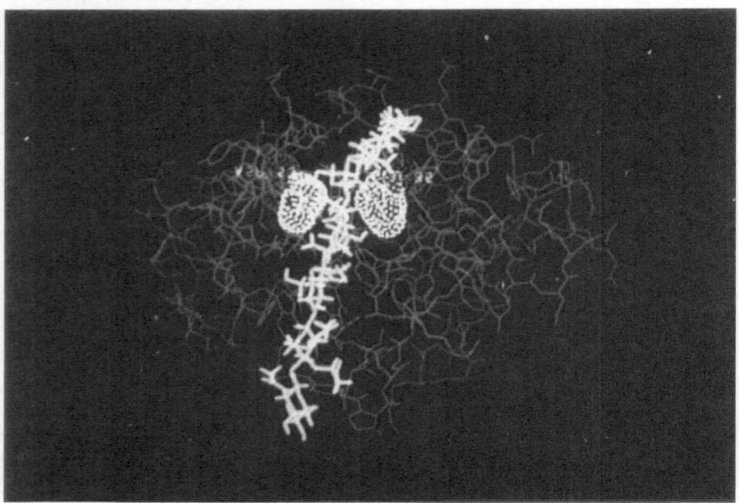

Computer Graphics Analysis of Modified Human-Lysozyme

Global Environmental Issues

We must accomplish our New Chemistry's goal without endangering human health or our natural environment. The challenge of global protection we face today, is undoubtedly a task for the New Chemistry.

We have many ploblems to be challenged and solved. The Japanese chemical industry has worked on solving a variety of domestic problems as well as saving energy. We must now extend those efforts to global environmental protection. We have many useful experiences, technology and ideas to develop and apply for this purpose, such as carbon-one chemistry, biotechnology, photochemistry, catalytic chemistry, polymer chemistry and organic synthesis.

For the global environmental protection, it is necessary to establish new research projects under strong support of governments and with collaboration of universities and industries worldwide.

Conclusion

The 21st century is considered by many to become the age of polymer science. I am firmly convinced that the future of polymer science is bright. The polymer industry should consistently meet with human needs and respond to challenge of human needs in order to continue to grow. The driving force of growth is basic research based on a truly long term perspective of human needs.

The polymer industry should be an industry with intensive Research and Development. Polymer science is the most exciting and promising science, as there remains many unknown inter-disciplinary areas to be explored.At the sametime, globalization of polymer technology will progress further. The importance of global cooperation with industries, academic and administrative sectors will increase steadily.

I would like to propose the following as management issues, that will grow in importance as we advance towards the 21st century;
(1) Establishing international basic research projects, such as the Protein Engineering Research Institute.
(2) Encouraging and enhancing the fused collaboration of researchers with interdisciplinary specialties.
(3) Emphasizing the importance of nurturing and supporting the young researchers. This is because the brightness and success of our future lies with them.

Polymer Science and the Arts

Panel Discussion

POLYMER SCIENCE AND THE ARTS

Organizer and Editor: Otto Vogl

Moderator: Hartwig Hoecker

A panel discussion has been organized at the First Pacific Polymer Conference to try to bridge the interests of artists and art restorers involved in using modern and well defined materials, mainly polymeric or oligomeric materials, with the polymer scientists interested in various aspects of art, especially art restoration. When the concept of organizing a microsymposium or a panel discussion on Polymer Science and the Arts first aroused my interest, in its initial stage of planning, I was in close contact with Professor Seizo Okamura. I have shared in the past many hours of discussion with Professor Okamura on the arts, humanity and specifically, on Japanese art; we have common interests in the paintings of the Kyoto Kano school, and the block prints of Hiroshige and other Japanese artists of the 19th century.

For the restoration of art objects, the use of modern and well defined materials, mainly polymeric materials rather then the use of traditional materials for art restoration, is now widely recognized. Many scientists are interested in the arts, visit museums and appreciate the beauty of the artistic expression of our heritage. The interests are rather diffuse and unorganized and it was thought important to focus the interests and knowledge into what is needed in art restoration and where we polymer scientists can contribute to the permanence of art objects through their maintenance and preservation.

We have brought together 4 scientists from Europe, Japan and the U.S. who have different interests in the use of polymeric materials in conservation, restoration, and generally, the appreciation of art objects.

VOGL: Welcome to the panel discussion on Polymer Science and the Arts. The idea of having a panel discussion at the First Pacific Polymer Conference occurred to us about two years ago. It had become obvious that we polymer scientists and chemists, in general, have

B. C. Anderson · Y. Imanishi (Eds.)
Progress in Pacific Polymer Science
© Springer-Verlag Berlin Heidelberg 1991

some responsibility not only to the environment but also to our
heritage and tradition. I had started discussions initially with
Professor Seizo Okamura about this proposal and found a very enthusiastic response. I then had the privilege of talking with Professor
Hoecker who is on my right and found another very enthusiastic person interested in art preservation. Professor Hoecker is involved in
a committee whose responsibility is the restoration of stone buildings, particularly cathedrals. I had the pleasure of talking with
Professor Hoffmann who is very generally interested in the arts.
Professor Chujo from Tokyo was very happy to join us on the panel;
he has agreed to cover the art of Japan for the panel.

I have asked Professor Hoecker to act as the moderator of this
panel discussion.

HOECKER: Thank you very much for giving me the honor and the
pleasure to chair this particular panel discussion. I think this is
an historic event for two reasons. First, it is the First Pacific
Polymer Conference and second, I think it is the first time that
polymer science shows the proper attention to art within a polymer
science conference. I would like to make a few comments in the beginning and one comment is the following: you will probably remember
that the keynote speaker emphasized the importance of environmental
problems and I think in the meantime everyone is aware of these particular problems and when we think of ancient art we face, in my
opinion, a very similar problem. The problems associated with the
conservation of objects of ancient art is as sensitive as the problems of the environment. Objects of ancient art are even more progressively decomposing as the environment is progressively being destroyed through influences of the most recent past. The environment
as well as objects of arts, of course, are the basis of our life.
The objects of art are the documents of our tradition and our
history and I think we should pay similar attention to the historic
and traditional artifacts as we do to the concern of our healthy
environment.

The problems we face in preserving art objects are of very high
complexity and interdisciplinary work is necessary to solve at least
some of these problems. I think that polymer science plays and should
play a key role here. I think further that it would be a very good
idea to face this general problem and see what we collectively can do
to provide some solutions.

I would like to emphasize that Professor Vogl was here particularly very much on the right track; this particular topic of discussions is needed for preserving objects of art. Or in more general terms, the application of polymer science to art is so important that this topic should be established as the subject of a symposium or at least of a discussion, at meetings of polymer science with high distinction.

Let me finally make one short remark; there are governments in various places and countries which require that a certain percentage of money be spent for the acquisition of contemporary art objects when a building is erected. On the other hand, I think that in the future we should also ask for a certain percentage of money to be spent for conservation of objects of ancient art when we do not wish to lose ancient buildings, paintings, or artifacts in general; at the same time we should ask for the devotion of a certain percentage of time to be alloted to the disussion of problems of art, of ancient art, and maybe also modern art, and their relevance to the scientific themes during scientific meetings.

The problems of keeping objects of art in a state of good repair are manifold. I touched briefly on the problems associated with the maintenance of buildings, paintings, ceramics and many other art objects. We have the problem of "inside art" and "outside art", the exposure of art objects to the indoor or outdoor environment.

We cannot offer any solutions and we do not even want to talk too much about possible ways of preserving art objects or trying to find solutions in todays panel discussion. But I think we four here on the panel are going to talk about the problem of conservation and polymer science in general. Each one of us will make a statement from his particular area of interest and his particular point of view. The aim should be to bring specific problems to your attention and to interest our fellow chemists in some of these problems and point out where problems exist. When I look around the panel I think that Professor Hoffmann is the best one to start; he will make a few general remarks concerning international science and the arts.

HOFFMANN: When I began to think about what to say I ran into a problem; I am not an expert in art preservation, as most of us are not, and neither am I, really a polymer chemist. Before I left home I looked at some of the slides that I have around for other reasons; I found that I had slides for a lecture that happened to be on the unity of arts and science. I brought the slides with me because they present a little bit of the cultural prespective of what we have, and what we need and want to preserve. It is very difficult to talk about these things unless one sees them. They also make a cultural point; because for this talk I had assembled a set of slides of Japanese ceramics.

The ceramic art, of course, has been a constant of civilization. Wherever there have been human beings there are pieces of pottery found; but in Japanese culture, the ceramic art reached a particularly high point of development. This is in part the result of Chinese and especially Korean influences, but also of native religions. It culminated in the remarkable, philosophical setting of the tea ceremony. Whatever object that has been associated with the tea ceremony acquired a greater degree of sophistication than almost anything else that we see. What you see here for instance is a classic bowl by Koetsu at the end of the Momoyama period, around the end of the 16th century. The chronology becomes important here. This bowl is a rather small object. It begs to be picked up in your hands and handled. One of the wonderful things about going to Japanese homes is that you will, as the guest, be occasionally asked to hold such objects which are centuries old.

Just a little bit later in the time prespective; this is a bowl by Ogata Kenzan who is the brother of a famous maker of screens, Ogata Korin. This ceramic bowl shows a combination of roughness and smoothness with the decorative elements that are so characterstically Japanese. The remarkable thing about Japan is not that Japan can make VCRs, and polymers better than we can, but that they can do it while preserving a tradition of respect of the arts and craft which is documented, f.i., by the continued existence of the ceramic tradition. These are some contemporary Japanese ceramics; one by Tagaki Shin which is from a collection of the Johnson Museum at Cornell. From the point of view of technique, you can see here a resist pattern of the cranes. This has something to do with the making of chips. The glaze was applied through a blow pipe. In the next piece, by Furukawa Takehise, you will notice again the elements of rough and smooth surface as the decorative element. Still another object is a dish which shows its beautiful pattern only in a certain light. It is by a woman potter from Shizuoka, Yamamoto Fukio.

Ceramics are inorganic three-dimensional inorganic polymers, so are various kinds of biopolymers that are used for the preparation of objects of art. As in the making of a western painting in this portrait of Federico de Montefeltro by Piero della Francesca which is in Florence. A typical portrait of the renaissance; a portrait like this could not have been painted 150 years before then when the representations were iconic. From a little bit to the north comes a work of Albrecht Duerer, a portrait of a serious man. These works of art are painted on canvas; there are polymers there, there are polymers in the paint, there are polymers in the varnishes and layers of restorations that such paintings have undergone.

These portraits are a part of our cultural heritage, just as much as this rather different portrait in which the light seems to emanate almost from the inside by Rembrandt von Rijn of his wife. Or in this multiple portrait, the greatest "representation of representation" ever painted, Las Meninas by Velasquez. If you are anywhere within a thousand miles of Madrid you should go to see it because this incredible portrait has just been cleaned.

The details of this portrait with its sculptural space with recognizable portraits in the background are remarkable. Note the gigantic canvas, Velazquez to the left of it; the Infanta Margarita in the middle surrounded by the ladies in waiting. And why are they all turned our way? Well, perhaps their Catholic Majesties, the King and Queen of Spain have come onto this sitting of Margarita for a portrait. Their image is reflected dimly in a mirror in the back of the room. An incredible series of images going back and forth; the canvas wih its backdrop, a mirror, people looking out in all directions. I go in for a close-up on the Infanta, so close that you can see the brush strokes of Velazquez' paint. We see very small emotional differences which exist between this portrait of a little girl by Velazquez and the work of Renoir, of a little girl three hundred years later.

The Velazquez portrait was painted chronologically in between the times that the two first Japanese bowls that I showed you were made. And we come more directly to the objects which more obviously deserve preservation. This is a portrait of the cathedral of Cologne. It looks to be in good shape but it isn't; we will hear more about this in just a while.

All of these objects, from a cathedral, to the ceramics, to the paintings, are part of our spiritual tradition. In many ways we could not be human beings, the kind of scientists that we are, if we did not value these cultural artifacts which cut across cultures. In our cultural heritage, the cathedral of Cologne has its place, and so

does Angkor Wat, and so do the Aztec ruins. These are objects that
are part of our humanity and worth preserving.

VOGL: My interest in arts preservation started when I came to
New York several years ago and became associated with the Metropolitan Museum of Art. A colleague of mine was involved in the restoration and preservation of paintings at the Museum The need for
interaction between scientist and conservationist became clear to
me about five years ago after Jane and I went "back" to the restoration section of the Metropolitan; I went to an easel and picked up
one of the pictures that was being readied to be restored. I said to
my colleague: "Isn't it this a nice picture! "This particular picture
appeared dreary and grayish with some cracks in the canvas, somewhat
torn in the back: Not anything one would look at when one is used to
the beautiful pictures on display in a museum. So my friend said:
"Yes, you are holding an 8 million dollar Rembrandt that has to be
restored." It is amazing how almost displeasing a run-down picture
can look when you are accustomed to seeing paintings in a excellent
condition in the picture gallery.

The problems of restoring Old Masters is rather complicated;
it involves many problems. As you know, many pictures are painted
on canvas, some of them on wood. Most frequently, the restoration
of the painting as well as that of the canvas is needed. Most of the
materials used are polymers, whether it is the materials of which the
canvas is made and or the adhesive or glue that has been used for preparation of the canvas prior to painting. In the last 10 or 20 years,
substantial changes have taken place from the use of restoration materials and varnishes from natural sources to the use of synthetic polymeric materials. Important problems in varnishing and revarnishing of
oil paintings have arisen in not only the cleaning and the removal of
the old varnish, but actually but also in the application of the new
varnish.

The varnish is not placed permanently on an oil painting, it is
applied as a very thin relatively uniform layer over the picture.
Important considerations for the properties of a "good" varnish application is the refractive index of the varnish layer as it gives the
right luster to the painting. After a few decades, for the next restoration, it must be possible to remove the varnish again in a safe
way. One of the favored material candidates right now is a cyclohexa-
none resin. The resin should be of low molecular weight; oligomers
with a molecular weight of about 500 to 1500 are used; acrylic resins

such as polymethacrylates are looked upon as potential candidates for this purpose. Paintings that have crevasses from the heavy application of the paint pose a special problem; in such cases the uniform application of the varnish is more problematic.

As mentioned earlier it is customary to clean oil paintings every few decades; this implies removing the varnishes and applying new ones. It is done preferably with mineral spirits and gentle rubbing. Over the years, especially coatings from natural products have had the tendency to "age" or to yellow. This is the result of autooxidation and photooxidation. This problem of aging has become more serious, since some of the display techniques require more elaborate lighting techniques. Modern light fixtures might have a light output with a higher percentage of ultraviolet light in the light sources; experimentations are underway to use ultraviolet stabilizers in the varnish, hopefully ultimately nonfugitive ultraviolet stabilizers, to protect the painting by having an ultraviolet screen in the varnish layer.

So much about some problems associated with the conservation and restoration of oil paintings.

To focus what is being done in the community of of the conservation of art, let me read from the communication of the International Council of Museums, you may find of interest: "It has been said that a number of synthetic resins produced by industry find now application in the conservation of artistic and historic works; although considerable research is carried out in the conservation field, the field could profit from the increased exchange with scientists working in industry and academia. It is also important to persuade the researchers in these environments to become involved in some of these special groups. There is a resins working group of the committee of conservation that is part of an organization which is called the International Council of Museums. This group has been working quite actively and its principles are to collect all information that is pertinent to natural and synthetic resins found in historic and artistic works. Of particular interest are those which are used or potentially can be used in coatings, adhesives, varnishes and other applications during the conservation and the restoration of historic and artistic works."

This committee of conservation then gives recommendations. There was a very important Congress in 1989, actually held in Kyoto; it was on the conservation of Far Eastern Art. It was the 12th International

Congress of the International Institute of Conservation of Historic
and Artistic works, that was held last year. What I am trying to say
is we are having actually at least one specific committee whose responsibility seems to be to coordinate the activities of restoring in the
art community and these people need the materials that we are producing or could produce if we knew the need.

One of our problems that we have in art restoration is that relatively small amounts of materials, varnishes, glues, of very specific
characteristics are needed. Ideally speaking it would be nice to have
available one type of material for each type of painting and each period. From the commercial point of view, it is not justifyable to
produce many types of small amounts of material. Consequently, the
cooperation of industry for the preparation of these specialty polymers, polymers that would in most cases have to be tailor made, is
limited. In the case of polymeric materials for electronic applications, it is much easier to arouse interest for polymers with low
volume but with larger returns. It is one of the objectives of this
discussion to call attention to the need of specific materials in
art restoration and preservation and to alert the scientists and
restorers to the importance of a bridge of knowledge in this field
and for industry to become aware of the opportunities.

CHUJO: I would like to introduce three items on Polymer Science
and the Arts. The first one is "Polymer Science and Conservation of
Art in Japan", the second is "Polymer Science and the Conservation
of Art on the Silk Road, and the third is "Polymer Science and Simulation of Art". If I were an expert in this area, I should point out
Polymer Science and the Creation of Art. However, I cannot comment
on this item.

On the first item: There are many efforts to conserve art made
in Japan. For instance, Mr. Arai is studying the mechanism of how
foxing of paper occurs. Japanese paper is subject to foxing. Foxing
describes a phenomena which produces brown spots on paper, similarly
to spots produced in the development techniques in paper chromatography. In order to prevent the foxing, Mr. Arai analyzed the mechanism of foxing and he found that Aspergillus penicilloides and
Eurotium herbaviorium are the primary cause for the formation of
the spots that are called foxing.

Mr. Oka is trying to apply a dry method to prevent the removal of the backing paper from paintings. He tried this method on silk and pigments. Traditionally, a wet method was used for removing the backing; however, according to his experience, the dry method is superior. He found that the pigment should first be consolidated with solutions of animal glue and then the paintings are faced using rayon paper and furunori (furunori is an traditional Japanese paste). After this treatment preservaton of the silk goods of very high quality was achieved.

Mr. Aoki tried to conserve excavated archaeological art objects. For instance, an iron sword is at first desalted by the Soxhlet extraction method. After that it is treated with an organofunctional silane followed by impregnation with acrylic resin emulsions; the analysis of polymeric material used for these preservations is very important, as in general, polymeric materials are important for the preservation of the art objects.

Now I would like to go on to a second item, "Polymer Science and Conservation of Art on the Silk Road". As you may know there are many frescos preserved on the Silk Road. The Silk Road is equivalent to a dry area. The restoration of these frescos must be done from a different point of view from the general preservation of Japanese art. The Japanese government has supported the conservation of the frescos on the Silk Road and offered financial and technical support to preserve the frescos in Dunhuang. This city is called Tonko in Japanese. This preservation program was started in 1983 under the auspices of the Ministry of Education; now the Foundation of Cultural Heritage is responsible for this project. This program is executed by the Tokyo University of Art; the President of this University is Professor Ikuo Hirayama and he is the most famous painter of frescos of the Silk Road. He is very active and the best person to promote this program. Special problems exist because of the local climate. Good adhesion must be achieved in order to provide good bonding between clay and pigment. To achieve this goal, in the last century a varnishing was used in attempts to cause proper bonding, but in th long run, the varnishing method is not very good because of the oxidation of the polymeric materials used in he varnish.

Next, an injection method was used. This method is also not very good; consequently, a backing with clay and pigment is applied on the backside. In order to prevent the hydrolysis of the polymeric

materials to occur, we recommend the use of the copolymer of vinyl
acetate and ethylene instead of poly(vinyl alcohol) as diluent.

The third item of my discussion is "Polymer Science and the
Simulation of Art". Simulation of the beautiful color of the butterfly is done in a development of a special polyester cloth developed
by Kuraray in Japan. The Morpho butterfly in South America shows
very beautiful luster. In the scanning electron microscope pattern
of the wings of the Morpho butterfly, highly regular ladder like
structures can be seen. This structure was simulated by the twist-
ing of a great number of filaments and a lustrous color can be obtained in a polyester fabric made from fibers made by this principle.
Comparison of the electron microscope patterns of the cloth and the
butterfly wing show a very similar picture. (Chujo showed several
samples of fabric made from this kind of polyester fibers).

HOECKER: Ladies and gentlemen, after having had such a nice
introduction to the more general aspects of our topic today I would
like to go a little bit more into details concerning the preservation
of buildings of natural stone. If you have never seen dead trees
you would not recognize the problems of dying forests as real problems.
The same thing is true with buildings; if you have never looked closely at a building which is falling apart you would not recognize the de-
terioration.

Therefore I would like to use this chance to give you an idea
of what may happen to artifacts upon degradation. This particular
degradation has been exponentially increasing during the last years.
I am showing you a picture of a figure at a church in a small city
of Germany, Marbach, built in 1511. You can see the figure as it was
in 1900 on the left hand side and on the right hand side you see the
same figure as it is today; the picture was actually taken in 1981.
The figure has completely lost its face; this is so true for not only
figures but also for many buildings such as cathedrals and, of course, for all kinds of buildings of natural stones: many public buildings, post offices, railroad stations, and even private houses.
Here you see a similar picture taken in 1900 and in 1969, sculptured
about 400 years ago and now loosing its face within 50 or 100 years.

In the next slide you see the morphology of the stones showing
up with increasing erosion. A very particular kind of erosion is
called "honeycomb structure"; soft parts of the stones, minerals and

the glue material between the different minerals in the stones are
dissolved. A particular event which sometimes is called scaling, or
conturscaling, where thin layers of different thickness are "eaten"
away from the stone. These are different signs you may see at buildings and sculptures, signs of slow disintergration of the material,
decomposition or weathering.

From afar, the cathedral of Cologne looks quite alright but
when you go into detail, when you come closer you see that the supporting pillars or columns are heavily weather-worn and have to be
restored sooner or later; the degradation or disintegration of the
square stones very often starts from the joint. Sometimes, in particular in the pillars high-up in the cathedral, the stones are
weather-worn, "sanded-off" or worn away by the weather. From a chemical point of view ion exchange reactions apparently play an important role, e.g. the formation of gipsum. Besides that, chemical,
physical and biological processes play important roles and they all
require water for these processes to proceed efficiently. Thus, you
might think, that keeping water away would help in preserving the
buildings or statues made of stone.

In a garden of a Bavarian castle, the castle of Seehof, near
Bamberg, the owners have taken away the water from the sculpture by
building a tent with a roof; beneath this tent a well preserved
sculpture can be seen. The Bavarians do this every winter and they
are very successful with this technique. On the other hand when the
stone has already decomposed and the single sand grains are insuffi-
ciently glued together then this method certainly would not be useful
and another way of saving the scuptures has to be used. A well known
method is the treatment with silicic acid esters which then form a
substitute within the stone which solidifies the stone upon chemical
reaction and elimination of alcohol; the material remains, however,
hydrophilic. To keep the water away, the surface has to become hydrophobic. Therefore a second method of preserving stones was considered:
namely, to apply polymers which are by definition hydrophobic, for
instance, silicones.

However, people did not have the best experience, particularly
at the Cathedral of Cologne, with polymers like silicones. In time
they became weather-worn themselves and, as a consequence, the surface became again hydrophilic. Therefore, at the Cathedral of Cologne, old weather-worn stones, not only the simple ones, but also

complicated, sculptured stones are replaced. This is a very expensive way of preserving the building; the substitution of a single square stone costs about $ 4,000. Presently, there is no polymers which could be applied or that the conservators would be willing to apply because they have no proof that they would work.

In cases of single statues, a full impregnation may be performed with methyl methacrylate in a vacuum oven; then MMA is radically polymerized to form PMMA in the entire inside of the sculpture; this indeed is a very efficient way of preserving sculptures, at least in many selected cases. However, a full impregnation cannot be done with the entire Cathedral of Cologne. The method of choice would be to impregnate a stone, preferably with an aqueous emulsion of a material which penetrates very deeply into the stone, adheres to the inner surface of the stone and leaves the pores open. The liquid water is kept away, the material remains open for water vapor transport. By means of scanning electron microscopy it may be shown that a mineral is fully covered with a very thin layer of a polymer.

A stone, after having been freshly broken and treated with the polymer, shows rounded edges covered with a polymer film. The glass transition temperature of the polymer has to be sufficiently low in order not to become brittle in winter time.

VOGL: Professor Ranby has been a member of a committee whose responsibility was to preserve the famous Vasa ship; Professor Ranby has received a medal of honor for his service. Would you please comment on the preservation of the Vasa; please tell us what is the Vasa ship, and why was the Vasa ship preserved and how was it preserved ?

RANBY: The Vasa ship is a large war ship, made from oak wood. It was built in Stockholm and made its maiden voyage in 1628. It sailed for 10 minutes and then it overturned and sank, due to a hard gust of wind. In a way, this was a good thing because the ship sank in a deep hole of the Stockholm harbor and became rather well preserved there. At that time and well into the 1900's, Stockholm had no sewage treatment and the ship was gradually covered and buried in dirt and silt from the raw sewage outlet. As a result, very little oxygen had access to the ship. The oak wood did not decay very much, the iron metal rusted away, but the cannons, cast from brass, remained virtually unchanged. The Stockholm water is brackish and no wood shipworms can live there. The Vasa shipwreck was located in 1956 and was raised to

the surface in one piece in 1961, 333 years after it had sunk. It
first appeared to be in rather poor shape. The surface of the oak
wood had deteriorated to a soft layer to a depth of about one inch
and all the wood was extensively water-soaked. The wooden hull contained more water than wood substance, 150 parts to 100 parts dry
wood. When a piece of the Vasa oak wood was dried, it cracked up
and crumbled from the surface. Some preservation treatment had to
be done. After a careful study the Vasa committee decided to use
poly(ethylene glycol) (PEG) in aqueous solution to preserve the wood.
Spraying of the big ship was started in 1961, first with a dilute
solution of PEG of low molecular weight (around 1400), then gradually
with higher concentrations of PEG of higher molecular weight (about
4000). The spraying was first done manually but after two years automatically with electronic control. After a total of 18 years of spraying the PEG had penetrated deep into the surface-decayed wood to an
average concentration of about 25% PEG which had replaced most of the
water. During the last 10 years the Vasa ship has been treated with
PEG melt on the surface and dried slowly to prevent cracking and coll-
apse of the wood. As a polymer PEG has no measurable vapor pressure
and stays in the wood. PEG is water soluble and can be extracted out
from the wood at any time. But the Vasa ship will not be launched into
the water any more. It will stay in its dry dock and is preserved for
a long time to come.

The Vasa ship and its surrounding exhibits now form a complete museum of Sweden and Swedish life as it was in 1628 when Vasa suddenly sank.
If you come to Stockholm next summer, the museum will be completed and
opened to the public on June 15, 1990. The ship stands in a huge hall,
moored at a stone key in its dry dock in full splendor, even with some
of the sails up. The Vasa sails which were raised when Vasa sank had
deteriorated but a spare set of sails were found in a wooden chest on
the ship. These sails were preserved on thin glass fiber fabric, impregnated and glued with a special polyacrylic resin which was prepared at the Royal Institute by my students. Resin and glass have about
the same refractive index which makes the glass fibers invisible. This
treatment made the sails strong. They look like new, they are shiny
like wet and very durable.

This is the story of the preservation of the Vasa ship. It is by
far the biggest experiment I have ever taken part in and also the most
extensive in time. The ship is 60 meters long, the hull in the stern
is 20 meters high and the masts originally 40 meters tall. The ship

now weighs 1200 metric tons of which presently about 25% is poly-(ethylene glycol) and about 1% sodium borate which is added as preservative. As it looks now, the preservation of the Vasa is a complete success.

HOECKER: Thank you very much Professor Ranby. We have learned that poly(ethylene glycol) has been and may be used very successfully for the preservation of materials, particularly of wooden objects.

HOFFMANN: I want to comment on the preservation of objects that cannot be replaced and the courage that is required to deal with the irreversibility. I don't know how easy it is to get out poly(ethylene glycol) from the Vasa. Is that the preception of the possible failures of such projects resides much longer in public opinion than the successes. Let me tell you a story about technology transfer. This has to do with aniline dyes in oriental carpets. Within about three years of the synthesis of the mauve dyes and the first replacement of natural indigo they reached Persia; on the first few synthetic dyes, there was no research done at all, particularly the black. So it you look at carpets of a certain period you can actually feel the black has gone away. The dye damaged the wool. There was issued an imperial edict in Persia prohibiting synthetic dyes, on the pain of death. To this day the use of dyes in Persian communities, by tradition, is fairly restricted while those tribes which work in the Soviet area, more commonly use pinks and greens which are difficult to get in natural, not synthetic colors. It is necessarily to recognize that quick fixes are dangerous.

CHUJO: I suppose the combination between the technique of application and the type of material used are the two most important factors in the preservation of art. Another important factor is our attitude trying to know in detail the polymeric materials, that we are using; our general knowledge is limited and often insufficient. Even if, for example, a polymeric material is the best because of its adhesive properties, we should not use it without knowing what its solubility characteristics are and what solvents can be used. If we find that some damage culd be done to an object of art because of the use of an inappropriate polymeric materials we must be able to remove the material quickly.

HOECKER: It is certainly not a field in which quick money can be made. A very strict quality control of products to be used is necessary in this business and the quality has to be guaranteed. This means the effect must be proven before the material can be applied, otherwise conservators who do not know how to apply the material properly can be the cause of bad experiences.

Are there any further remarks from the audience ?

CHUJO: I would like to make two points. The item of destruction. The nature of the polymeric materials and the procedure of their application is very important. I suppose the combination of procedure and the material are most important for the preservation of art. The second problem is the type of polymer used, because after using polymeric material we have the preconcieved notion that the damage is due to the use of polymeric materials. If we find some damage we must look out quickly for instance, for a second polymer is a different lack of solvent if the second polymer is the best in adhesive we don't use exchange for the preservation of arts.

Let me just say the procedure is very important, the preparation is very important; the preparation is the first thing that has to be taken into consideration. The purification of the material once grown with algae and things like that has to be removed. It is very important to have to consider that deep penetration and reapplication must be possible when it becomes necessary.

RANBY: During the experiment with the preservation of the Vasa ship with PEG, which took about 20 years of spraying, we were well aware of the importance of having selected the right polymer. The poly-(ethylene glycol) could be removed by extraction with water any time if it was found to be the wrong polymer. But I think PEG was the right choice. PEG is sensitive to degradation with light. But the ship is now exhibited in a museum under controlled conditions of humidity and temperature an not exposed to sunlight. We have to limit the amount of UV light on the ship to avoid photodegradation of PEG and wood.

VOGL: Of course, poly(ethylene oxide) is very readily autooxidized.

HOECKER: I hate to finish before I have additonal comments from the audience.

ANDERSON: I don't know whether you consider fishing rods as objects of art but some of us do. Bamboo has been a material of con-

struction for several hundred years amd there has been considerable
work to preserve the functionality of the bamboo against decay by
impregnation with resins. For example, The Orvis Company of Manchester,
VT, sells impregnated rods which are in excellent condition after 50
years or more. I believe the impregnation is done with a phenol-
formaldehyde resin system using vacuum-pressure cycles to achieve
complete impregnation. The appearance and function of the bamboo
is preserved and the durability greatly enhanced.

HOECKER: Thank you. Any further comments from the audience.
It is in our great interest to talk about what is going on in the
world and where the progress is being made. If somebody has any idea
and can comment I would like to invite him to make his comments.

VOGL: I would like to comment on another item that is not directly related to the preservation of art objects but has to do the construction of string instruments using polymers and composite materials.
As you know, the Amati and Stradivarius violins are world renowned.
Famous violinists like Isaac Stern are playing instruments of the
Cremona school rather than instruments of more modern construction.
In recent years the understanding of the individual components of
construction materials that are essential to make a good violin has
improved considerably. Scientific investigations have been undertaken
to identify the basic characteristics of the materials of construction, of the type and quality of the wood and of the lacquers. Characterization techniques that are commonly done in polymer science such
as the determination of modulus, tensile strength of these basic construction materials, have become of interest; interest has also been
focused on the application technique of the lacquers. The type of
materials and their application is essential for the superior performance of these instruments.

HOFFMANN: I would like to mention some special problems which
still need solutions. In general, textiles have special problems which
require attention for their preservation. If you think about it we
don't have many wool and silk objects that have survived certain
periods. There is a need for a method of reproduction of textiles
or an impregnation method that lasts for a long time. We have very
few oriental carpets; we have one remnant that has come from Russia,
but otherwise we have relatively few samples that have survived from
the years 400/500 A.D.

Modern materials, as you all know, present special problems and
there are art forms that have evolved using some modern materials.
Isn't it interesting that we have more trouble preserving films than
we are having preserving stones or pictures made with oil paints? You
know the problems with color photography; even the problems of preserving black and white photographs are severe. In part they have to do
with the acidity of the papers that are used. Magnetic tape now is
the medium of choice for preserving images both on video tape and for
tape recording presenting problems. The solutions are, of course,
electronic and it is now often possible to redigitize the information
and preserve most of it.

HOECKER: I would like to make one final point. In Germany there
is a project which is sponsored by the Federal Ministry of Research
and Technology of the F.R.G.. Some 20 University Institutes in the
F.R.G. are involved and some important chemical companies of the
Federal Republic of Germany, like Bayer, Wacker (a subsidiary of
Hoechst) and many others are actively cooperating. Industry provides
and develops products and Universities investigate the performance of
these products in combination with different stones. Furthermore, they
look for reasons why stones disintegrate. We are confident that we
will find a solution before too long. As long as we do not have solutions, however, it is highly desirable and recommended to conduct discussions on polymer science and the art regularly during polymer con-
ferences of high visibility. This will allow to keep a "Standing Committee" of polymer scientists active and make sure that other scientists are kept aware of the problems of art restoration and our responsibilities for our heritage.

The Violin

William J. MacKnight

Polymer Science and Engineering Department
University of Massachusetts
Amherst, Massachusetts   01003

The modern violin family consists of only three members: the violin itself, the viola, and the violoncello. (The doublebass or contrabass is a member of the related but distinct family of viols.) The violin was developing during the 17th century and achieved its present form during the l8th century. In fact, it is commonly held that the greatest examples of the violin family were produced in the Italian town of Cremona in the eighteenth century and are associated with names such as Amati, Guarneri, Bergonzi, Guadagnini, and Stradivari, among others. The name of Stradivari or Stradivarius is the one which has most captured the attention of the general public.

The figure shows some of the parts of the violin. The belly, or table, is usually made of pine and the back is made of maple. Sound is produced by the communication of the strings, which are set into vibration with the bow, with the body of the instrument. This is accomplished first through the bridge and then by means of the sound post. There is a great deal of mythology and pseudo-science extant concerning the means by which a "great" violin is made or identified. For example, the superiority of the Cremonese makers was at one time attributed to the quality of their varnish, but it is generally recognized today that the varnish has little or nothing to do with the tone of the instrument. Of course, the most important single factor in achieving a beautiful sound from a violin is the player. Great masters can make a rather mediocre instrument sound wonderful while, at the other end of the spectrum, the amateur player most often plays out of tune and produces a thin, unpleasant tone whether he is playing on a Guanerius del Gesu or a violin mass-produced in the workshops of Bohemia. The violin is an extremely difficult instrument to play and is very demanding physically since the position in which it must be held is quite strained and unnatural. A number of physical afflictions tend to beset violinists, including back, shoulder, and arm problems. Often, the neck becomes irritated where the violin is held against it under the chin and, on occasion, surgery may be required to correct the problem.

It is the case that world class concert artists, almost without exception, choose to play on an instrument made by one of the Cremonese masters. It may be asked if it is really possible that artisans working 200 years ago were capable of producing a product superior to anything that can be made today. The general answer given by most violinists to this question would be yes. It is certainly true that the Cremonese instruments possess an

B. C. Anderson · Y. Imanishi (Eds.)
Progress in Pacific Polymer Science
© Springer-Verlag Berlin Heidelberg 1991

412

amazing combination of brilliance, power, and beauty of tone. This is, of course, traceable to the mixture of fundamental and harmonics produced in each note. It is also true that these instruments are very "even". That is, the quality of the sound remains quite constant over the whole range of the instrument (about four octaves) and on the different strings. From the figure it can be seen that the violin is a complex instrument and, in modern technical parlance, might best be regarded as a "system" for producing musical sounds. It seems clear that the sound will be affected by the quality of the wood (a "materials" problem), the overall shape of the body, the shape of the sound holes (or "f" holes), and the positioning of the sound post and base bar, among other things.

There are a number of modern violin makers who produce instruments with many of the characteristics that make the old Italian violins so desirable. If they are not prized as highly this may be due to perception rather than reality. Recently, there have been a number of efforts on the part of scientists, including chemists, to reproduce or equal the products of the Cremonese masters. In this endeavor they are making use of a battery of techniques including electron microscopy and fourier analysis of the sounds produced. Computer-aided design and manufacture are also utilized.

It is unlikely that the 18th century Italian instruments will be supplanted anytime soon, not only because of the beauty of the sound which can be produced from them, but because they are also "objects d'art" in their own right. Great connoisseurs spend lifetimes studying the characteristics of these violins and become capable of immediately identifying their makers and even the date at which they were made. Many collectors exist around the world who become fascinated by fine examples of the violin maker's art.

When one hears a great violinist in concert it is perhaps worth reflecting on both the astonishing skill of the performer and the almost unbelievable beauty and complexity of the instrument on which he is playing.

From the polymer science point of view, the most obvious relationships which exist with the violin are through the strings. In former times the strings were made from natural products such as gut from the intestines of sheep. Today they are either metal or metal wound on a polymer, such as nylon. The nylon strings have the advantage that they are relatively insensitive to temperature and humidity variations and so remain "in tune" under a variety of conditions. However, a number of violinists find the tone produced by these strings to be rather harsh and prefer the older gut strings over them for that reason.

Of course the wood from which the body of the instrument is made is built up from the polysaccharide cellulose. One might be tempted to conclude from this that the production of sound from the instrument reduces to the dynamic mechanical properties of cellulose in the kiloherz region. Naturally this is greatly oversimplified inasmuch as the violin in reality is a complex structure and the wood from which it is made is an anisotropic composite material. Thus, as alluded to previously, the secret of the tone quality of the instrument is unlikely to reside in any one component such as the varnish, or the quality of the wood, or the shape of the "f" holes, or whether the back is or is not made from a single block, etc. Modern instrumentation and analytical techniques do make possible a microscopic examination of the constituents of a fine violin. Whether the results of such an examination are really capable of leading to the routine production of violins "equal to Stradivarius" is another question.

The Violin.

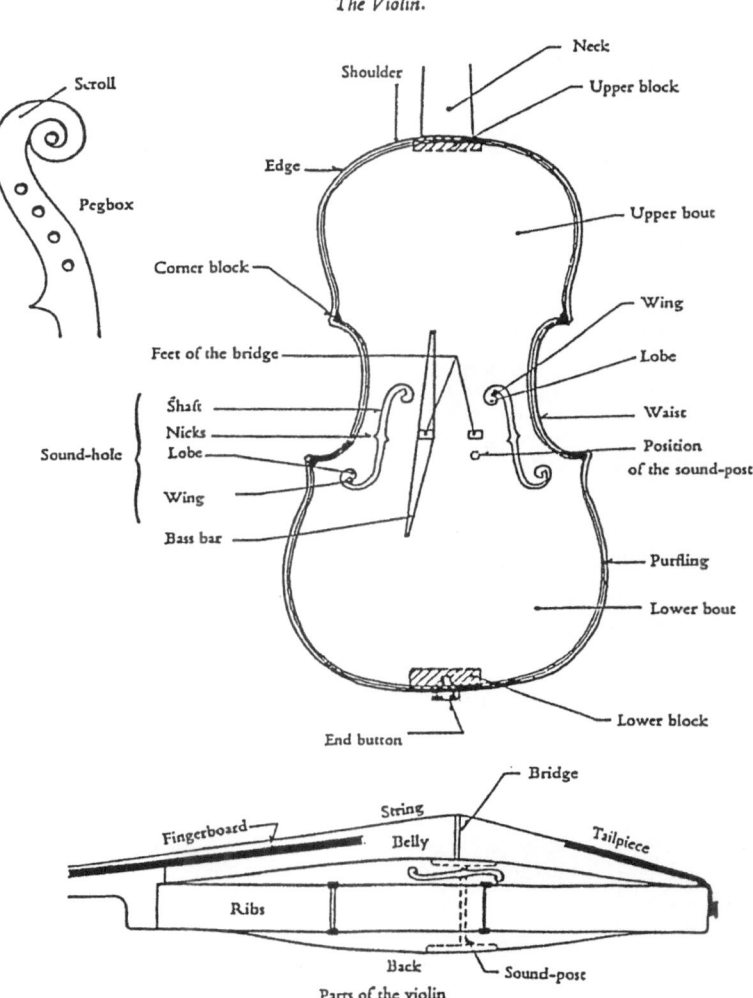

Parts of the violin

Bibliography

The Violin, Its Physical and Acoustic Principles, by Paolo Peterlongo, a Crescendo Book, Taplinger Publishing Company, New York, 1979.

Lining Adhesives Used in Paintings Conservation, An Introduction

Christopher W. McGlinchey

Department of Paintings Conservation

Metropolitan Museum of Art, New York, New York

Abstract: The lining process is described as a method by which a new cloth material is adhered to the back of a painting whose canvas is decayed to such an extent that it can no longer support itself. Due to the close relationship of the paint film to the canvas, the activation parameters employed in the lining process are at risk of changing the appearance of the paint film. The constraints the adhesive and adhering process must be activated under are discussed.

1 Introduction

Older paintings on canvas are often found adhered to an additional backing canvas. This auxiliary support, known as the lining canvas is necessitated by the incapacity of the painting's deteriorated canvas to support the reflective ground and paint layers. Lining processes have existed for centuries, current techniques are derived from 18th Century lining adhesives using starch and animal based glues [1] and more recently developed synthetic adhesives based on either ethlyene-vinyl acetate type hot melt [2] or solvent activated Polyvinylacetate [3] adhesives.

The three elements that control the activation of lining adhesives are pressure, temperature and solvents (organic or aqueous); in a properly executed lining these factors never exceed the tolerance of the painting. Knowing the tolerance of paintings to these parameters requires extensive study, working at conditions beneath them requires great skill. Exceeding the tolerance of these three factors in lining processes, however, can lead to their deformation, melting, cracking and change in optical appearance. However, as more becomes known about the chemical, physical and mechanical properties of paintings the technique of lining can thus become more specialized when used on the broad variety of paintings requiring treatment.

B. C. Anderson · Y. Imanishi (Eds.)
Progress in Pacific Polymer Science
© Springer-Verlag Berlin Heidelberg 1991

Since artist's painting techniques have in the past, as well as promise to evolve into some ways that could be adversely affected by current activation parameters of conservation adhesives, there exists a persistent demand for new materials that are safer to use. It is hoped that with an understanding to the constraints of adhesives used in paintings conservation, recent advances in polymer science can assist further understanding of the lining problem. The purpose of this article is to familiarize the polymer scientist with lining adhesives used in paintings conservation in order to facilitate the transfer of knowledge between scientist and conservator.

2 Esthetic Considerations

Paintings conservators have the responsibility of maintaining the intrinsic values of paintings, not only to prevent art historians and curators from being misguided, but also the general public. This is clearly no simple task that can be accomplished without careful thought and open discussion. Discussion with scholars and scientists to further understand the meaning of the picture, nature of the changes that may have occurred with time and prior restorations frequently occurs before any treatment or intervention occurs. While it is understood that fugitive pigments and yellowed media cannot be returned to their original state, it is however hopeful that a conservator's materials and techniques can be employed so that a significant portion of the elements established by the artist can be read by the viewer.

The lining process is not immediately apparent to those admiring pictures, if properly lined the lining canvas should never enter the viewers mind (except for the curious conservator). There are varying degrees to which the lining process can cause harm to pictures, ranging from the obvious to those that are more subtle in nature. For example, brushwork whose details can only be scrutinized from close will cause a painting to reflect light in a certain way. If these subtle undulations in surface texture are even slightly modified, the painting will reflect light differently, altering how it appears from a viewer's distance. Distinguishing these diminutions from an artists working methods merely depends upon the experience of the viewer. If the condition of the painting prior to the lining has been studied carefully, such adverse changes in state can be noted easily after a faulty lining; if, however, the painting's condition prior to lining is unknown, it is usually just a matter of time before the cause for the seemingly aberrant nature of the painting is discovered.

Using an adhesive that is capable of penetrating too vigorously, or over time capable of migrating or bleeding into the paint surface will drastically alter the optical saturation of the painting. For example, using an adhesive that contains a large proportion of wax on a painting which has a porous structure will provide more medium to absorb light so that less is scattered. If a painting that was intended to appear matt is lined in this manner it will become a glossy, saturated image that appears over all much darker. Such processes have altered many impressionist paintings by replacing the visual effect of matt whiteness caused by diffusely scattered light with one of glossy, specularly reflected light. In fact, due to the entropic migration of the low molecular weight wax into the visible elements of the paint structure this devastating change is considered to be irreversible since the wax cannot be removed without taking pigment particles or binding components along with it.

3 Constraints Imposed Upon Lining Adhesives

The adhesional constraints that lining adhesives must operate under distinguish these materials from almost any other class of adhesive in use today. In most other adhesional applications, the materials being joined, or laminated have a higher tolerance for the activation method employed and commonly permit more rigorous surface preparation than lining applications allow. On the other hand, the bond strength required of a lining adhesive is quite considerable, though its viscoelastic nature has yet to be agreed upon. The author's opinion is that the adhesive should permit the natural expansion and contraction of the paint film. The author thinks it unwise and possibly unethical to attempt to restrain this movement in an attempt to reduce cracks that may occur if no intervention were needed. As noble as these efforts may be, unresolved stresses as a result of a new adhesive may cause the paint film to crack in a manner telling of the nature of the lining adhesive and not the nature of the paint film.

Any conservation material or technique used should produce results that are reversible and not permanently bonded to the artist's original work; this allows for the removal of conservation treatments that later no longer serve their intended purpose. This constraint alone eliminates all thermosetting type adhesives because of their thermal irreversibility. It also excludes adhesives that are applied as solutions of low molecular weight material, due to their tendency to penetrate and dry in unreachable areas. Lining adhesives have therefore been limited in principle to stable, thermoplastic materials. One important exception is the use of starch and animal type glues discussed in section 4.1. These adhesives are not readily reversible by thermal means and are usually removed by mechanical scraping. The negative ramifications of strongly hydrogen bonded adhesives are serious and their use for many types of lining applications has recently been discouraged. The most serious of these is that hydrogen bonded materials tend to develop large strains that depend upon moisture content of the air. But for many paintings that already include these materials in their ground structure the only caveat is from the initial stages of the adhering process which requires water, heat and pressure and may adversely react with the ground layer.

The peel strength of lining adhesives has been recommended by Phenix and Hedley to be no less than 300 g/2.5cm [4]. This minimum value has been established to avoid not only adhesive failure, but to avoid cupping out of the canvas plane that is a result from weightier regions of the painting that attempt to redistribute their weight dependant stress patterns and often appear as ripples in more thinly painted regions that neighbor more heavily painted areas [5]. It is worthwhile to note that according to Phenix and Hedley some degree of solvent flocking is even necessary for certain heat seal adhesives in order to obtain the minimum peel strength. Katz has shown that sizing of the original canvas with a dilute solution of a methacrylate greatly improves interfacial adhesion [6].

4 Lining Adhesives
4.1 Lining Adhesives of Natural Origin.

Recipes of starch paste glues mixed with polysaccharides have been frequently used as early lining adhesives. The basic principle behind this type of adhesive is that they are warmed in a double boiler for sufficient time to achieve the gel state, once applied to the canvas the adhesive is then activated by ironing the water out of the adhesive. For paintings whose canvas fibers have not been saturated with oil (i.e. have been properly sized and gessoed) a strong bond is formed between the lining canvas and the original with starch paste glues. In spite of their misgivings, these glues have been found as linings on paintings that were applied over 300 years ago and are still in good condition.

Water soluble proteinaceous glues of animal origin such as hide glue, rabbit skin glue and fish glue have been used as lining adhesives as well. The basic gelling agent in these glues is collagen, while not gelling as much as the starch/polysaccharide adhesives, upon evaporation of the water it is possible to achieve a high degree of tackiness to set the adhesive. One setback of these and other natural glues is that they are subject to microbial attack.

Because of the setbacks of these hygroscopic materials, wax resin linings were developed in the 19th Century. Wax/resin adhesives are based essentially on mixtures of beeswax and dammar, a low molecular weight, high softening point resin from Indonesia. Since the wax resin mixture is not hygroscopic these adhesive layers and all that they impregnate are no longer dependant on fluctuations in humidity. The method does have the benefit of adhering more strongly with canvases whose fibers have some fat content. Since the materials are thermoplastic, assuming insufficient oxidation has occurred, this adhesive in theory can be thermally reversed. Unfortunately the materials do oxidize as well as densify (because of their thermal properties), thus rendering the ultimate reversal more difficult.

There is a more immediate threat to paintings from the wax resin lining of paintings that have either no ground preparation or a cracked and porous ground. In this instance, the wax is capable of flowing into the paint layers, saturating the colors, thereby making them appear darker. In the case of a cracked ground the wax only migrates in these regions underscoring the crackelure. Wax behaves as a highly viscous material at ambient temperatures; regardless of the stresses and strains involved, upon removal of a stress the material experiences no "recall" or elastic behavior and hence continues to diffuse into less saturated domains. It is because of this nature some have referred to wax/resin linings as was/resin impregnation [7,8].

Because of the hygroscopic nature of starch and fish glues and the wicking effect of wax resins, conservators have been prompted to search further for new materials to function as lining adhesives. Fortunately in the 20th C., polymer science has provided many new raw materials and processes to develop alternate techniques.

4.2 Synthetic lining adhesives

The earliest synthetic materials applied as lining adhesives were either solvent cast or emulsified copolymers or homopolymers of vinylacetate or methacrylate polymers. These materials are admired because they do not respond to changes in relative humidity because of their reduced amount of functional groups capable of forming hydrogen bonds with moisture present in air.

Mehra has used Plextol B-500, an acrylic methacrylate copolymer in dilute solution along with a thickener. He first used hydroxyethyl cellulose to gel the adhesive to keep it at the lamina interface, later he used microspheres to apply the film as a slurry. These lining methods were easily activated in the absence elevated temperatures and only slight pressures. Hacke has used a thickener based on methacryl acid in small concentrations with plextol D360 as another adhesive system for a low pressure activated adhesive [9].

Recently, thermally activated, heat seal or hot melt type adhesives have been modified for paintings conservation. These materials offer the benefit of containing no solvent that needs to be dried off. The first hot melts were commercially available to industry in 1965, and use of hot melts in conservation was first suggested in 1968 by Lodewijks [10]. Ethylene-vinylacetate copolymers, being internally plasticized, have been welcomed by conservation since there is no chance of low molecular weight material migrating into the paint film.
Berger has developed, especially for conservation a hot melt or heat seal adhesive based on EVA copolymers [11]. It includes other materials to act as wetting agents and viscosity modifiers to improve the handling properties of the adhesive. Its activation temperature is about 60-65 C, which is a temperature range that some paintings can tolerate without any visual changes to the paint film if carefully lined. This lining adhesive is activated by both pressure and temperature, though as mentioned previously some claim that solvent flocking is necessary to achieve adequate adhesional strength.

5 Lining Procedures

The actual process of adhering the lining fabric to the original is needless to say a delicate operation, and has recently undergone refinements to reduce the dangers that are involved. Earliest methods used heated irons, in some instances the heat was only crudely controlled, producing often hazardous circumstances. Some treatments were so reckless scorch marks could be seen from the back of the canvas, along with a melt distorted paint surface on the front. Properly used though the technique does posses some benefits; since the ironing is local one has greater control of focusing on areas that need more pressure or less temperature and can be modified according to the needs of the painting.

Today a variety of specially designed tables exist with vacuum systems for controlled pressure and heaters for temperature regulation. Most applications first developed these tables for ambient temperatures and only employed low pressures- because of the nature of the adhesive being used at the time, slight pressure was all that was needed. Later, to incorporate thermally activated adhesives, tables with evenly heated surfaces were developed. These hot tables are capable of obtaining temperatures in excess of 70 C, which is enough to jeopardize most paint films. Pressure can be applied in one of two ways; 1) a mylar envelope surrounds the entire arrangement of canvas and lining canvas and a vacuum is pulled, or 2) the original canvas functions as part of this seal and a vacuum is drawn between the lining canvas and the original. The first method presses from the top of the painting, often threatening the flattening of impastos that protrude upwards due to concentration of localized pressures. The latter technique reduces the risk of flattening impastos since lining canvas and the original are actually pulled together, however even distribution of pressure is somewhat difficult to control.

Discussion

Synthetic materials can eradicate some problems natural adhesives have by using thermoplastic polymeric materials of sufficiently high molecular weight. Such materials should be elastic enough to prevent cold flow from occurring, and lack polar functionalities or crystallizability caused by hydrogen bonding so they would not change dimensionally during humidity cycles. At this point it is important to mention that the effect of elasticity, or more exactly the magnitude of the modulus of elasticity for such materials which has yet to be thoroughly evaluated. The question should be posed as thus: Should an adhesive essentially fix the strain to zero at the back of the canvas, effectively preventing hygroscopic materials in the structure of the paint film from straining while developing stresses, or should the adhesive be flexible enough to allow the strains to occur with little additional stress development.

The stress components that a painting experiences in its native state are a result of gradients of strain response initiated by changes in atmospheric moisture content which cause hygroscopic material to expand against material that is hydrophobic or unsaturated. After high humidities under the proper thermal conditions, brittle material not responding to moisture may crack if the stresses involved are large enough. This phenomena applies to all paintings that have any degree of heterogeneity, which due to the creativity of painting is difficult to avoid. It has been thought that such processes that occur in paintings requiring a lining should be minimized by restraining these stresses with a lining adhesive that limits the strain responses to changes in humidity. Considering such an option will however only restrain at the adhesive interface, leaving hygroscopic paint films at the surface of the painting to respond in their unrestrained manner. Any materials acting under such a constraint prompt the development of stresses within these materials, which if sufficient will cause them to fracture.

Adhesives should be elastic in the sense that upon removal of the environs inducing a stress, strains are once again brought back to nil, i.e. creep phenomena or cold flow should be avoided. Though the following serious question remains unanswered: what is the ideal Youngs Modulus an adhesive material must posses, or in other words, what should the permissible stress response to a strain be? This no doubt is highly dependant upon the painting and is unlikely to be a simple quantified answer. Adhesives that permit even small degrees of creep should always be avoided. This is because samples that have undergone some form of long term strain without a restoring force exist at a new dimension of strain upon removal of the stress. Then, if an event such as a sudden temperature drop induces the reversal of the strain below the time frame of the creep response, it is possible that the stress response may be excessive thereby inducing failure under either compression or tension depending upon the circumstances.

It is the authors opinion that these physical responses are best controlled by providing a stable environment through the careful monitoring of temperature and humidity. Therefore one may say that the final constraint (paradoxical as it may seem) is to use an adhesive that does not alter the paintings previous response to climatic variations, though act as a structural support. In order to understand just exactly what these responses are requires not only an understanding of the chemical and physical properties of artist's materials but how their chemical and physical aging effects these properties over time.

Conclusions

Due to the complex, heterogeneous nature of paintings the formulation of adhesives that can be used in their linings include a large number of constraints. While highly trained conservators have relied on their intuition to evaluate all of these complexities, it seems possible that scientific instrumentation can begin to answer some hitherto unanswerable questions. Due to the specialization of instrumentation and lack of conservators with a highly specialized technical background, it is necessary for scientists and conservation scientists to get involved in devising these questions and answering them.

REFERENCES

1 Percival-Prescott, W.: "Report on the Greenwich Lining Conference." ICOM Committee for conservation, 4th Triennial Meeting, Venice, 1975, 75 pp. 1-9.

2 Berger, G.A.: "A New Adhesive for the Consolidation of Paintings, drawings and textiles." Bulletin IIC-AG, 11 no. 1, pp.36-38 (1970).

3 Mehra, V.R.: "Comparative study of conventional relining methods and materials and research towards their improvement." Interim Report to the ICOM Committee for the Care of Paintings, Madrid (1972) Offprint 22pp.

4 Phenix, A. and Hedley, G.A.: "Lining without heat or moisture." ICOM 75 no.10, sect. 2, pp.38-44. (1975)

5 Mecklenburg, M.F.: "Some Aspects of the mechanical behavior of fabric supported paintings", Report for Science and technology in the service of conservation. IIC, (1982)

6 Katz, K.B.: "The quantitative testing and comparisons of peel and lap/shear for Lascaux 360 H.V. and Beva 371." JAIC 24 pp.60-68 (1985)

7 Hedley, G.A.: "The effects of Beeswax/Resin Impregnation on the Tensile Properties of Canvas." ICOM 75 no. 11 sect 7, pp.1-13. (1975)

8 Berger, G.A. and Zeliger, H.: " Wax impregnation of cellulose: an irreversible process." Conference on Comparative Lining Techniques, National Maritime Museum, 31 no. 7 pp 3-7 (1974).

9 Hacke, B.: " A low pressure apparatus for treatment of paintings." ICOM no. 12 section 2 pp.1-17. (1975)

10 Lodewijks, J. "Heat Sealing for relining paintings." ICOM (1967)

11 Berger, G.A.: "Application of Heat Activated Adhesives for the Consolidation of Paintings." Bulletin of the IIC. 11 no. 2 pp. 124-128.

Author Index

Subject Index

430

Member Organizations of the Pacific Polymer Federation

Founding Members:

The Society of Polymer Science, Japan
The Division of Polymer Chemistry, American Chemical Society
The Polymer Division, The Royal Australian Chemical Institute

Members:

Macromolecular Science and Engineering Division, Chemical Institute of Canada
The Society of Polymer Science, Korea
Polymer and Industrial Section, Malaysian Institute of Chemistry
The Polymer Group, New Zealand Institute of Chemistry
Polymer Division, Chinese Chemical Society